6th International Conference on Computers Management & Mathematical Sciences (ICCM 2020)

IOP Conference Series: Materials Science and Engineering
Volume 1020

Moscow, Russia
27 - 29 October 2020

Editors:

Rajesh Kumar
Gaurav Trivedi
N. Kishorjit Singh
M.M. Dixit
Nishtha Kesswani
Frank Lin
Ajit Kumar Singh Yadav
Shibabar Choudhury
Manmohan Mall

ISBN: 978-1-7138-3323-9
ISSN: 1757-8981

Printed from e-media with permission by:

Curran Associates, Inc.
57 Morehouse Lane
Red Hook, NY 12571

proceedings.com

Some format issues inherent in the e-media version may also appear in this print version.

Institute of Physics
Dirac House, Temple Back
Bristol BS1 6BE UK

Phone: 44 1 17 929 7481
Fax: 44 1 17 920 0979

techtracking@iop.org

Additional copies of this publication are available from:

Curran Associates, Inc.
57 Morehouse Lane
Red Hook, NY 12571 USA
Phone: 845-758-0400
Fax: 845-758-2633
Email: curran@proceedings.com
Web: www.proceedings.com

TABLE OF CONTENTS

IOP Conf. Series: Materials Science and Engineering 1020 (2021) 011001 doi:10.1088/1757-899X/1020/1/011001

Preface

6th International Conference on Computers, Management & Mathematical Sciences (ICCM) 2020

We are pleased to present this volume of the selected contributed articles, submitted and presented in the 6th International Conference on Computers, Management & Mathematical Sciences (ICCM) 2020 which held on November 22 & 23, 2020 at Department of Electronics and Communication Engineering, North Eastern Regional Institute of Science and Technology (NERIST), Arunachal Pradesh, India, in collaboration with International Association of Academicians (IAASSE), USA. (Conference link: http://iccm.iaasse.org/.)

The International Conference on Computers, Management & Mathematical Sciences (ICCM 2020 is the 6th in the series) is a platform for researchers, engineers, academicians as well as industrial professionals from all over the world to present their research results and development activities in Computers, Management, Mathematical Sciences.

The event (ICCM 2020) took place in hybrid model and thereby allowing some authors to present their papers virtually because of Covid–19 and the government's restriction on any gathering. And by that, organizers fulfilled their commitments with researchers and supports the scientific research.

The eminent professors and experts in their fields were invited to deliver their speech atICCM 2020. The Speakers were

- Dr. Alexander Pfeiffer-The MIT Education Arcade, USA-Danube University Krems, Austria;
- Dr. Joseph G. Vella, Faculty of ICT, University of Malta;
- Dr. Gaurav Trivedi, Associate Professor, EEE, IIT Guwahati, India;
- Dr Lalit Garg, Faculty of ICT, University of Malta)

The papers presented in the ICCM 2020 went through strict refereeing and examination resulting in a current acceptance rate of 43.9%. All papers were selected for oral presentation in the conference after an initial review. Every submitted full length paper was sent for peer review to at least two potential reviewers of related areas of expertise once it was passed by the program committee. We are delighted to say that this is in no small part due to the hard work the editorial board and reviewers, in not only refereeing the papers submitted but raising the standard of the quality of papers that are to be published.

IOP Conf. Series: Materials Science and Engineering 1020 (2021) 011001 doi:10.1088/1757-899X/1020/1/011001

This volume of the proceedings from ICCM 2020 will be a very useful contribution for the researches, educators and scientists for enthusing new vistas of research and development in the emerging areas of computers, management and mathematical sciences.

ACKNOWLEDGMENT

The guest editors warmly thank and record their heartfelt deep appreciation to all the eminent guests including the keynote speakers, plenary speakers and other delegates who made their valuable contributions towards the success of the ICCM 2020. It is a great opportunity for us to thank all the reviewers for sparing time from their busy schedule and provide timely and valuable comments to improvise the original versions of the papers towards these extended papers. We would like to thank representatives of IOP also for their collaboration and full cooperation.

Last but not the least, as the guest editors of ICCM 2020 proceedings, we are thankful to people who have worked with us in planning and organizing both technical arrangements. We also thank all learned authors for their kind cooperation and contribution. Hope that the quality research work published in this issue will be able to serve something new to Humanity and Science.

Disclaimer: The authors are solely responsible regarding the authenticity and ethical issues related to their article contained in this volume.

Guest Editors:

Dr. Rajesh Kumar, Associate Professor Department of ECE, NERIST, Arunachal Pradesh
Prof. Gaurav Trivedi Department of EE, IIT Guwahati
Dr. N. Kishorjit Singh Department of CSE, IIIT, Manipur
Dr. M M Dixit Department of Mathematics, NERIST, Arunachal Pradesh
Dr Nishtha Kesswani, CURAJ, Rajasthan
Dr Frank Lin, CSUSB, USA
Mr. Ajit Kumar Singh Yadav Department of CSE, NERIST, Arunachal Pradesh
Dr. Shibabarta Choudhury Centre for Management Studies (CMS), NERIST, Arunachal Pradesh
Dr. Manmohan Mall Centre for Management Studies (CMS), NERIST, Arunachal Pradesh

IOP Conf. Series: Materials Science and Engineering 1020 (2021) 011001 doi:10.1088/1757-899X/1020/1/011001

6th International Conference on Computers, Management & Mathematical Sciences (ICCM) 2020

Organized by

Department of Electronics and Communication Engineering
NERIST, Nirjuli

In association with

International Association of Academicians (IAASSE), California, USA

22nd and 23rd November 2020

Venue: ET Cell Seminar Hall, NERIST

पूर्वोत्तर क्षेत्रीय विज्ञान एवं प्रौद्योगिकी संस्थान
North Eastern Regional Institute of Science and Technology
Deemed-to-be-University under Ministry of Education, Govt. of India
Nirjuli - 791109, Arunachal Pradesh

ICCM 2020 in pictures:

IOP Conf. Series: Materials Science and Engineering 1020 (2021) 011001 doi:10.1088/1757-899X/1020/1/011001

IOP Conf. Series: Materials Science and Engineering 1020 (2021) 011002 doi:10.1088/1757-899X/1020/1/011002

Peer review declaration for ICCM 2020

All papers published in this volume of *IOP Conference Series: Materials Science and Engineering* have been peer reviewed through processes administered by the Editors. Reviews were conducted by expert referees to the professional and scientific standards expected of a proceedings journal published by IOP Publishing.

- Type of peer review:
  A double blind peer review process was adopted with a minimum of 2 reviews for each paper. Reviewer selection was carried out by the General Chair and technical chairs from a classified list of invited and consented reviewers, who are widely acclaimed, national and international authorities on the subject and registered in the database through the submission management system. In case of a split verdict, a 3rd review was also sought.

- Conference submission management system: **Easy Chair**

- Number of submissions received: **91**

- Number of submissions sent for review: **91**

- Number of submissions accepted: **40**

- Acceptance Rate (Number of Submissions Accepted / Number of Submissions Received X 100): **43.95%**

- Average number of reviews per paper: **2**

- Total number of reviewers involved: **110**

- Any additional info on review process:

- Contact person for queries: **Dr Rajesh Kumar, General Chair, ICCM 2020**

IOP Conf. Series: Materials Science and Engineering 1020 (2021) 012001 doi:10.1088/1757-899X/1020/1/012001

Voice Pathology Detection Based on Deep Neural Network Approach

Jeberson Retna Raj, J Jabez, S Senduru Srinivasulu, S Gowri, and J S Vimali

Department of Information Technology, Sathyabama Institute of Science and Technology, Chennai, India

Abstract. The advancement of technology offers solution to the complex problems faced by the society and brings the wellbeing of the individuals. Smart healthcare is prominent nowadays for diagnosis, treatment and constant monitoring which reduces visitation of hospital, transport cost and waiting time. Voice pathology is a decease which affects the person vocal cord so that one who facing difficult in speech. If the decease not identified in time, it leads to permanent loss of voice for an individual. Traditionally, the decease is identified through oral examination or manual procedures. Due to the advent of smart phone, one can record the voice and send it to the cloud server for processing. Our system classifies the voice data and provides the decision to the user. This greatly reduces the transport cost, and waiting time for oral examination at medical center. The mobile phone recorded the patient voice data and it will be stored into the cloud. The voice data is synthesized to signals and with the help of deep Neural network the voice pathology can be identified. The system has been tested with the data set and the test result shows promising.

1. Introduction

Integration of wireless network with cloud technology allows seamless interaction of applications and message exchange. Mobile healthcare is famous due to its application helps in diagnosis of deceases, training the medical practitioners, testing the deceases with mobile based applications. The health industry and the products are reaching out the customer based on the credibility and accuracy of solving the patient problem. Smart health care is imperative for countries and smart cities integrate the patient, medical attendant and physician. This will highly reduce the patient overhead by visitation of hospital manually. Early diagnosis, preventive advices can be remotely handled with the available technology. Voice pathology is one of the decease affected worldwide. Survey shows that over 7 million people affected in US for voice pathology decease. The consequence of the decease is broken voice, change in voice pattern, and degradation of voice. The decease affected by the small growth of fuss, tumor in the path of vocal cord. Prompt and timely identification and treatment is required. Otherwise, one who loss the voice and in the worst case operation is necessitated. The decease may common for those who involves in teaching, music and related profession. Furthermore, there is less muscle movement at the vocal cord is also victims of the decease. In essence, the voice pathology can be identified with two procedures namely subjective and objective methods. The subjective method deals with oral examination of doctor with the patient. The voice test will be examined physically and the doctor can identify the level or stages of the decease. The objective method is based on supported or assistive tools with the help of software for identifying the decease. The software based applications employed with the different algorithms to identify the decease. However, the accuracy if the objective based methods for detecting voice pathology is a challenge due to its accuracy. Oral examination sometimes better for conclude the decease and its pattern. The objective procedure works based on the algorithm given and the dataset taken for training and testing. The objective method provides better accuracy if better the algorithm has taken. In our work, a CNN algorithm based voice pathology detection model has been introduced to identify the disease in the voice samples. With the help of smart phones, the voice data is recorded and stored into the cloud. The voice data is preprocessed before applying the CNN algorithm. The training and testing data is given to the algorithm and the algorithm efficiently classifies the test data with positive or negative remarks.

1.1. Literature Survey

Numerous work has been done with voice pathology identification has been discussed in literature. In [1], an LBP and ELM based classifier has been used to classify the voice samples. The Local Binary Pattern (LBP) used to extract the feature description and Extreme Learning Machine used for binary classification. In [2], a detail survey of voice pathology detection system has been discussed. The comparison of the machine learning algorithm is shown with their accuracy. Furthermore, the author presented the issues and challenges pertained in detecting voice pathology. In [3], a voice pathology detection system has been introduced. They are using transfer learning based CNN algorithm is used to detect the decease. The method is tested with SVD database and the accuracy is shown. Al-Nasheri et al developed a voice pathology detection system for detecting early stages of decease. The system divides the voice frame and detect the peak signal valued frame is extracted for classification. The SVM classifier is used to classify the voice samples [4]. Hossain et al proposed big data base voice pathology detection system for classifying the decease. The voice samples of user from hospital has been collected in the MPEG-7 audio features and IDP features are normalized and fusion. With the help of SVM, ELM and GMM algorithms, the voice samples are classified as either pathological or non-pathological voice [5]. In [6], a Dysphonia Detection Index (DDI) calculation for testing the quality of voice has been introduced. The input voice data is taken and the four acoustic parameters are defined. With the help of tree regression model the relationship between these four parameters measured and a threshold has been fixed. If the quality of the voice exceeds the threshold, then the system reported as DDI which means pathological voice otherwise it will be a healthy voice. In [7], a pathology voice detection system has been introduced to distinguish between the defect voice and healthy voice. A mobile health care unit is framework of modules such as voice acquisition, authentication, signal processing, classification, storage and reporting modules. The Boost tree algorithm is combination of decision tree is used to distinguish between pathology voice and healthy voice. Verde et al introduced the machine learning classifiers applied with the voice signals and the performance is compared. The voice signal main parameters such as fundamental frequency, jitter, shimmer, harmonic to noise ratio and cepstral coefficients are used for clinical evaluation. The algorithms which includes decision tree, logistic model tree, Bayesian classifier, instance-based learning are applied with the SVD data set and the performance is compared [8]. In [9], a deep neural network based algorithm for identifying multi-vowel has been discussed. In [10], a multidimensional voice program (MDVP) featuring with 33 voice parameters experimented with the SVM algorithm and voice pathology is detected. In [11], characterization of pathological voice signal is presented. Three different approaches which include classification, divergence criterion and entropy based computation for detecting pathological voice decease. All these existing systems are limited in accuracy of source estimation in time domain, high estimation error, and low estimation accuracy. Furthermore, some of the existing systems provide theoretical way of solution to the voice pathological decease and not in practical. The existing algorithm incurs high in computational cost and complexity in processing.

The proposed method equipped with a smart healthcare framework which integrates patient, doctor and medical attendants with cloud and wireless technologies. The patient smart phone records the voice data as an input and it will be stored in a cloud server. The voice data is preprocessed by removing silence and make it as window file format. The voice signal parameter like jitter, shimmer, fundamental frequency is extracted for evaluation. The convolutional neural network algorithm is applied with the jitter parameter and classified with the given input is pathology voice or healthy voice.

2. Voice pathology Detection process

Detection of voice pathology is a challenging process as the voice data is complex in nature. It is very difficult to differentiate the healthy voice and pathological voice. The survey shows that the pathological voice decease may happen at least once in a person's lifetime. There are 15 percent people affected in Arab countries regarding voice related deceases. It may be slur voice due to paralysis or other related deceases. The growth of cist, gland or small tissue in the voice cord leads to the decease. If the decease not identified at the right time, one can lose the speech permanently. In order to detect the voice pathology, the software based model to distinguish the healthy and patient voice. The proposed system employed with CNN algorithm for efficiently detecting the voice pathology. The complete framework integrated with smart phone, wireless network, and cloud computing technologies. The detailed architecture is shown in figure 1.

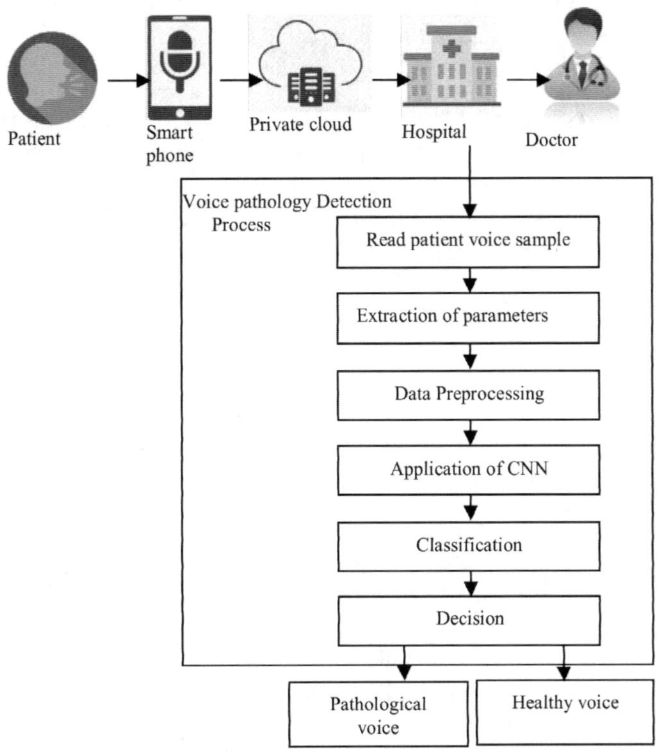

Figure .1 Voice pathology detection process

3. Working Methodology

The patient recorded the voice through the smart phone and the audio data is stored into the private cloud server. The cloud server is connected with the hospital, doctor and medical staff. The hospital equipped with our voice pathological detection system for detecting voice pathology. The doctor feed the input voice data to the system and the system reports the given voice is pathological voice or healthy one. The proposed voice pathological detection system employs with the following modules:

- Data acquisition
- Extraction of parameters
- Data pre-processing

- Application of CNN
- Classification
- Report Decisions

3.1. Data acquisition

The voice data is captured through the smart mobile and store it in the cloud server database. The collected data will be converted into the required format for further processing. The captured data will be in .wmi file format and it will be converted into vector representation for the application of algorithm.

3.2. Extraction of parameters

The important parameters such as fundamental frequency, jitter, and shimmer, harmonic to noise ratio and cepstral coefficients are extracted for process the data. Jitter is an important parameter due to its pitch and frequency helps in accurately classifies the data.

3.3. Data preprocessing

Data preprocessing is important in removing the noise and cleaning the data. The silence in between voice signals will be removed. This help in increase the accuracy.

3.4. Application of CNN

Convolutional neural network algorithm is efficient for classifying the complex data. Here, the binary classification labeled with voice pathology and healthy voice is the two classified data. The system reports the decision and it will be stored into the cloud server and eventually the doctor conveyed to the patient. This will greatly reduces the visitation time by the patient, transport cost, waiting time etc.

4. Experimental Results

The voice pathology system is tested with the sample voice data. The dataset SVD and voice data from Gaggle is considered for processing. The pathology detection process implemented using python framework and the voice samples are tested successfully. The open source dataset is considered and stored it in a database which can be in the .csv format. In our work, the dataset of a collection of 3,168 voice samples of both pathological and non-pathological voice which are equally present in its Datasets. In the data acquisition phase, the images of this data set are taken for processing. The data is preprocessed and the noise free data is in the form of an image. The CNN algorithm is applied with the preprocessed data and the pathology voice is classified. The split function used to feed the train and test voice data for validating and predicting the answer (i.e...Pathology or not). In x train some values are given to predict either the parameter is pathology or not, for that we use y train. Similarly x test is tested with some more different voice samples and predict it, for that we used y test. Therefore, the predicted values are both of pathology and non-pathology. Numpy is used for vector representation. The label encoder by fit transforms to change the Pathology Data as 1 and non-pathology Data as 0 which are binary values. Then we uses pipeline algorithm, here we are taking an array of n number of features (voice samples) in x test. so, from that we do predict the answers of the pipe_svc _predict (X_test). As we are use an label Encoder by fit transform , the result will be like array of both 1's and 0's format which are pathology and non pathology respectively. For validating the work, we used a cross validation check by taking a dataset from voice sample, we checked it ten times and from that we are getting the result as 97% of accuracy on an average. Figure 2 shows the plotted graph blue labelled as healthy voice and yellow labeled as pathological voice.

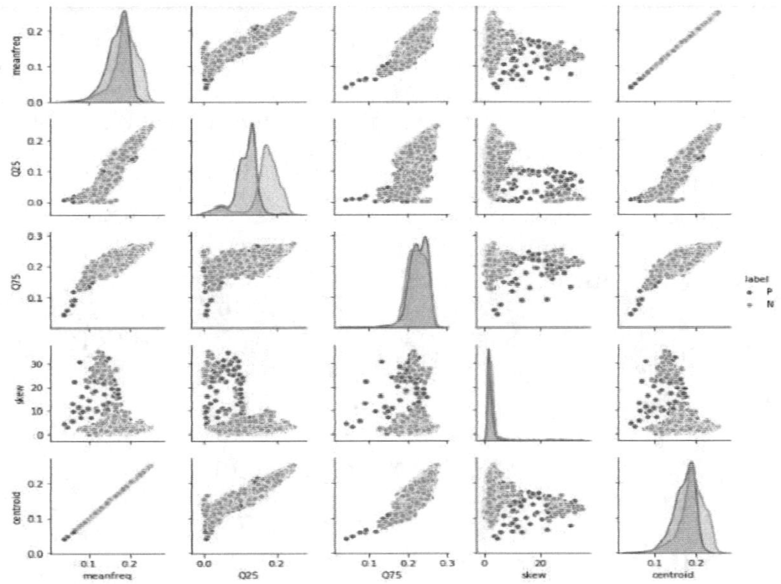

Figure 2. The plotted graph blue labelled as healthy voice and yello lebeled as voice pathology

Two models are implemented for predicting the pathology and non-pathology decease. They are Pipeline SVC model and Grid search Model. Both models give a 97.6% Accuracy Results. Both are quite efficient only when we have taken in thousand numbers of Datasets. The difference we observed between these models when we have given a million of datasets. Figure 3 shows the classification accuracy of the algorithm

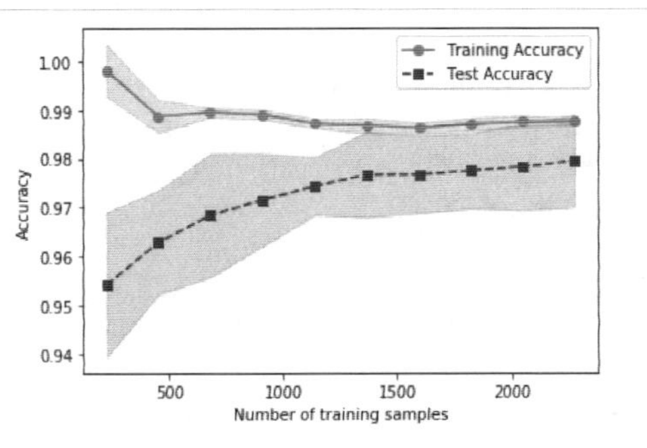

Figure 3. performance of the algorithm

5. Conclusion

Voice pathology can be concluded in two observations. First using medical way by using highly expensive equipment to check and another is by system check by using Deep learning approach. The system successfully distinguished between pathology and normal voice with the given dataset. The voice data of the user is considered as an input and the system extracts the features shimmer, jitter, harmonic noise ratio and ceptrol coefficient. The CNN algorithm works well in the given data set and identifies the pathology and non-pathology decease. The presented model can work well against low band speech analysis for pathology identification with continuous signals. This model is one kind of observation for identifying the pathology decease. However, additional observation is required to confirm the voice pathology decease.

References

[1] G. Muhammad, S. M. M. Rahman, A. Alelaiwi and A. Alamri, "Smart Health Solution Integrating IoT and Cloud: A Case Study of Voice Pathology Monitoring," in IEEE Communications Magazine, vol. 55, no. 1, pp. 69-73, January 2017.
[2] F. T. Al-Dhief et al., "A Survey of Voice Pathology Surveillance Systems Based on Internet of Things and Machine Learning Algorithms," in IEEE Access.
[3] M. Alhussein and G. Muhammad, "Voice Pathology Detection Using Deep Learning on Mobile Healthcare Framework," in IEEE Access, vol. 6, pp. 41034-41041, 2018.
[4] A. Al-Nasheri et al., "Voice Pathology Detection and Classification Using Auto-Correlation and Entropy Features in Different Frequency Regions," in IEEE Access, vol. 6, pp. 6961-6974, 2018.
[5] M. S. Hossain and G. Muhammad, "Healthcare Big Data Voice Pathology Assessment Framework," in IEEE Access, vol. 4, pp. 7806-7815, 2016.
[6] L. Verde, G. De Pietro, M. Alrashoud, A. Ghoneim, K. N. Al-Mutib and G. Sannino, "Dysphonia Detection Index (DDI): A New Multi-Parametric Marker to Evaluate Voice Quality," in IEEE Access, vol. 7, pp. 55689-55697, 2019.
[7] L. Verde, G. De Pietro, M. Alrashoud, A. Ghoneim, K. N. Al-Mutib and G. Sannino, "Leveraging Artificial Intelligence to Improve Voice Disorder Identification Through the Use of a Reliable Mobile App," in IEEE Access, vol. 7, pp. 124048-124054, 2019.
[8] L. Verde, G. De Pietro and G. Sannino, "Voice Disorder Identification by Using Machine Learning Techniques," in IEEE Access, vol. 6, pp. 16246-16255, 2018.
[9] T. Zhang, Y. Wu, Y. Shao, M. Shi, Y. Geng and G. Liu, "A Pathological Multi-Vowels Recognition Algorithm Based on LSP Feature," in IEEE Access, vol. 7, pp. 58866-58875, 2019.
[10] A. Al-nasheri et al., "Voice pathology detection with MDVP parameters using Arabic voice pathology database," 2015 5th National Symposium on Information Technology: Towards New Smart World (NSITNSW), Riyadh, 2015, pp. 1-5.
[11] C. D. Maciel, J. C. Pereira and D. Stewart, "Identifying healthy and pathologically affected voice signals [Lecture notes]," in IEEE Signal Processing Magazine, vol. 27, no. 1, pp. 120-123, Jan. 2010.Computing, 104 (2020) 104046

IOP Conf. Series: Materials Science and Engineering 1020 (2021) 012002 doi:10.1088/1757-899X/1020/1/012002

FLOT : Location Privacy preservation in Internet of Things

Nishtha Kesswani

Central University of Rajasthan, Ajmer, India 305801

E-mail: nishtha@curaj.ac.in

Abstract. Internet of Things includes sensor enabled devices that are capable of doing a wide variety of tasks. As these devices capture huge amounts of information including the location of the user, these generate privacy concerns. In this paper, Fuzzy Logic Based Location Obfuscation Technique (FLOT), fuzzy logic based Location privacy preservation technique has been proposed. The technique was compared to other existing technique and the results indicate that the original location of the user is obfuscated to a great extent in the proposed technique.

1. Introduction

With the connected world and technologies provisioning the connections, sensor enabled Internet of Things (IoT) devices have revolutionized the world. IoT is a system of interconnected sensor enabled devices with unique identifiers and are capable of transferring data without human intervention. These devices are used in varied applications such as automation, smart cities, smart grids etc.

Augmenting the user experience with sensors and actuators enhances the capabilities of the devices, but at the same time, the devices are collecting huge amounts of data about the user. This data includes the personal details such as the date of birth of the user, her telephone number, and email address etc. which raises privacy concerns. Location-based services also collect users' location which may be used by third party applications. Fundamental characteristics of IoT devices that make it difficult to ensure privacy are heterogeneity, inter-connectivity, dynamic networks and enormous scale. Ensuring privacy is vital not only for the safety of the users, but also to avoid intrusive interference in the personal space. In such a scenario, ensuring privacy of the users' data and location is of utmost importance.

In this paper, a Fuzzy Logic based privacy preservation technique has been suggested. The main contribution of this piece of work is to ensure Location privacy using Fuzzy logic based obfuscation technique. The rest of the paper is organized as follows. Section 2 discusses the Background; Related Work is mentioned in Section 3. Section 4 contains the proposed Fuzzy Logic based Obfuscation Technique (FLOT). Experiments and results are given in Section 5. Section 6 concludes the paper.

2. Background

Ensuring privacy is important from the point of view of the user. There are several devices in an IoT based network that are connected to the Internet and these devices need to be protected. For instance, the user may wish to anonymize her identity, location or restrict access to her data from third party services. In order to facilitate this, there are several different techniques that are used for privacy preservation. These include Obfuscation based techniques, Noise based techniques, and Anonymization based techniques. Figure 1 shows broad categorization of privacy preservation techniques and methods.

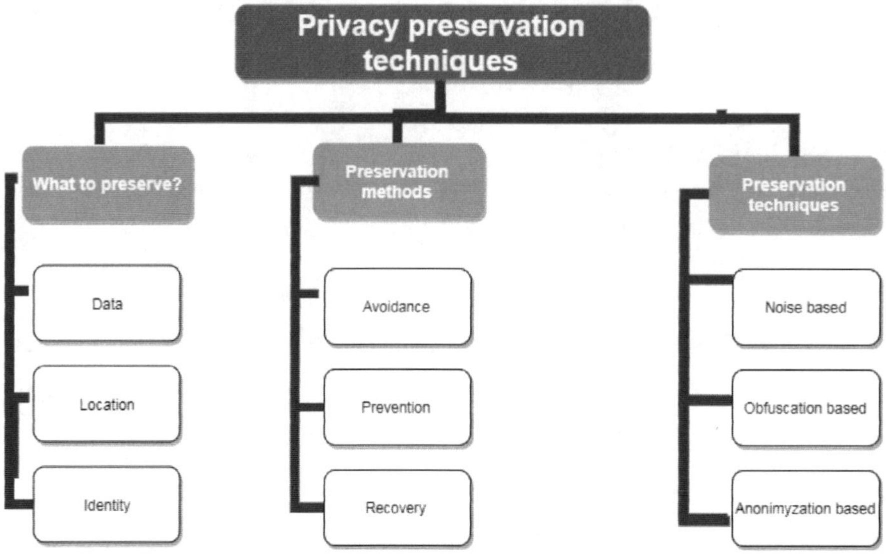

Figure 1. Privacy preservation techniques and methods

3. Related Work

Due to its importance, many researchers at different points of time tried to find methods of preserving privacy. For this piece of work, only Location privacy preserving techniques have been mentioned. Major works in this field are listed in Table 1.

Though there are several techniques available in literature, but most of the techniques are not suitable for the resource constrained IoT devices. An ideal privacy preservation technique should be able to ensure privacy and at the same time should be simple to implement. The technique proposed in this paper, tries to fill this research gap.

IOP Conf. Series: Materials Science and Engineering 1020 (2021) 012002 doi:10.1088/1757-899X/1020/1/012002

Table 1. Related works in Location privacy

Reference	Privacy Preservation technique	Domain	Details
Beresford and Stajano[1]	Anonymization	Pervasive computing	Used mix zones to ensure privacy
Campbell *et al.* [3]	Anonymization	Pervasive computing	Communication infrastructure Mist that separates location from identity
Duckham *et al.*[4]	Obfuscation	Context aware system	Formal method based location obfuscation
Yingjie *et al.*[5]	Anonymization	Mobile crowd sourcing systems	Two-stage auction algorithm based on trust degree and privacy sensibility
Xi *et al.* [6]	Anonymization	Mobile online social networks	Centralized privacy preserving location sharing system
Agostino *et al.*[2]	Obfuscation	Cellular networks	Obfuscate the location of the user by increasing the radius, shifting the center and reducing the radius of the circle.
Han[7]	Obfuscation	IoT	Confused arc based location obfuscation
Han[8]	Obfuscation	IoT	k-means based approach
Gope *et al.*[9]	Obfuscation	IoT	Privacy preserving RFID authentication scheme
Riahi *et al.*[10]	Anonymization	IoT based transportation systems	Game theory based approach
Jain and Kesswani [16]	Obfuscation	IoT	Random variance added for privacy preservation
Kaur *et al.* [17]	Randomization	IoT	Random path and random neighbor used for preserving location privacy
Minghui *et al.*[11]	Anonymization	Healthcare IoT	Reinforcement learning based Privacy aware offloading with energy harvesting
Guan *et al.*[12]	Anonymization	Fog enhanced IoT	Privacy preserving data aggregation
Jain *et al.*[15]	Data splitting and obfuscation	IoT	Used Storage as a Service and Key Management as a Service

4. Fuzzy Logic based Obfuscation technique (FLOT)

4.1. System Model

Privacy preservation can be done in three ways:

- Avoiding privacy leakage by adopting the best practices that preserve privacy.
- Preventing privacy leakage by protecting points that may cause privacy leakage in the network.
- Recovery after the privacy is breached.

One such technique of preventing privacy leakage is to restrict access and reducing the points which are open to the outside network. But restricting access may not be feasible in the modern days' connected world where all the devices are connected to each other and to the Internet.

In this paper, a Fuzzy logic based obfuscation technique has been proposed to prevent privacy leakage in IoT based networks. Let us assume the scenario given in Figure 1 in which IoT nodes are deployed. For the current piece of work, it has been assumed that the nodes are deployed randomly. In such a scenario let us assume that the attacker tries to detect the location of the user from the mobile phone shown in Figure 2. The system model consists of three components:

- IoT Network that comprises of several kinds of connected sensor enabled devices.
- User, who decides the level of privacy that she wants.
- Fuzzy Inference system that utilizes Fuzzy rules to obfuscate the location.

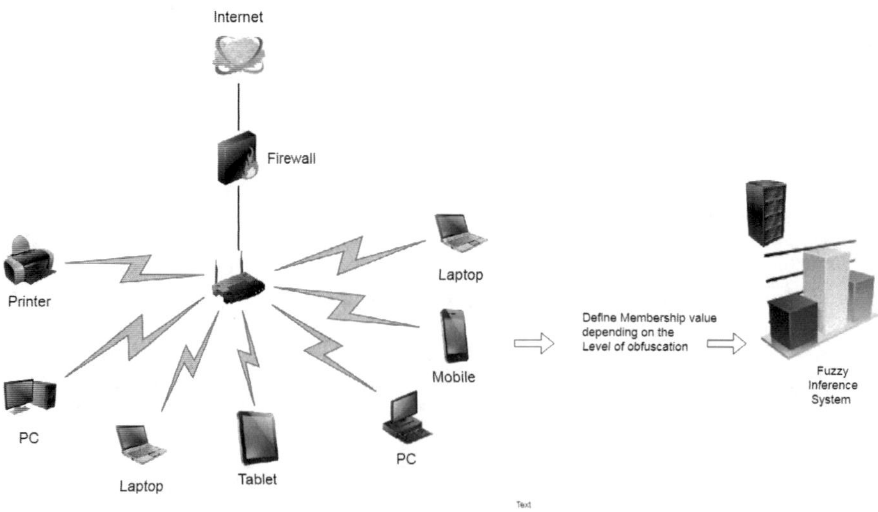

Figure 2. System model for FLOT

4.2. Adversary Model

The proposed system has three main entities; IoT devices, users and Fuzzy Inference system. The Fuzzy inference system is assumed to be a trusted entity that has complete information about the user but does not affect the privacy of the user. The attack model is as follows:

(i) Other devices in the IoT network may try to capture private information about the user.

(ii) Third parties and Location based services may use users' location to increase user experience, but at the same time raise privacy concerns.

(iii) An intruder may enter the system and try to get users' location and make inferences.

4.3. System Design

The location of the user is stored in a fuzzy set $\mathbb{A} = \sum_{x \in X} (\mu_A(x))/x$ Where X is the Universe of all the locations used for the purpose of developing the system. And the membership of x in Fuzzy set A is denoted by $\mu_A(x)$. The strategy is to create 10 locations in the same fuzzy set with membership value in the range [0-1] .As the distance increases the membership value would be governed by the Gaussian membership function given in Equation 1.

$$Gaussian(x; c, \sigma) = e^{-1/2} \left(\frac{x - c}{\sigma} \right)^2 \tag{1}$$

For instance, let us assume that the user having the Mobile phone shown in the Figure 2 wants to obfuscate her location, and then at this particular location her membership would be defined by Equation 1. As we move away from the location of the user, the membership value would decrease. The user can obfuscate the location using 3 levels: Level 1 indicating low level of privacy, Level 2 indicating medium privacy and Level 3 indicating high level of privacy. More the level of privacy, difficult it is to get the location of the user. The Algorithm 1 and 2 for FLOT show the calculation of the obfuscated location, given the Original Location of the user. The procedure uses every 10 nearby locations and places them in the same fuzzy set whose membership is governed by Gaussian membership function mentioned in Equation 1. Larger the set, difficult it is to make inferences about the location of the user. The summary of Notations used in the Algorithm are given in Table 2.

The user is supposed to give her Privacy preference from 0 to 2. This level decides the calculation of the final obfuscated location. As indicated in the procedure, the k-nearest sets are taken into consideration with k taken from the domain (3, 5, 7) which is just indicative of increasing distance. That is, as the value of k increases, the distance between the Original Location (Loc in the Algorithm) and the Obfuscated Location (Obf) increases. The final obfuscated location is a Random point of the third, fifth and seventh set of locations. Random point is taken for the sake of convenience so that it is difficult to detect the obfuscated location.

Table 2. Summary of Notations

Term	Description
Loc	Original location of the user
Obf	Obfuscated Location
LocMF	Membership value at Original location i
ObfMF	Membership value at Obfuscated location
LocFS	Set of all Fuzzy sets of all locations

In Algorithm 2, Original location is in fuzzy set A and level indicates the privacy levels in fuzzy set B where $x \in A$. The obfuscated location is calculated in Algorithm 2 using Mamdani Fuzzy Inference System. The Original location and the Privacy Preference Level act as inputs and the output is final Obfuscated location. Since Mamdani Fuzzy Inference system returns a fuzzy set as output, final precise Obfuscated location is decided using Centroid de-fuzzification. Fuzzy rules used in the inference system are shown in Algorithm 2 .

5. Experiments and results

The Experiments were conducted on Shared Cars Location Dataset [13]. The Dataset has five columns; Timestamp, Latitude, Longitude, TotalCars and CarsList. A snapshot of the dataset is shown in Table 3. It was pre-processed and only the relevant fields i.e. Latitude and Longitude were taken into consideration.The dataset containing 1090 records was taken for this piece of

Algorithm 1 Fuzzy Logic based Location Obfuscation Technique FLOT

Input: *Original Location Loc*
Output: *Obfuscated Location Obf*

$ObfuscateLoc() = FunctiontoObfuscatetheLocation$
$Loc_{MF} \leftarrow Gaussian(x; c, \sigma) = e^{(-1/2)\frac{(x-c)}{\sigma}^2}$
$Levelprivacy \leftarrow (0, 1, 2)$ //Privacy Level selected by the user 0:Low, 1:Medium, 2: High
$k \leftarrow (3, 5, 7)$ //k-Nearest Fuzzy set
For each fuzzyset in LocFS
For each Loc to be obfuscated
if $Levelprivacy = 0$ then Move to a Random point of the k nearest set **then**
  $Obf = ObfuscateLoc()$
else if $Levelprivacy = 1$ then Move to a Random point of the k+1 nearest set **then**
  $Obf = ObfuscateLoc()$
else if $Levelprivacy = 2$ then Move to a Random point of the k+2 nearest set **then**
  $Obf = ObfuscateLoc()$
end if
End For
End For
Return Obf

Algorithm 2 Procedure ObfuscateLoc() for Obfuscation using Mamdani Fuzzy Inference system

FuzzyRules:
If x in original location A and level in B then obfuscated location in A +3
If x in original location A and level in B then obfuscated location in A +5
If x in original location A and level in B then obfuscated location in A +7

$Loc \leftarrow Original\ location\ of\ the\ user$
$Obf \leftarrow Obfuscated\ Location$
$Levelprivacy \leftarrow (0, 1, 2)$ //Privacy Level selected by the user 0:Low, 1:Medium, 2: High
$k \leftarrow (3, 5, 7)$ //k Nearest Fuzzy set
For each set in k do
if Levelprivacy = 0 **then**
  $Obf \leftarrow Random(k)$
else if Levelprivacy = 1 **then**
  $Obf \leftarrow Random(k + 1)$
else
  $Obf \leftarrow Random(k + 2)$
end if
End For
Return Obf

work. Every 10 successive coordinates from the dataset were taken into one fuzzy set. Euclidean distance from the Median value was calculated using Equation 2.

$$\sqrt{(x_1 - median)^2 + (x_2 - median)^2 ... (x_n - median)^2} \quad (2)$$

Snapshot of the intermediate results is shown in Table 4. As seen from the results that most of the Co-ordinate locations are near the crossover points with membership value around 0.5. This is because the dataset contains co-ordinates from the same city and the locations are nearby. This problem can be solved by increasing the size of the fuzzy set to lets say 100 coordinates to

Table 3. Snapshot of the dataset

Timestamp	Latitude	Longitude	Totalcars	Carlist
2019-01-10 11:45:55.070781 UTC	32.09995	34.78794	1	[182]
2019-01-10 11:45:55.070781 UTC	32.06567	34.79612	1	[268]
2019-01-10 11:45:55.070781 UTC	32.06465	34.80322	1	[106]
2019-01-10 11:45:55.070781 UTC	32.05978	34.81034	1	[180]
2019-01-10 11:45:55.070781 UTC	32.05133	34.75089	1	[16]
2019-01-10 11:45:55.070781 UTC	32.04223	34.7742	1	[72]
2019-01-10 11:45:55.070781 UTC	32.04156	34.77128	1	[160]
2019-01-10 11:45:55.070781 UTC	32.12373	34.81346	1	[210]
2019-01-10 11:45:55.070781 UTC	32.11874	34.83406	1	[136]
2019-01-10 11:45:55.070781 UTC	32.03351	34.75509	1	[27]
2019-01-10 11:45:55.070781 UTC	32.14288	34.79361	1	[75]
2019-01-10 11:45:55.070781 UTC	32.14306	34.79729	1	[132]
2019-01-10 11:45:55.070781 UTC	32.083175	34.776552	0	[]
2019-01-10 11:45:55.070781 UTC	32.088379	34.775111	0	[]
2019-01-10 11:45:55.070781 UTC	32.074877	34.773515	0	[]
2019-01-10 11:45:55.070781 UTC	32.098603	34.778565	0	[]
2019-01-10 11:45:55.070781 UTC	32.09478	34.79728	0	[]
2019-01-10 11:45:55.070781 UTC	32.098032	34.798089	0	[]
2019-01-10 11:45:55.070781 UTC	32.12047	34.800318	0	[]
2019-01-10 11:45:55.070781 UTC	32.04409	34.80421	0	[]

incorporate more variability and increasing the distance between the Original and Obfuscated location. Since the Obfuscated location is randomly chosen from the third, fifth and seventh consecutive sets, larger the sets are, more is the distance between the Original and obfuscated location. Thus, by regulating these parameters, the privacy level can also be regulated.

The results show that there is variation in the Original location and Obfuscated location. Table 5 shows the Original location and the location after obfuscation.

The results were also compared to ESOT [14]. For instance for the same location, 51.50477 -0.1155168 (Sutton Wall, London, UK) used in ESOT gives the Obfuscated Location at 51.502966 -0.117233253(Chicheley St. London, UK), the proposed technique FLOT gives the Obfuscated location as33.39826 4.13252 (Bou Ikerkourene, Morocco). As seen from the results that in the proposed technique FLOT, more obfuscation is achieved by increasing the distance between the Original and obfuscated location. Both the algorithms were compared on the basis of Entropy which is a measure of unpredictability. For all the location data points d_i in the entire set of Locations D, Entropy (E) is measured as shown in Equation 3.

$$E = \sum_i \frac{d_i}{D} log_2 \frac{d_i}{D}$$

The results indicate that the proposed approach has more unpredictability as compared to the existing approach. As the level of privacy increases, the Entropy of FLOT increases.

6. Conclusion

Ensuring location privacy is an important aspect in the modern days' connected world. In this paper, a Fuzzy Logic based Location Obfuscation Technique was proposed and the results

Table 4. Snapshot of the Intermediate results

Latitude	Longitude	X1-Median	X2-median	Euclidean Distance	Membership
32.03025	34.74837	1.3614	1.35672	1.922003933	0.520290298
32.03025	34.74837	1.3614	1.35672	1.922003933	0.520290298
32.03028	34.74837	1.36137	1.35672	1.921982683	0.520296051
32.03049	34.74791	1.36116	1.35626	1.921509233	0.520424249
32.03049	34.74791	1.36116	1.35626	1.921509233	0.520424249
32.03351	34.75509	1.35814	1.36344	1.924451323	0.519628628
32.035393	34.75873	1.356257	1.36708	1.925705267	0.519290266
32.035393	34.75873	1.356257	1.36708	1.925705267	0.519290266
32.035393	34.75873	1.356257	1.36708	1.925705267	0.519290266
32.035393	34.75873	1.356257	1.36708	1.925705267	0.519290266
32.035393	34.75873	1.356257	1.36327	1.925691392	0.519294007
32.035393	34.75873	1.360067	1.36327	1.925691392	0.519294007
32.0369	34.7625	1.35856	1.36704	1.927299571	0.518860698
32.0369	34.7625	1.35856	1.36704	1.927299571	0.518860698
32.03761	34.751066	1.35785	1.355606	1.918703794	0.521185189
32.03761	34.751066	1.35785	1.355606	1.918703794	0.521185189
32.03761	34.751066	1.35785	1.355606	1.918703794	0.521185189
32.03985	34.77473	1.35561	1.37927	1.93392456	0.517083252
32.03985	34.77473	1.35561	1.37927	1.93392456	0.517083252
32.041052	34.769815	1.354408	1.364125	1.929526856	0.518261768
32.041052	34.769815	1.364638	1.364125	1.929526856	0.518261768
32.041052	34.769815	1.364638	1.364125	1.929526856	0.518261768
32.041052	34.769815	1.364638	1.364125	1.929526856	0.518261768
32.041052	34.769815	1.364638	1.364125	1.929526856	0.518261768
32.041052	34.769815	1.364638	1.364125	1.929526856	0.518261768
32.041052	34.769815	1.364638	1.364125	1.929526856	0.518261768
32.04156	34.77128	1.36413	1.36559	1.930203799	0.518080008
32.04156	34.77128	1.36413	1.36559	1.930203799	0.518080008

Table 5. Experimental results after applying FLOT

Original location	Original location coordinates	Obfuscated location	Obfuscated location coordinates	Privacy Level	Distance (kms)
Bat Yam, Israel	32.03025 34.74837	Gunbad, Pakistan	33.39165 68.93284	2	3247
Aviv Yafo, Israel	32.041052 34.769815	Haven Park, Western Australia	33.40569 117.22228	3	111000
Middleburg, Netherlands	51.50477 -0.1155168	Bou Ikerkourene, Morocco	33.39826 4.13252	1	2106

indicate that the Original location of the user is obfuscated to a great extent.In the proposed technique FLOT, as the Level of obfuscation increases, the location privacy increases. The advantage of the proposed technique is its simplicity.

References

[1] Beresford, Alastair R., and Frank Stajano. "Mix zones: User privacy in location-aware services." IEEE Annual Conference on Pervasive Computing and Communications Workshops, 2004. Proceedings of the Second. IEEE, 2004.

[2] Ardagna, Claudio Agostino, et al. "Location privacy protection through obfuscation-based techniques." IFIP Annual Conference on Data and Applications Security and Privacy. Springer, Berlin, Heidelberg, 2007.

[3] Campbell, Roy, et al. "Towards security and privacy for pervasive computing." International Symposium on Software Security. Springer, Berlin, Heidelberg, 2002.

[4] Duckham, Matt, and Lars Kulik. "A formal model of obfuscation and negotiation for location privacy." International conference on pervasive computing. Springer, Berlin, Heidelberg, 2005.

[5] Wang, Yingjie, et al. "Truthful incentive mechanism with location privacy-preserving for mobile crowdsourcing systems." Computer Networks 135 (2018): 32-43.

[6] Xiao, Xi, et al. "CenLocShare: a centralized privacy-preserving location-sharing system for mobile online social networks." Future Generation Computer Systems 86 (2018): 863-872.

[7] Han, Guangjie, et al. "CASLP: A confused arc-based source location privacy protection scheme in WSNs for IoT." IEEE Communications Magazine 56.9 (2018): 42-47.

[8] Han, Guangjie, et al. "KCLP: A k-means cluster-based location privacy protection scheme in WSNs for IoT." IEEE Wireless Communications 25.6 (2018): 84-90.

[9] Gope, Prosanta, et al. "Lightweight and privacy-preserving RFID authentication scheme for distributed IoT infrastructure with secure localization services for smart city environment." Future Generation Computer Systems 83 (2018): 629-637.

[10] Sfar, Arbia Riahi, et al. "A Game Theoretic Approach for Privacy Preserving Model in IoT-Based Transportation." IEEE Transactions on Intelligent Transportation Systems (2019).

[11] Min, Minghui, et al. "Learning-based privacy-aware offloading for healthcare IoT with energy harvesting." IEEE Internet of Things Journal (2018).

[12] Guan, Zhitao, et al. "APPA: An anonymous and privacy preserving data aggregation scheme for fog-enhanced IoT." Journal of Network and Computer Applications 125 (2019): 82-92.

[13] Shared Cars location, Kaggle, Available: https://www.kaggle.com/doit-intl/autotel-shared-car-locations/version/1?

[14] Ullah, Ikram, et al. "ESOT: A new privacy model for preserving location privacy in Internet of Things." Telecommunication Systems 67.4 (2018): 553-575.

[15] Jain, Shelendra Kumar, Nishtha Kesswani, and Basant Agarwal. "Security, privacy and trust: privacy preserving model for internet of things." International Journal of Intelligent Information and Database Systems 13.2-4 (2020): 249-277.

[16] Jain, Shelendra Kumar, and Nishtha Kesswani. "IoTP an Efficient Privacy Preserving Scheme for Internet of Things Environment." International Journal of Information Security and Privacy (IJISP) 14.2 (2020): 116-142.

[17] Kaur, Rajwinder, et al. "Efficient Routing Protocol for Location Privacy Preserving in Internet of Things." International Journal of Information Security and Privacy (IJISP) 13.1 (2019): 70-85.

IOP Conf. Series: Materials Science and Engineering 1020 (2021) 012003 doi:10.1088/1757-899X/1020/1/012003

MADM based Optimal Nodes Deployment for WSN with Optimal Coverage and Connectivity

Prince Rajpoot[1], Pragya Dwivedi[2]

Assistant Professor, REC Ambedkar Nagar, Akbarpur, UP, India[1]
Assistant Professor, MNNIT Allahabad, Prayagraj, UP, India[2]

E-mail: princeraj2904@gmail.com

Abstract. It is the digital era that provides importance for the research in Wireless Sensor Network (WSN). Different issues with different solutions are suggested by many authors. The realistic deployment having restrictions for cost, area coverage, Cluster Head (CH) coverage, and sink connectivity for WSN that demands an estimation of total count of sensors for deployment with the conditions for CH coverage and sink connectivity. Area coverage means how much portion of total area is sensed by deployed sensors nodes. CH coverage means the sensor node is able to transfer the sensed data to its CH without amplification and sink connectivity means that the CHs are connected with the sink and able to send the data directly without amplification. AHP is a Multi Attribute Decision Making (MADM) method that is used for the selection of best choice having less cost and efficient coverage and connectivity.

1. Introduction

WSN is a group of small sensors that do sense the environment for particular reason and provide the information to sink. One node may have one or more sensors. As the applications increased, area of deployment is also maximized toward remote areas. This creates the un-reachability of humans frequently, so the replacement of battery is also not possible. This leads to the temporary network that works until the battery discharged. To maximize the data collection time, we need to optimize the energy in such a way so that data can be sensed and transfered for long period of time. Clustering provides comparative larger lifetime to network. CH coverage and connectivity plays an important role in clustering. Total count of nodes and area coverage are also considerable factors for cost efficiency and maximum area coverage respectively.

So, here we are considering total four factors for network deployment. Our aim is to provide optimal solution for network deployment having 100% area coverage with optimal count, optimal CH coverage, and optimal Sink connectivity. Optimal count helps to reduce the total cost of deployment and we need to put small amount of sensors. So this count should be minimized, due to this, cost is a non-beneficiary factor. Maximum CH coverage reduces the energy required for transferring the data from neighbour nodes to CH. So we need higher value for this factor, due to this, CH coverage is a beneficiary attribute. Finally, maximum BS connectivity helps CHs to transmit the data to the sink with minimum power requirement. So high value is needed for this factor, due to this, BS connectivity is a beneficiary factor. So finally among four factors, we need 100% area coverage with minimum nodes having maximum CH coverage and sink connectivity. The layout of next sections are as follows: Section II represents the literature present on the WSN communication and layout. Section III illustrated our approach in detail, where as, the results of simulation are provided and compared in section IV. Conclusion is illustrated in section V.

2. Related Work

As we discussed before that WSN is a lime light field in research. Many works are being carried out by many authors and delivered a lot of methods for optimal data collection and deployment in WSN. It can be represent in following manner:

LEACH [2] is the base for efficient protocols that is based on probability with TDMA method for forming the schedule for packet passing. It selects the CHs based on a threshold value and probability to select as a CH and use signal intensity to connect with other nodes. It also rotates the authority of CH among all nodes and assigns a time schedule to transmit the data. But it uses some randomness in the phases, due to this, unbalancing occurs during clustering and power consumption.
LEACH-C [2] reduces some limitations of LEACH and provides constrain on minimum energy. It verifies that the nodes are of high power, so that they may able to process the data. H-LEACH enhances the clusters by partitioning the network in optimal number and LEACH verifies the CH selection with rotation in each part. Aaditya et al. verified the optimal CH authority in each part of H-LEACH and verifies the optimal clusters.
HADCC is an approach to determine a path in clustering form suggested by Aslam [5]. In it, the nodes are placed in two levels: inner and outer. The inner level contains the circle form for advance sensors positions. Sink is having the responsibility of choosing the CHs. Many factors have been used for this like: Total and remaining power, distance etc. CSMA method assigns schedule for data transfer.
Another process uses the sink in the mid of the deployed area and the area is partitioned into optimal parts such that every part is having equal amount of energy [6]. After this, total 5% of total nodes are chosen as CHs. Probability and transmission power are used for clustering. Now the CHs communicate with other CHs with any partition restriction.
Another approach uses the concept of multi-hop [7], in which the maximum power nodes are placed near to sink and low power nodes placed far from the sink. This concept uses to reduce the effect of black hole problem since every packet pass through nearest nodes.
Subhasis et al. provided a hierarchy based node deployment in which each node has two parents except the root. This algorithm starts with finding the level for each node with its residual energy in next phase. After this, the nodes are divided into active node or inactive node. Active node finds two parents to connect and inactive node remains in sleep mode. Last phase is of data transmission.
Delaney et al. [9] proposed a metrics based route finding approach and balance the distance in routes. It uses and takes the support of routing matrices of nodes and find efficient route.
Halder et al. [10] provided routing and clustering process using the model of Archimedes spiral. This spiral is start from the center and make a view of turn till the end point. This spiral is used to calculate the distance between two nodes and also helps to maintain about equal distance among the nodes. Gujman et al. [11] provided ON/OFF capacity nodes for lifetime increment with low and high energy nodes.
Lee et al. [12] suggested an approach that uses semi-distribution method in which four levels are used as upper and lower. Upper level CHs are chosen in centralized manner, whereas, CHs are chose in distribution form in lower level. First sink is there then after grid head level 2, after it, level1 CHs are there and at last level 0 nodes are placed. Level 0 nodes sense the surroundings and transfer it to the level 1 and forwarded to sink via grid head.
Sanjay [13] proposed clustering for integrated network in which the area is partitioned into some regions with centralized sink. Now, CHs are chosen using its weight and also change the

responsibility in continuous fashion.
PRRP method [14] is suggested by zoman et al. in which the grid form is used for area division. Due to this, every node position got known and reduces the energy needed in searching process. Krishna et al. [15] provided two layered deployment approach in which they compared the two deployment approach. One is when inner deployment be in circular form and outer is in hexagonal form whereas in other the outer form is of pentagonal. Both the layers is having different CHs and hierarchical transmission takes place. Shubham et al. [16] suggested optimized clustering to this paper using divisional method.
Some surveys [17], [18] are also performed regarding node deployment strategies. Many other authors also suggested many methods for efficient communication [19], [20], [21], [22], [23]. Some approaches by [24], [25], [26] also suggested intra cluster transmission using many efficient routes. Matrix method is also proposed by [27] to select best population for next generation. Many conflicting attributes are also suggested and applied with MADM approaches for load balance by [28] and for energy balanced by [29]. But these techniques can be applied with high computational sensor networks. Some algorithms for low computation network with lower attributes and simple calculations are also suggested by [30], [31], [32], [33], [34], [35], [36], [37], [38]. Neetu et. al [39] provides efficient communication to supervise the gas pipeline using clustering. Fuzzy tool can also be a option taken by [40], [41], [42], [43] for selection of optimal CHs that are able to perform data transfer with minimum use of energy.
Here, we have proposed a method that uses AHP protocol that chose best alternative solution for node deployment with 100% area coverage, low cost, higher CH coverage and higher sink connectivity. The approach is explained in detail in next section.

3. Proposed Approach

We have proposed an AHP based method that provides a way to chose best node deployment with maximum area coverage, connectivity and less cost among many options. The Figure ?? shows the working of proposed algorithm in step by step process as a flow chart of complete procedure.
Table 1 shows the parameters to generate network and deploy the nodes.
Here, first we see whether the network is covering whole area or not. For this, first we calculate

Table 1. Factor Values for network deployment

Factor	Values
Size	200m x 200m
Sensor Position	Variable
Prime Power of sensor	1J
Sensor Nodes	Variable

the area coverage using each position of area and the area covered by any one node. It starts from (0,1), (0,2), and process to (n,n-1), (n,n) and check the coverage node for each and every point. It can be given as:

$$Cover\_percent = \frac{\sum_{x1=0}^{x_{max}} \sum_{y1=0}^{y_{max}} SC(x1,y1)}{x_{max} * y_{max}} * 100 \tag{1}$$

and

$$SC(x1, y1) = \begin{cases} 1 & \text{if } \exists \text{ p} \in (1..N),\ D_{(x_p,y_p),(x1,y1)} \leq R_{Sense} \\ 0 & \text{otherwise} \end{cases}$$

where SC(x1,y1)- supervise condition of point (x1,y1)
We find the alternatives with different count and position of nodes. Now, our aim is to chose the best alternative having minimum cost with maximum CH coverage and sink connectivity. For this we have used AHP algorithm that ranks the alternatives and provides best solution among them that would satisfy all the constraints with better conditions. Per unit cost contains sensor cost, LIPO battery cost, and GPS device cost for each sensor node, so when we calculate the cost of sensor, it comes around 7350 Rs. per unit. There are following steps that can determine the complete approach.
Step I: Generate the alternatives and calculate the values of factors-

Table 2. Alternative solutions with respective values for factors

S.No.	Nodes	Cost	Area Cover	CH Cover	BS Connect
1	100	735000	98.2	100	50
2	100	735000	98.2	95	90
3	100	735000	100	95	80
4	100	735000	100	92	60
5	100	735000	100	94	50
6	100	735000	100	92	50
7	100	735000	100	95	80
8	100	735000	100	97	80
9	120	882000	100	100	40
10	120	882000	100	97.5	60
11	120	882000	100	100	50
12	120	882000	100	100	60
13	120	882000	100	97.5	80
14	130	955500	100	100	50
15	130	955500	100	99.23076923	60
16	130	955500	100	99	80
17	130	955500	100	98.46153846	70
18	150	1102500	100	100	60
19	150	1102500	100	100	50
20	150	1102500	100	99.33333333	80

Table 2 shows the values for different alternatives having different node count with the cost, area coverage, CH coverage and sink connectivity. Now we will apply the AHP algorithm and rank the alternatives, finally we will select the best alternative and the corresponding CHs for the final transmission of data. The next step explains the complete AHP process.
Step II: Apply MADM Approaches for rank the solutions:

3.1. Apply AHP method to rank the solutions
Step A.a: Normalization- Normalized matrix (NM) is calculated as:
For beneficiary factors:

$$Norm\_value_i = \frac{Value\_Factor_i}{Max\_value\_Factor\_in\_All\_CH} \tag{2}$$

For Non-beneficiary factors:

$$Norm\_value_i = \frac{Min\_value\_Factor\_in\_All\_CHs}{Value\_Factor_i} \tag{3}$$

Step A.b: Compute Relative Importance Matrix RW**-** It shows the relative importance of the factors with each other. In our proposed method, it is formed as:

Table 3. RW Matrix for attributes

cost	coverage	connectivity	
cost	1	1.7	1.3
coverage	0.59	1.00	0.91
connectivity	0.77	1.10	1.00

Step A.c: Compute Geometric Mean- It is calculated as row wise and is calculated as:

$$G_{Meank} = (\prod_{k=1}^{N} RW_{pq})^{(1/N)}, p - 1, 2, ., m \ \& \ q = 1, 2, ., m \tag{4}$$

where RW is the relative weight importance matrix and m is the number of factors.

Step A.d: Enumerate Weights in Matrix Form- Matrix contains weight of all factors is enumerated as:

$$Weight_{Mk} = \frac{G_{Meank}}{\sum_{j=1}^{N} G_{Meanj}} \tag{5}$$

It should verify that $\sum_{k=1}^{N} Weight_{Mk} = 1$

Table 4. $Weight_{M3*1}$ Matrix for attributes

0.425661937
0.265248859
0.309089204

Step A.e: Consistency Verification of Weights- Following steps are used to verify the consistency of our calculated weights:

Step A.e.1: Enumerate Consistency Matrix1 as-

$$(CM_1)_{N*1} = (RW)_{n*n} * (G_{Mean})_{n*1} \tag{6}$$

Step A.e.2: Enumerate Consistency Matrix2 as-

$$(CM_2)_{N*1}= \frac{(CM_1)_{N*1}}{(Weight_M)_{N*1}} \tag{7}$$

Enumerated consistency matrix2 for proposed approach is:

Table 5. CM_{23*1} Matrix for attributes

3.003324589
3.007364497
3.004310686

Step A.e.3: Enumerate λ_{max}-

$$\lambda_{max}= \frac{\sum_{p=1}^{N} (CM_2)_P}{N} \tag{8}$$

Enumerated λ_{max} for proposed approach is: 3.004999924
Step A.e.4: Enumerate Consistency Index-

$$Cont\_Index = \frac{(\lambda_{max}-N)}{N-1} \tag{9}$$

Enumerated Consistency Index for proposed approach is: 0.002499962
Step A.e.5: Enumerate Consistency Ratio-

$$\text{Const\_Ratio}= \frac{Cont\_Index}{RI} \tag{10}$$

where RI = Random Index.
Const_Ratio should be less than 0.1 for error less than 10%.

Enumerated Consistency Ratio for proposed approach is: 0.00480762. Means total error is 0.48%.

Since it is less than 0.1, so our enumerated weights are acceptable.

Step A.f: Enumerate Efficiency Value for Each Alternative- Following two methods are used to enumerate efficiency value for each alternative. Simple Additive Weighting (SAW) or Weighted Product Method (WPM) uses these weights to compute the values of P_i.

1) SAW Method: It is known as Simple Additive Weighting method. It enumerates efficiency value for alternative as:

$$EV_k = \sum_{k=1}^{N} (Weight_M)_k * (NM)_{jk}, j = 1, 2, ., m \tag{11}$$

2) WPM Method: It is known as Weighted Product Method. It enumerates efficiency value for alternative as:

IOP Conf. Series: Materials Science and Engineering 1020 (2021) 012003 doi:10.1088/1757-899X/1020/1/012003

$$EV_k = \prod_{k=1}^{N} (Weight_M)_k * (NM)_{jk}, j = 1, 2, ., m \tag{12}$$

In proposed approach, we have used SAW method to enumerate efficiency value.

Step A.g: Rank the Solutions using Efficiency Value- After ranking the final rank for each solution or alternative in proposed approach is shown in Table 6:

Table 6. Ranking of attributes using SAW

Pop Num	1	2	3	4	5	6	7	8	9	10	11	12	13	14	15	16	17	18	19	20
Rank	8	1	3	6	9	10	4	2	18	14	16	12	5	17	15	7	11	19	20	13

3.2. TOPSIS method to rank the solutions

Rank the patients as higher value of condition C_{worst} using TOPSIS is shown in Table 7.

3.3. PROMETHEE method to rank the solutions

Rank the patients using PROMETHEE is shown in Table 8.

Step III: Take the top rank situation and deploy the nodes accordingly: In our proposed approach, second population is having the best efficiency. But it is not able to provide 100% coverage to the area so we will not select this population as the final best. Now we proceed to the next population, i.e. population is P8 having Make the chosen nodes as final CHs.

Step IV: Start the transmission and gather the sensed data from the network. So we choose the CHs selected in P8 and other nodes join the nearest CH to make the the clusters. All normal nodes start the sensing of environment and transmit the data to the sink via CHs.

4. Results and Analysis

We have simulated our approach on WSN modeled by using MATLAB tool that gives realistic support to perfect vision of data transfer and energy consumption with death of nodes. We have applied our approach on homogeneous network having equal amount of prime power with equal characteristics of nodes for energy depletion. A node is known to be dead of the energy becomes less than the threshold value this threshold value is equal to the energy required to send the packet to a distance where no amplification is needed for packet transmission. We have compared among Proposed algorithm, Base Algorithm [16], SNPCM [15], and LEACH [2].

Case 1: Sensors=500; Sink=0,0

Case 2: Sensors=500; Sink=150,50

Table 7. Ranks assigned to various pop by TOPSIS

Pop Num	1	2	3	4	5	6	7	8	9	10	11	12	13	14	15	16	17	18	19	20
Rank	12	1	3	9	13	14	4	2	20	11	18	10	5	19	15	6	8	16	17	7

Table 8. Ranks assigned to various populations by PROMETHEE

Pop Num	1	2	3	4	5	6	7	8	9	10	11	12	13	14	15	16	17	18	19	20
Rank	5	1	3	8	14	16	4	2	11	13	10	6	7	19	17	9	15	18	20	12

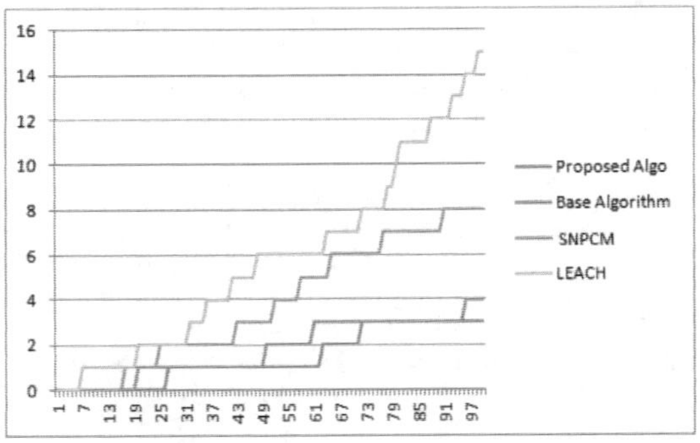

Figure 1. Dead sensors graph for case1.

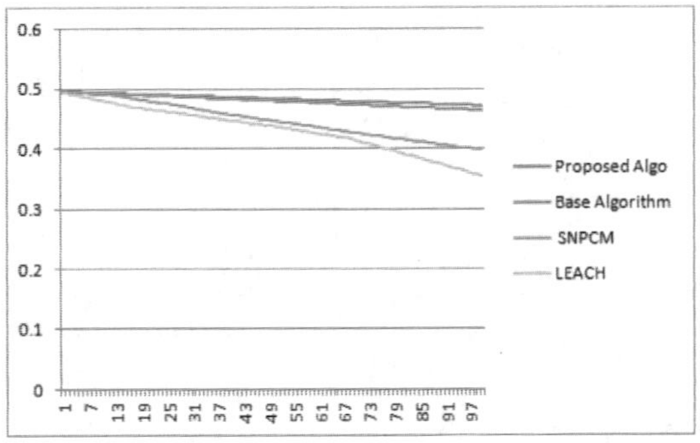

Figure 2. Remaining energy graph for case1.

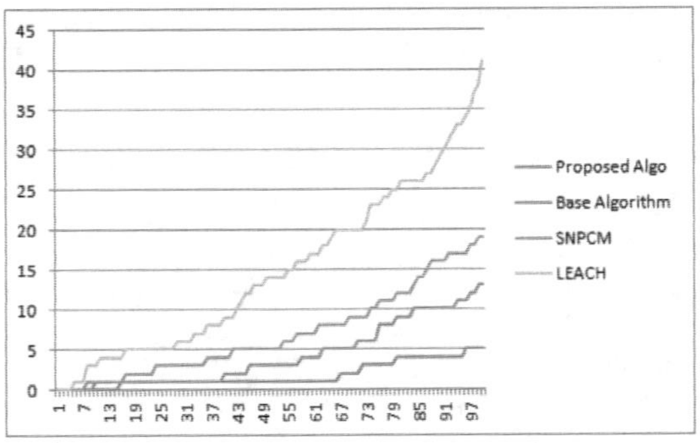

Figure 3. Dead sensors graph for case2.

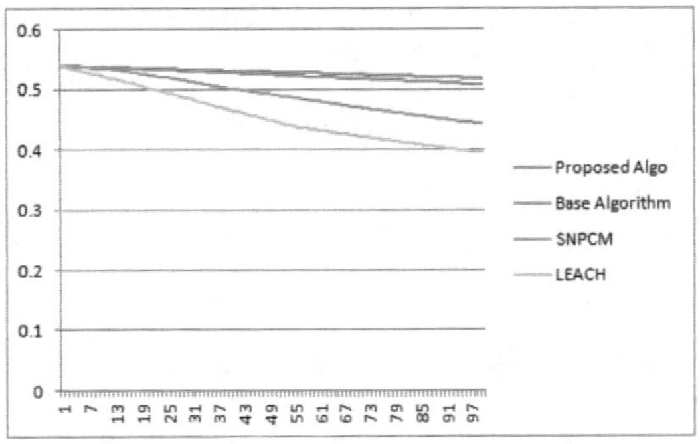

Figure 4. Remaining energy graph for case2.

5. Conclusion

Proposed algorithm provided efficient deployment of sensor in the field in such a way that the whole area can be sensed by at-least one sensor means 100% area coverage must be provided to the network. It also gave the possible smallest amount of nodes for the coverage, that minimizes the cost with the constraint of effective data collection. The effectiveness of data collection is confirmed by the maximum CH coverage and sink connectivity. Maximization of CH coverage confirms the minimization of the distance to CHs from all the respective nodes that minimized the power consumption from nodes to CHs. Whereas, the maximization of sink connectivity confirms the minimization of distance from CHs to the sink that reduces the power consumption for data transfer from CHs to the sink. Finally, the results also support the effectiveness of the proposed approach.

6. References

[1] Kumar R, Dubey K, Chaudhary H, Rajpoot P, Mishra S and Yaduvanshi R. 2019. Energy Distribution based Cluster Head Selection to Enhance the Lifetime of WSN, *International Conference on Advances in Electrical, Electronic and System Engineering (ICAEESE 2019)*, IEEE. (In press)

[2] Heinzelman WB, Chandrakasan AP and Balakrishnan H, 2002. An application-specific protocol architecture for wireless microsensor networks. *IEEE Transactions on wireless communications*, **1**(4), pp.660-670.

[3] Kumar R and Rajpoot P, 2018, May. Optimized H-LEACH algorithm for clustering to improve lifetime of WSN. *In 2018 3rd IEEE International Conference on Recent Trends in Electronics, Information & Communication Technology (RTEICT)* (pp. 2393-2398). IEEE.

[4] Rajpoot P, 2017. Data aggregation and distance based approach to boost life span of WSN. *International Journal of Engineering Technology Science and Research, IJETSR* **4**(11).

[5] Aslam M, Munir E U, Bilal M, Asad M, Ali A, Shah T, and Bilal S, 2014. HADCC: hybrid advanced

distributed and centralized clustering path planning algorithm for WSNs, *In 2014 IEEE 28th International Conference on Advanced Information Networking and Applications*, pp. 657-664. IEEE.

[6] Chen C, Liu X, Qi H, Zhao L, and Ren Z, 2015. A security enhancement and energy saving clustering scheme in smart grid sensor network, *IEEE 16th International Conference on Communication Technology (ICCT)*, pp. 848-855. IEEE.

[7] Cheng B C, Yeh H H, and Hsu P H, 2011. Schedulability analysis for hard network lifetime wireless sensor networks with high energy first clustering, *IEEE transactions on reliability* **60.3**: 675-688.

[8] Dash S, Mallick S S, Hansdah R C, and Swain A R, 2015. A distributed approach to construct hierarchical structure for routing with balanced energy consumption in wsns, *In 2015 IEEE 29th International Conference on Advanced Information Networking and Applications*, pp. 382-388. IEEE.

[9] Delaney D T, Higgs R, and OHare G M P, A stable routing framework for tree-based routing structures in WSNs. *IEEE Sensors Journal* **14.10**: 3533-3547.

[10] Halder S, and Ghosal A, 2015. Lifetime maximizing clustering structure using archimedes' spiral based deployment in WSNs, *IFIP/IEEE International Symposium on Integrated Network Management (IM)*, IEEE.

[11] Guzman-Medina C A, Rivero-Angeles M E, and Orea-Flores I Y, ON/OFF protocol for the life extension of low energy level nodes in wireless sensor networks, *International Conference on Computing Systems and Telematics (ICCSAT)*, IEEE.

[12] Lee J S, and Kao T Y, 2016. An improved three-layer low-energy adaptive clustering hierarchy for wireless sensor networks, *IEEE Internet of Things Journal* **3.6**: 951-958.

[13] Kumar S, 2015. Energy efficient clustering algorithm for WSN, *2nd International Conference on Signal Processing and Integrated Networks (SPIN)* (pp. 990-993).

[14] Zaman N, Low T J, and Alghamdi T, 2015. Enhancing routing energy efficiency of wireless sensor networks, *17th International Conference on Advanced Communication Technology (ICACT)* IEEE.

[15] Krishna R K, and Ramanjaneyulu B S, 2018. A Strategic Node Placement and Communication Method for Energy Efficient Wireless Sensor Network. *Proceedings of 2nd International Conference on Micro-Electronics, Electromagnetics and Telecommunications. Springer, Singapore.*

[16] Singh S H, Verma K R and Rajpoot P, 2018, May. Partition based strategic node placement and efficient communication method for WSN. *3rd IEEE International Conference on Recent Trends in Electronics, Information & Communication Technology (RTEICT)* (pp. 1807-1812). IEEE.

[17] Thakur A, Prasad D, and Verma A. 2017. Deployment Scheme in Wireless Sensor Network: A Review. *International Journal of Computer Applications* **163.5**.

[18] Saha S, 2018. Sensor Node Placement Methods Based on Computational Geometry in Wireless Sensor Networks: A Review. *International Research Journal of Engineering and Technology (IRJET)* **5**: 799-804.

[19] Aderohunmu F A, and Deng J D, 2009. An enhanced stable election protocol (sep) for clustered heterogeneous wsn, *Department of Information Science, University of Otago, New Zealand.*

[20] Naeimi S, Ghafghazi H, Chow C O, and Ishii H, 2012. A survey on the taxonomy of cluster-based routing protocols for homogeneous wireless sensor networks, *Sensors* **12, no. 6**: 7350-7409.

[21] Almalkawi I T, Zapata M G, and Al-Karaki J N, 2011. A secure cluster-based multipath routing protocol for WMSNs, *Sensors* **11, no. 4**: 4401-4424.

[22] Saleh A M S, Ali B M, Rasid M F A, and Ismail A, 2012. A self-optimizing scheme for energy balanced routing in wireless sensor networks using sensorant, *Sensors* **12, no. 8**: 11307-11333.

[23] Tufail A, Qamar A, Khan A M, Baig W A, and Kim K H, 2013. WEAMRA weighted energy aware multipath reliable routing mechanism for hotline-based WSNs, *Sensors* **13, no. 5**: 6295-6318.

[24] Yaqoob M M, Israr I, Javaid N, Khan M A, Qasim U, and Khan Z A, 2012. Transmission delay of multi-hop heterogeneous networks for medical applications, *In 2012 Seventh International Conference on Broadband, Wireless Computing, Communication and Applications*, pp. 428-433. IEEE.

[25] Javaid N, Qasim U, Khan Z A, Khan M A, Latif K, and Javaid A, 2013. On energy efficiency and delay minimization in reactive protocols in Wireless Multi-hop Networks, *In 2013 Saudi International Electronics, Communications and Photonics Conference*, pp. 1-4. IEEE, 2013.

[26] de Freitas E P, Heimfarth T, Pereira C E, Ferreira A M, Wagner F R, and Larsson T, 2009. Evaluation of coordination strategies for heterogeneous sensor networks aiming at surveillance applications, *In SENSORS, 2009 IEEE*, pp. 591-596. IEEE.

[27] Rajpoot Pand Dwivedi P, 2018, January. Matrix Method for Non-Dominated Sorting and Population Selection for Next Generation in Multi-Objective Problem Solution. *8th International Conference on Cloud Computing, Data Science & Engineering (Confluence)* (pp. 670-676). IEEE.

[28] Rajpoot P and Dwivedi P, 2020. Optimized and load balanced clustering for wireless sensor networks to increase the lifetime of WSN using MADM approaches. *Wireless Networks, 26(1)*, pp.215-251.

[29] Rajpoot P and Dwivedi P , 2019. Multiple parameter based energy balanced and optimized clustering for WSN to enhance the Lifetime using MADM Approaches. *Wireless Personal Communications*, **106**(2), pp.829-877.

[30] Rajpoot P, Singh S H, Verma R, Dubey K, Pandey S K and Verma S, 2020. Multi-factor-Based Energy-Efficient Clustering and Routing Algorithm for WSN. *Soft Computing: Theories and Applications* (pp. 571-581). Springer, Singapore.

[31] Sharma V, Rajpoot P, Gupta A, Dubey K, Pandey N and Verma N, 2019. Heterogeneous Clustering for Energy Optimization in Wireless Sensor Networks, *In 2019 9th International Conference on Cloud Computing, Data Science & Engineering (Confluence)* (pp. 587-592). IEEE.

[32] Dubey K, Kumar P, Pathak P, Kumar S, Rajpoot P and Pandey S K, 2019, March. A Survey on Fundamental of Wireless Sensor Network with Various Issues in Optimization. *3rd International Conference on Computing Methodologies and Communication (ICCMC)* (pp. 288-293). IEEE.

[33] Dubey K, Sharma V, Gupta A, Srivastava A, Mishra S and Rajpoot P, 2019, March. Load Balanced Clustering protocol for lifetime enhancement of WSN. *3rd International Conference on Computing Methodologies and Communication (ICCMC)* (pp. 282-287). IEEE.

[34] Yadav A K, Rajpoot P, Kumar P, Dubey K, Singh S H, and Verma K R, 2019. Multi Parameters Based Heterogeneous Clustering Algorithm for Energy Optimization in WSN, *In 2019 9th International Conference on Cloud Computing, Data Science & Engineering (Confluence)* (pp. 587-592). IEEE.

[35] Rajpoot P, Dwivedi P and Dubey K, 2019, July. Power Balanced Efficient Clustering Algorithm for WSN. *International Conference on Communication and Electronics Systems (ICCES)* (pp. 585-589). IEEE.

[36] Rajpoot P, Dubey K, Kumar V, Prajapati A, Kumar S and Mishra S, 2019, March. An enhanced approach for energy optimization using protocol based on multi factors. *3rd International Conference on Computing Methodologies and Communication (ICCMC)* (pp. 300-305). IEEE.

[37] Dubey K, Yadav A K, Kumar P, Shekhar P, Rajpoot P and Kumar S, 2019, March. Power Optimization Algorithm for Heterogeneous WSN using Multiple Attributes. *3rd International Conference on Computing Methodologies and Communication (ICCMC)* (pp. 294-299). IEEE.

[38] Prajapati A, Kumar V, Rajpoot P, Dubey K, Singh S H and Verma N, 2019, January. Multi-Objective Heterogeneous Clustering Approach for Efficient-Energy Optimization in WSN. *9th International Conference on Cloud Computing, Data Science & Engineering (Confluence)* (pp. 85-91). IEEE.

[39] Verma N, Pandey N and Rajpoot P, 2018, May. Energy Efficient Protocol to Supervise Overground Pipelines using WSN. *2018 3rd IEEE International Conference on Recent Trends in Electronics, Information & Communication Technology (RTEICT)* (pp. 1818-1823). IEEE.

[40] Anjali, Nidhi, Dubey K, Rajpoot P, Singh A K, KumarA and Yaduvanshi R, 2019, July. Fuzzy based Technique for Nodes Coverage with Load Balancing Data collection using Multiple Conflicting Factors. *International Conference on Communication and Electronics Systems (ICCES)* (pp. 579-584). IEEE.

[41] Dubey K, Verma K, Mishra S, Rajpoot P, Yaduvanshi R and Kumar S, 2019. Optimization of Data Collection process in Heterogeneous WSN using Fuzzy and Multi-Attribute based Load Balancing Approach, *International Conference on Advances in Electrical, Electronic and System Engineering (ICAEESE 2019)*, IEEE. (In press)

[42] Maurya AK, Dubey K, Pandey A, Rajpoot P, Mishra S and Yaduvanshi R, Relative Distance Based Fuzzy Clustering Approach for Wireless Sensor Networks, *International Conference on Advances in Electrical, Electronic and System Engineering (ICAEESE 2019)*, IEEE. (In press)

[43] Saini R, Dubey K, Rajpoot P, Gautam S and Yaduvanshi R, 2020. Lifetime Maximization of Heterogeneous WSN using Fuzzy-based Clustering. *Recent Advances in Computer Science and Communications* **13**: 1. https://doi.org/10.2174/2666255813999200824123409

Acknowledgment

This work would not have been possible without the financial support of TEQIP-III, Rajkiya Engineering College, Ambedkar Nagar.

Axial Ratio Improvement of Spiral Antenna with Resistive Loading

Moirangthem Edison Singh[1], Bhaskar S Naik[2]

[1]ECE Department, NERIST,Arunachal Pradesh, India 791109
[2] SAMEER ,Maharastra, India 400076

em@nerist.ac.in

Abstract. Spiral antenna belongs to the class of frequency independent antenna. It radiates circularly polarized wave bidirectionally. Due to the finite size of the antenna, the performance at the lower frequency degrades. This could be accounted to the reflection of the current from the truncation. Wide band characteristics from 300MHz to 3.5GHz are able to achieve through the partial resistive loading of the antenna. The lower frequency 3dB axial ratio has been improved significantly from 600MHz for the unloaded case to 200MHz for the loaded case without much effect on the efficiency. At 300MHz, the efficiency is around 60% and axial ratio at within 3dB range.

1. Introduction

The Spiral antenna has been a good candidate for the Ground Penetrating Radar (GPR) application (10MHz-2.6GHz) due to its circularly polarized radiation. The higher frequency components are radiated at the inner part of the spiral and die out finally. But the lower frequency components are radiated at the outer part of spiral. If the spiral is truncated before the current correspond to the lower frequency dies out, then it is reflected from the end. At the lower frequency, the axial ratio deteriorates due to reflection from the end. In paper[1] ,it uses lumped resistor loaded in between the adjacent arm to match the antenna but axial ratio performance has not been shown. In paper [2] the lumped resistor are loaded along the end of the arms. But its impact on efficiency has not been discussed. Wu and King[3] came up with the idea of continuous resistive loading. The current towards the end of the antenna is slowly attenuated with the help of continuously distributed resistance so that there is no reflection from the end. Later, this idea was experimentally verified by Shen[4]. Due to the difficulty in the realization of continuously varying resistance, step impedance was used instead. M Kanda[5] also developed several tapered resistive loaded antenna for picoseconds pulse measurement. Kurt L. Shlager[6] also applied this resistive loading for the optimization of the bow tie antenna. He also used resistive sheet of different sheet resistance value in succession to reduce the reflection from the end of the bow tie antenna.

IOP Conf. Series: Materials Science and Engineering 1020 (2021) 012004 doi:10.1088/1757-899X/1020/1/012004

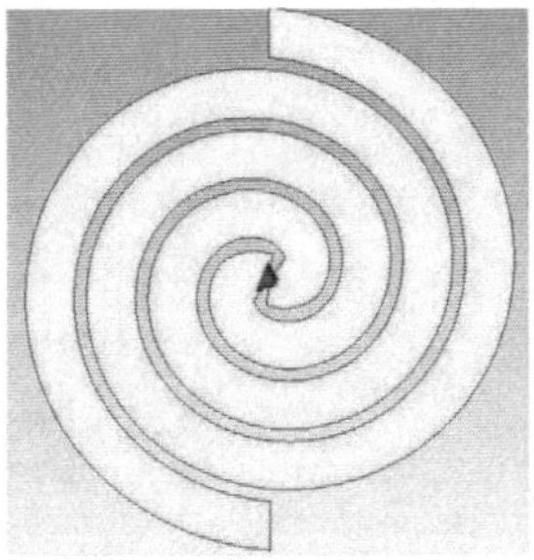

Figure 1. Unloaded Spiral Antenna

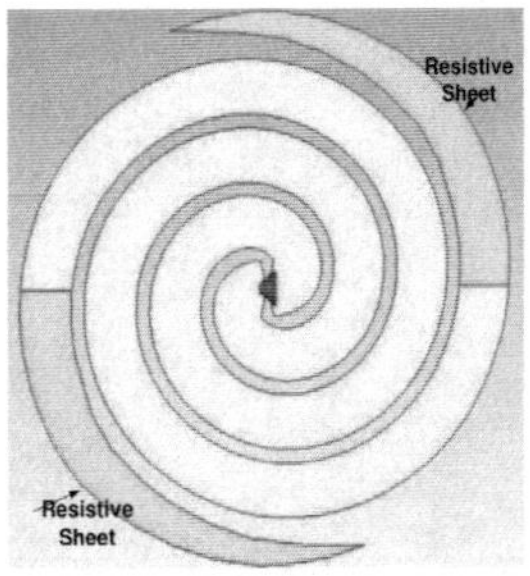

Figure 2. Resistive Loaded Spiral antenna

The challenges in the resistive loading are effect on the efficiency. Most of the papers of resistive loading, the loading is done on the whole antenna structure, which leads to a tremendous decreases in the efficiency. In order to reduce the effect of resistive loading on the efficiency, a new approach is applied. Instead of loading the whole antenna structure it is applied at the truncation only where the reflection occurs. Wide band characteristics are able to achieve through the partial resistive loading of the antenna. The lower frequency 3dB axial ratio has been improved significantly from 600MHz for the unloaded case to 200MHz for the loaded case without much effect on the efficiency.

2. Antenna Design

A spiral antenna is designed to operate in the UHF range of frequency using the equations below [7].

$$r = r_o + a\theta \qquad (1)$$

$$a = \frac{s + w}{\pi} \qquad (2)$$

Here r_o is the inner radius , a is the spiral growth rate, w is the width of the strip and s is thespacing width between the arm. The lower frequency is chosen at 200 MHz and higher frequency at 3.5GHz. The width of the metal strip is 40mm. The spacing between the arm is 10mm. The spiral growth rate is 15.9 mm/rad. The antenna structure is made up of copper metal. The size of the antenna in figure 1 is 4.2cm x 4cm. Some part of the metal part is removed and the resistive loading is done with the resistive sheet of 25 Ω/m^2 as shown in figure 2. The resistive sheet is tapered to increase the resistance along the end. The loaded and unloaded spiral in figure 1 and figure 2 are simulated in CST.

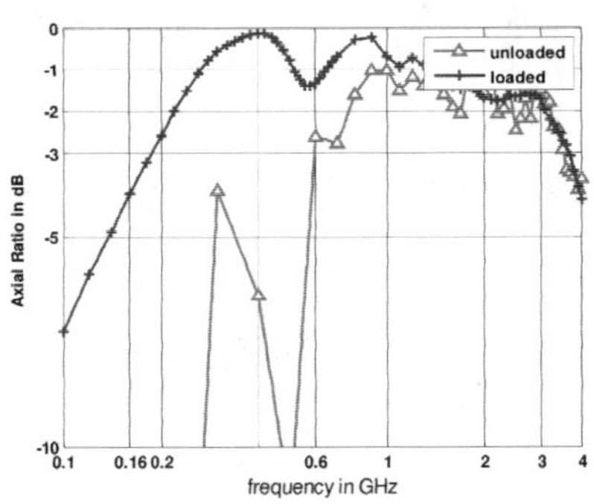

Figure 3. Axial ratio plot with frequency of unloaded and loaded antenna

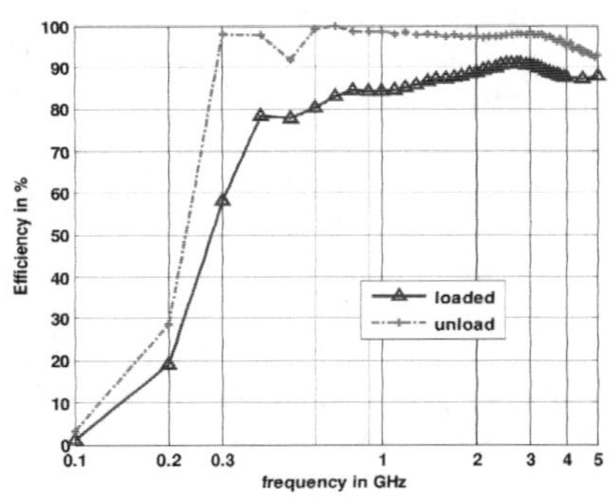

Figure 4. Efficiency plot with frequency of unloaded and loaded antenna

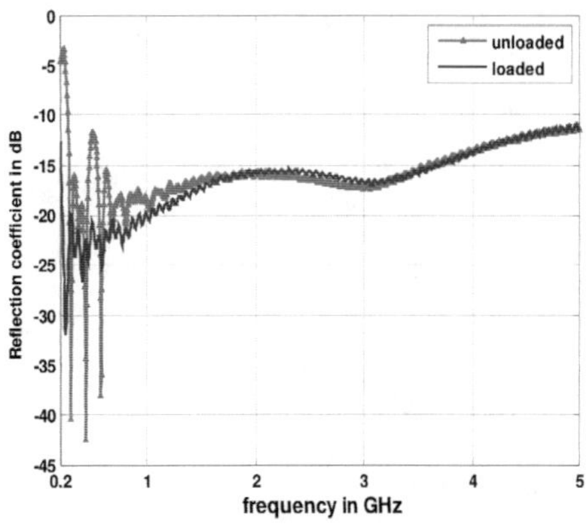

Figure 5. Reflection coefficient plot with frequency of unloaded and loaded antenna

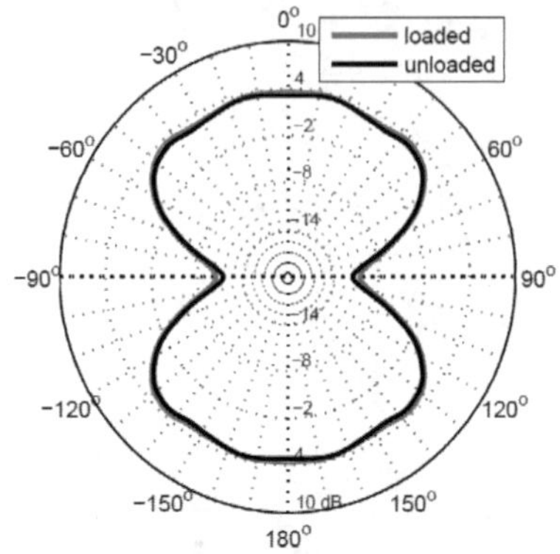

Figure 6. Radiation pattern of the antenna at 300MHz at $\phi = 0^0$ plance

3. Results

The performance comparison of the resistive loaded antenna and the unloaded antenna is compared as shown in the figure 3 , figure 4, figure 5 and figure 6. At the lower frequency the unloaded antenna performance is not desirable. It is due the reflection of the lower frequency components which could not be radiated due to finite size of the antenna. In order to avoid these reflections, resistive sheet with gradual increased in impedance replaced the terminal section of the antenna. This sheet attenuates the current which correspond to the lower frequency components thereby reducing the reflections. The 3dB axial ratio at the lower frequency has been improved from 600 MHz to 200 MHz . The effect of resistive load on the efficiency is quite less as compared to Wu King Profile where efficiency is always around 50% of the unloaded case. The impedance matching at the lower frequency in Fig.5 is improved in case of the resistive loaded antenna. The radiation pattern of the antenna is shown in Fig.6. The effect of the resistive load is minimal in radiation pattern.

4. Conclusion

A resistive loaded spiral antenna has been developed which performances cover the UHF frequency range. Its lower frequency of 3db axial ratio is improved from 600MHz (unloaded case) to 200MHz (loaded case) with resistive loading. At 300MHz, the efficiency is around 60% and axial ratio at within 3dB range.

References

[1] M. Grady and T. M. Weller, "Using resistive loading to control the radiation efficiency of a spiral antenna," *WAMICON 2014*, Tampa, FL, 2014, pp. 1-4, doi: 10.1109/WAMICON.2014.6857804.

[2] M. Zou, J. Pan, P. Li and L. Zuo, "Design and performance of a lumped resistively loaded spiral antenna," *Proceedings of the 2012 IEEE International Symposium on Antennas and Propagation*, Chicago, IL, 2012, pp. 1-2, doi: 10.1109/APS.2012.6349372.

[3] T. Wu and R. King, "The cylindrical antenna with nonreflecting resistive loading," in *IEEE Transactions on Antennas and Propagation*, vol. 13, no. 3, pp. 369-373, May 1965, doi: 10.1109/TAP.1965.1138429.

[4] Liang-Chi Shen, "An experimental study of the antenna with nonreflecting resistive loading," in *IEEE Transactions on Antennas and Propagation*, vol. 15, no. 5, pp. 606-611, September 1967, doi: 10.1109/TAP.1967.1139025.

[5] M. Kanda, "A relatively short cylindrical broadband antenna with tapered resistive loading for picosecond pulse measurements," in *IEEE Transactions on Antennas and Propagation*, vol. 26, no. 3, pp. 439-447, May 1978, doi: 10.1109/TAP.1978.1141853.

[6] K. L. Shlager, G. S. Smith and J. G. Maloney, "Optimization of bow-tie antennas for pulse radiation," in *IEEE Transactions on Antennas and Propagation*, vol. 42, no. 7, pp. 975-982, July 1994, doi: 10.1109/8.299600.

[7] J.Kaiser,"The Archimedean Two-Wire Spiral Antenna"IRE Transactions on Antennas and Propagation, Vol.AP-8,no. 3,pp. 319-325,May 1960

IOP Conf. Series: Materials Science and Engineering 1020 (2021) 012005 doi:10.1088/1757-899X/1020/1/012005

Analysing Benefits and Challenges of Information Communication Technologies: A Study on Nagaland

K.D. Vizo[a], M. Mall[b], R. M. Pant[c]

[a,b,c] North Eastern Regional Institution of Science and Technology, Nirjuli, Arunachal Pradesh, India

Email: vizo123@gmail.com

Abstract: This paper examines the status of Information Communication and Technology (ICT) in the state of Nagaland by using well-structured questionnaires and secondary data. The study had analysed the benefits and challenges of ICT covering the entire state. It is observed that although ICT had picked up well, there are many places especially in rural areas where ICT usage is still very low. There is no doubt that the use of ICT is contributing a lot to the uplift of socio-economic of the people but there is also a necessity of further improving the electronic public service delivery to citizens in a transparent, efficient, cost-effective and easy manner. In general, it is seen that the younger generation appears to be much more in favour of engaging ICT tools in their day to day work and while framing ICT policies more focus have to be given on the lower age groups so as to improve trust between citizens and government and between citizen and various institutions. The study also reveals that sincere efforts have been made to use the latest state of art technologies to take the state of Nagaland to the next higher level and the state is now fully aware of the benefits and challenges of ICT. The findings of this study shows that placing more emphasis on ICT can significantly speed up the overall inclusive socio-economic growth of the state and therefore more initiatives have to be taken to strengthen ICT so as to bring quick changes in governance and all other spheres of life.

Key Words: ICT, Nagaland, Benefits of ICT.

1. Introduction

Today, ICT is indispensable in all spheres of national life, in economic life, social and cultural development (Ali, 2020). ICTs can be used to interlink information technology devices such as personal computers with communication technologies such as telephones and their telecommunication networks (Akram et al.,2017). Bindu (2019) have defined ICTs as a range of electronic technologies which when converged in new configurations are flexible, adaptable, enabling and capable of transforming organizations and redefining social relations. ICT is rapidly changing our lives, the way we do business (Gaudio et al., 2020), access and communicate information with each other (Choudhury et al., 2019). Every government motto is to serve the citizens and ICT is now being effectively used by the government (state or central) in providing services to the citizens. Therefore, on the premise of citizen-centric approach-where citizens are being treated as customers, ICT enables the government to provide improved service quality, which in turn enhances citizen satisfaction. Citizen-centric service involves designing of services from user's point of view and the benefits have to be directly visible and translated into economic and social gains of the citizens. Thus, ICT can be used as a tool in enabling the government to reach all the citizens and act as a catalyst in development and apart from this, ICT can be used significantly as a means in combating poverty and fostering sustainable development through creating information-rich societies and sustainable livelihoods (Gangopadhyay et al. 2019). As per Jun, (2010), in order to develop a state especially in rural areas, we need to focus on poverty alleviation (Karanasios, et al., 2019), better livelihood (Kompella, et al.2017), improvement of basic infrastructure facilities and promotion of agricultural sector where rural economy is heavily dependent. Lee-Geille, (2019) emphasised upon the rising awareness amongst the citizens on government schemes and services, the demand for better delivery of services is increasing day by day and the time has now come to infuse Information and Communication Technology (ICT) to play a prominent role in strengthening such demand for better delivery of services. Information Technology (IT) today is becoming as important as water, bread, and air (Lalrochunga, et al., 2020). In the early part of 20th Century, people used to believe in the secrecy of information but in this 21st Century, the concept is totally reversed (Koutroumpis, et al., 2020). Now people like to share information and thus Information Technology is emerging as a powerful tool to improve socio-economic development. Experts in this field confidently predict that in future, poverty line will no longer be measured in terms of money, but in terms of information (Ghosh, et al., 2020). Therefore, emerging Information technologies can possibly be used to reduce information gap between the information rich and information poor to enhance productivity and improve quality and efficiency for all-round socio-economic development (Lee-Geille, 2019).

1.1 Objective

The following are the main objectives of this study:

- To analyse the present awareness level and benefits of Information Technology and Citizen Centric Services in the state of Nagaland.
- To study the possible role of Information Technology and other citizen centric services in improving socio economic development of the state.
- To ascertain the factors determining the role of Information Technology in socio economic transformation in the said region.

To suggest suitable policy measures to make Information Technology and Citizen Centric Services an effective tool for Socio Economic development of the study region.

1.2 Cooperative Farming and Renewable Energy

Both primary and secondary data has been used and the secondary data was collected from the publications of various organizations, and published journal papers of relevant topics. It is not possible to fulfill all the objectives of the study through secondary data alone and therefore, to fulfill all the aspects of the study, primary data have been collected by undertaking field study through two well-structured questionnaires and the status, benefits and challenges of ICT have been collected. For this study, the sample survey was conducted following random sampling. In this study, basic citizen-centric IT related services were selected for sampling survey. From each of this basic citizen-centric, IT related services, data has been collected from all districts in such a way that it captures the basic citizen-centric IT related services with variations and economic realities of the survey area.

The Data was collected through the common service center (CSC) District Managers, who are dealing with both urban and rural citizens on Citizen-centric related matters in each District. Data thus collected were processed, tabulated and analyzed using simple statistical tools.

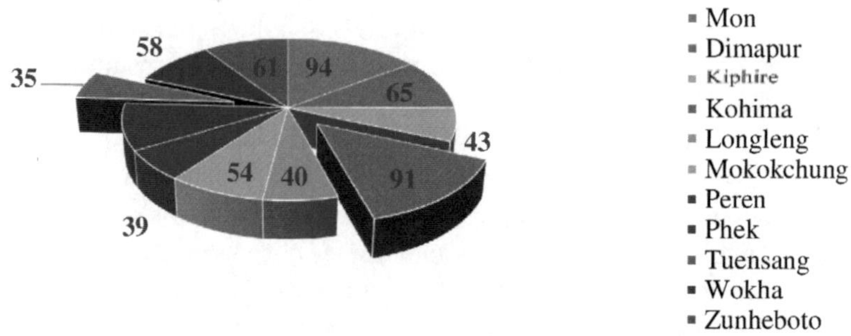

Fig 1.a :District wise sample survey of Nagaland state (Phase -I)

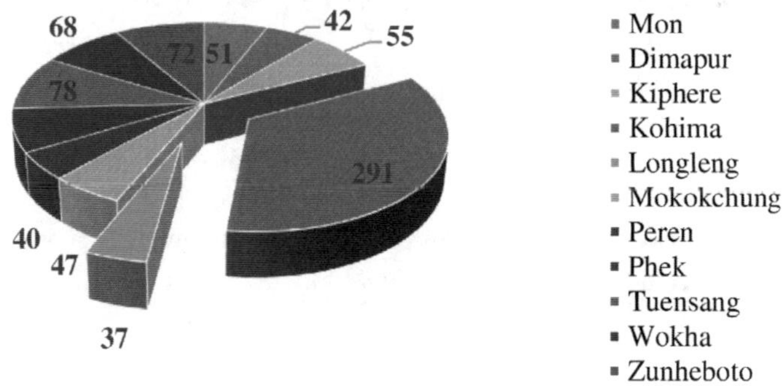

Fig. 1.b: District wise sample survey of Nagaland state (Phase -II)

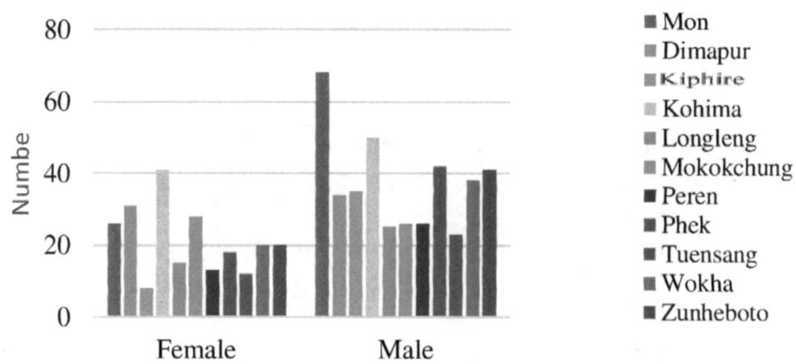

Fig. 1.c. District wise survey participation of Nagaland (Gender based)

The study of socio-economic scenario, the present awareness levels, benefits and challenges of ICT in the state of Nagaland has been studied. The survey data has been collected from all the districts of the state in two phases, the respective respondents is presented in graphical format in fig 1.a and fig 1.b. Similarly, the respondents were also maintained the gender equity while collecting the data from each districts of the state (fig 1.c).

2. Current Status of ICT in the state of Nagaland

In the state of Nagaland, e-governance is slowly growing and the Department of IT&C have started promoting effective use of e-governance tools. It is clear that the state of Nagaland is facing a set of complex, multi-faceted and interdependent challenges which includes challenges like overcoming poverty, inequality, restoring peace, improving security, improving health, roads, education systems and so on. However, there are no single actor—let alone single departments or units that can effectively deal with such challenges on their own and therefore effective collaboration among various departments/agencies across all levels of government is essential. As it is, with government and non-governmental actors, to ensure good governance and good development outcomes for the state of Nagaland, collaborative e-governance underpinned by a well-functioning public administration, is crucial to improving people's lives in the state. Herein, all the government departments must deliver, equitably and efficiently, essential services that meet citizen needs, provide opportunities for economic growth, as well as facilitate citizen engagement and participation in public policymaking and service delivery, so as to promote the empowerment and well-being of all the people in the state. In general, the penetration of ICT and citizen-centric services has not reached the national level and there is also a disparity in the penetration in urban and rural areas. Also, some districts have better penetration than other and therefore in this study, all the 11 districts were taken into consideration and it was ensured that Responses were collected from both urban and rural areas.

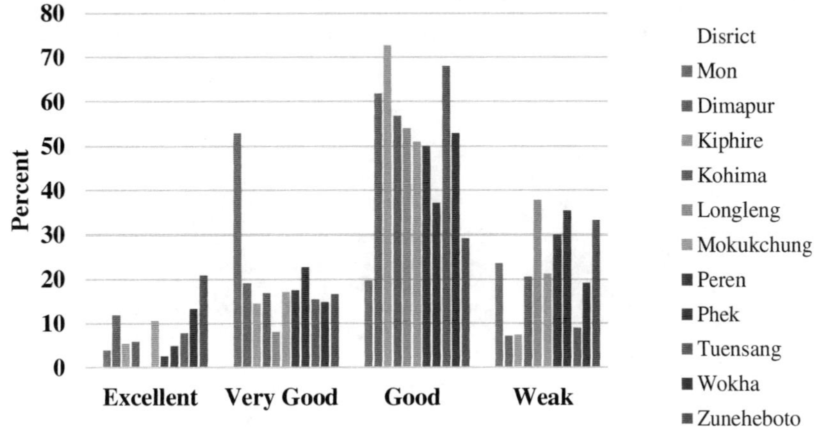

Fig.2.a: ICT Knowledge District wise

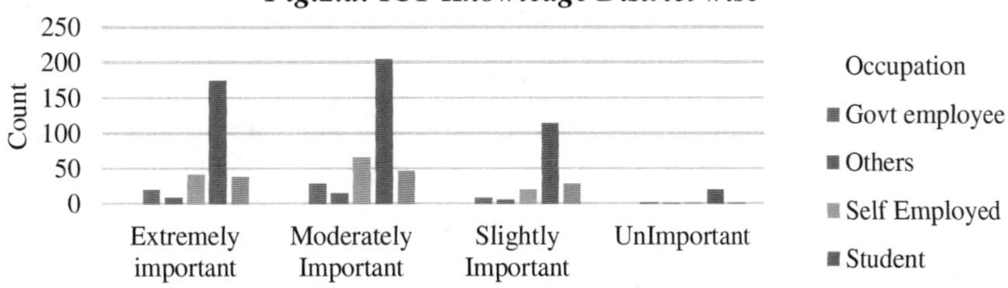

Fig. 2.b:Importance of ICT based on Occupation

The questionnaire was surveyed regarding the preference status amongst the citizens of the different districts of Nagaland. The preference was recorded on ordinal scales. The two parameters: ICT knowledge, and importance of ICT of different occupations were measured based on different questions from diverse segment of citizen. Fig. 2.a gives snapshot about the ICT knowledge of different districts. On the ICT knowledge, very few respondents given opinion about weak ICT knowledge. Majority have agreed upon having good ICT knowledge. Amongst all districts it was found that district like Mon, Dimapur have outperformed in this ranking. Similarly, fig 2.b has focused on importance of ICT knowledge over various occupations summarised.

It has also reflected very nominal group of people with negative response importance of ICT. However, amongst all occupations, students have prioritised their importance of ICT as compared to all other occupations. Professionals and self-employed personnel have also shown a positive trend in accepting ICT in their occupations. The view of all other occupations have also shown inclination towards adopting ICT knowledge in their fields. ICT has been perceived as indispensable tool for the holistic development of the individual citizen. Moreover, due to interdependency nature also they are also getting motivated in adopting the modern tool.

3. Benefits of ICT in the context of Nagaland

Palvia, et al., (2018) have emphasised upon the role of ICT in socio-economic development of the region. It can help in new process of manufacturing, social bondage, as well as in the distant learning for bridging the knowledge gap. Especially, utility services like the banking facility have been improved a lot due to introduction of ICT (Gaudio, et al., 2020). Financial transaction, and other supporting elements of production and distribution also foster the scope of entrepreneurship (Millán, et al.,2019).

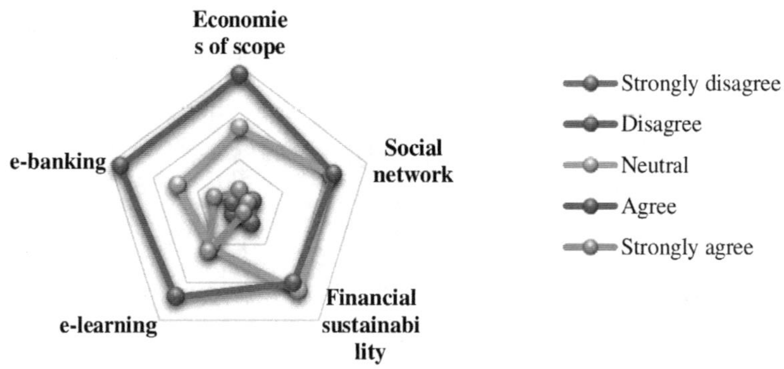

Fig. 3: Survey summarized results of Benefits of ICT in Nagaland

To measure the benefits of the ICT in the state of Nagaland, five different parameters were considered: economies of scope, social networking, financial sustainability, e-learning, and e-banking. The responses were recorded in the five scale tool, with lowest one as 'strongly disagree', and best one is 'strongly agree'. Majority of the respondents as shown in fig.3, has .given consent of 'agree' in all the five factors of benefits. However, financial sustainability and social bondage still need attention The respondents have mostly limited their response either 'agree' or 'neutral'. Respondents further negligibly preferred to 'disagree', 'strongly disagree', as well as 'strongly agree'. Hence, further confidence build up would definitely result in citizen migrating towards the preference of 'agree' towards 'strongly agree'.

4. Challenges of ICT and Initiatives of Nagaland government

Meijer, (2015) highlighted various barriers in adopting ICT as a result of which the full benefits of the e-governance is not harnessed by the citizens. The barriers of the ICT adoption is region specific, hence the appropriate appraisal of the scenario can be helpful for any government to overcome such hurdles (Moghaddam, et al., 2013). The socio-economic development in the present era is dependent on the customized adoption of ICT (Tambotoh, et al., 2015). Poncian, (2020) has also emphasized the role of ICT for the citizen empowerment in implementing all the schemes. In the developing country like India also emphasized in the different process of overcoming the same (Rao, 2009).

Fig4.a :Dr Arvo Ott, Executive Director, e-governance Academy of Estonia delivering lecture during Training on e-governance with state Departmental IT Nodal Officers.

Fig4.b :Dr Lüa Hanni, former Minister and Chairman Constitution Drafting Committee of Estonia delivering lecture on e-governance to state Cabinet.

Fig 4.c:Shri T.R. Zeliang, Hon'ble Chief Minister, flagging off eMarathon on 3rd July 2015 during Digital India Week.

Fig 4.d: Shri Pankaj Kumar, IAS, Chief Secretary, Nagaland during the launching of Biometric Attendance System for Nagaland Civil Secretariat.

Fig4.e: Students and participants during the 1st e-Naga Summit, 2015 at NBCC Convention Centre Kohima.

Fig.4.f:Hon'ble Minister, IT&C, Chief Secretary, Nagaland, Director General STPI, Commissioner & Secretary, IT&C during the signing of MoU for setting up of STPI at Kohima, Dimapur.

The department of IT&C of Nagaland has introduced many initiatives to overcome the same. The one point solution has been devised through the state portal www.nagaland.gov.in. The electronic service facility of the state is rendered by Common Service Centre (CSC), e-district project. The department has performed some of the benchmark activities as listed below to boost up the confidence amongst its citizens:

Formulating State IT Policy.

- Inviting world leading key persons e-governance institute of Estonia for successful imparting of training (fig 4.a).
- Besides government officials, state government ministers were also trained by the e-governance Academy of Estonia (fig 4.b).
- State government proactive in implementing flagship 'digital India' program of union government (fig 4.c).
- Initiation of biometric attendance system for civil secretariat of the state government (fig 4.d).
- To popularize ICT among school students e-naga summit organized (fig 4.e).
- Setting up of software technology park of India (STPI) in the state (fig. 4.f).

5. Conclusion

It is observed that the growth of ICT and IT enabled services are still in initial stage in Nagaland and if necessary thrust is given it has all the potential of growing at a very fast pace. In general, it is making considerable impact on the lives of the people and people feels that the use of ICT is resulting to an impartial delivery of services. The general citizens are of the opinion that there is more accuracy in IT enabled services and the cost of service delivery is getting reduced. It is observed that there is considerable reduction in the cost of obtaining civic and other services and more ICT intervention can go a long way in minimizing the cost of availing citizen centric services. In general ICT initiatives are increasing service delivery and productivity and the use of ICT tools will improve job satisfaction. The citizens wants the government to invest more in IT enabled services to improve their overall livelihood and there is lot of scope for further research in this area.

References

[1] Akram, W., Kumar, R., (2017) "A Study on Role and Applications of ICT in Development of Rural Areas," *IJSRM*, vol 5, issue 8, pp. 6758-6763.

[2] Ali, M.S. B., (2020). Does ICT promote democracy similarly in developed and developing countries? A linear and nonlinear panel threshold framework. *Telematics and Informatics,* 50, 101382.

[3] Bindu, N., (2019). From conventional governance to e-democracy: Tracing the evolution of e-governance research trends using network analysis tools. *government Information Quarterly*, 36(1), 385-399.

[4] Choudhury, S., Parida, A., Pant, R.M., Chatterjee, S. (2019). GIS augmented computational intelligence technique for rural cluster electrification through prioritized site selection of micro-hydro power generation system. *Renewable Energy*, 142, 487-496.

[5] Director of Census Operations, Nagaland - Nagaland Census 2011.

[6] Directorate of Economics & Statistics, government of Nagaland. Nagaland Economic Survey 2014-15.

[7] Directorate of Economics & Statistics, government of Nagaland. Statistical Handbook, 2014

[8] Department of Planning & Coordination, government of Nagaland. UNDP Report 2013 for Dimapur District.

[9] Gangopadhyay P. K., et. all. (2019). Spatial targeting of ICT-based weather and agro-advisory services for climate risk management in agriculture. *Climatic Change*, 154:241-256, doi.org/10.1007/s10584-019-02426-5.

[10] Gaudio, B.L. D., & et al (2020) How do mobile, internet and ICT diffusion affect the banking industry? An empirical analysis. *European Management Journal*, in press.
[11] Ghosh, S.K., Das, A.K., Choudhury, S., (2020). ICT interventions in rural development schemes: A case of Rupshi development block, Assam, *2020 8th IEEE International Conference on Reliability, Infocom Technologies and Optimization (Trends and Future Directions)(ICRITO),* pp. 349-354.
[12] Jun, X., (2010). Linking ICTs to rural development: China's rural information policy. *government Information Quarterly* vol. 27, No.2, pp. 187–195, 2010
[13] Karanasios, S., & Slavova, M. (2019). How do development actors do "ICT for development"? A strategy-as-practice perspective on emerging practices in Ghanaian agriculture. *Information Systems Journal*, 29(4), 888-913.
[14] Kompella, K,. (2017). E-governance systems as socio-technical transitions using multi-level perspective with case studies. *Technological Forecasting & Social Changes*, 123, 80-94.
[15] Koutroumpis, P., & et al., (2020) Small is big in ICT: The impact of R&D on productivity. *Telecommunications Policy*, 44 (1), 101833.
[16] Lalrochunga, D., Parida, A., Choudhury, S., (2020). Sustainability of UAVs in developing countries: Prospects and challenges. *Journal of Discrete Mathematical Sciences and Cryptography*, 23 (1), 237-248,
[17] Lee-Geille, S., (2019). Using government websites to enhance democratic E-governance: A conceptual model for evaluation. *government Information Quarterly*, 36(2), 208-225.
[18] Meijer, A., (2015). E-governance innovation: Barriers and strategies. *government Information Quarterly*, 32(2), 198-206.
[19] Millán, J.M., & et al.(2019) 'Digital divide' among European entrepreneurs: Which types benefit most from ICT implementation? *Journal of Business Research*, in press.
[20] Moghaddam, B. K , Khatoon-Abadi, A., (2013). Factors affecting ICT adoption among rural users: A case study of ICT Center in Iran. *Telecommunications Policy* vol.37, pp. 1083–1094.
[21] Palvia, P., Baqir, N., & Nemati, H., (2018). ICT for socio-economic development: A citizens' perspective. *Information & Management*, vol. 55, issue 2, pp. 160-176, Mar. 2018.
[22] Poncian, J., (2020) ICT, citizen engagement and the governance of extractive resources in Tanzania: Documenting the practice and challenges. *The Extractive Industries and Societies*, in press.
[23] Rao, S.S., (2009). Achieving millennium development goals: Role of ICTS innovations in India. *Telematics and Informatics,* 26(2), 127-143.
[24] Tambotoh, J. J. C , Manuputty, A. D, & Banunaek, F. E., (2015). "Socio-economics Factors and Information Technology Adoption in Rural Area," *Procedia Computer Science*, vol. 72, pp. 178–185.

Various Methods of Link Design for Transferring Data in Wireless Sensor Networks for different Applications

Richa Tiwari[1] Rajesh Kumar[2]

[1]Assistant Professor, Dept. of ECE, Vishveshwarya Group of Institution Gr.Noida(U.P) Indai.
[2]Associate Professor, Dept. of ECE, North Eastern Regional Institute of Science &Technology, Nirjuli, Itanagar, Arunachal Pradesh, India

Email: richavgi@gmail.com/rk@nerist.ac.in

Abstract. This paper offers the different type transmission schemes used in wireless sensor networks. The wireless sensor network used in a number of applications such as Agriculture, Military, Medical, Multimedia etc .According to the applications there are different quality of service parameters and designing requirements for different layers in network design. While designing any ad hoc network or sensor network many designing principals and challenges are taken into consideration. For the designing of protocols layering in ad hoc sensor network basically a five layer model is taken into consideration. These layers are application, transport, network, Medium access control and physical layer. In this paper different types of designing constraints associated with physical and medium access layer has been introduced for different applications in sensor networks. In that manner this paper helps us to select the designing of a link for particular application.

Keywords: bit error rate (BER);Quality of service(QoS);wireless sensor networks(WSNs);error control schemes(ECS); packet error rate (PER); forward error correction (FEC); Amplitude shift keying(ASK) modulation, Binary Phase shift keying(BPSK);signal to interference plus noise ratio(SINR);orthogonal space time block code(OSTBC)

1. Introduction

Wireless Sensor Networks (WSNs) is a combination of tiny nodes with sensing, computational and short distance transmission capability. Sensor network have potential for practical deployment in both consumer and military applications. These networks have a wide range of applications in target tracking or gathering intelligent information. Now a day's biological sensors are also very popular in medical fields. Home based applications for monitoring electricity, gas and water remotely via sensor network are also very popular in market. The link design more or less related to the application for which sensor network is going to be installed in particular area. In this paper the main focus is on the designing of physical layer and MAC layer protocols. Most of sensors are deployed with non- rechargeable battery so Energy optimizations are the challenging task in wireless sensor network design. There are many factors that are taken into consideration while calculating the life time of any ad hoc sensor network, Nodes have fixed transmit power, the link signal to noise ratio between two communicating nodes will typically decrease as the distance between them increases. Link signal to noise ratio or interference ratio determines the communication performance of any link. The energy consumption shall achieved at all layers of network design, mainly the power control depends on two layers, the physical layer and the MAC layer in addition of that it is also a part of network layer when we talk about resource allocation .The physical layer of wireless sensor network deals with transmission of bits or packets over wireless link. The Quality of service parameter that governs the quality of that link is data rate and bits or packet error rate that is supported by that layer. The designing parameters that are taken into consideration while designing the physical layer are the modulation, coding, diversity, spread spectrum etc. There are many ways to improve the bit error rate performance of any network such as multiplexing, diversity and error control coding. These techniques are employed for achieving high data rates low bit error rates. Multiple antennas are used to provide diversity in fading and increase the data rate of the link. The Diversity gain will reduce the PER leading to fewer retransmission and multiplexing will increase the link rate, which reduce the congestion and delay on the link. The advantages of diversity with high data rate are also achieved by using OSTBC coding.

After physical layer the Access layer designing comes into consideration this layer controls how the user share the available spectrum this layer control the functionality in terms of sharing of channel when the users are not accommodated by the system then this layer deny the access to users ,this layer also responsible for retransmission for providing error free communication. The link SINR varies randomly over the time this may leads fading and interference. The flexibility in link connectivity that comes out from varying link parameters such as power and data rates has major implications for routing. However it is much more challenging to support high data rates and low delays over a multi hop sensor network so this is one of the main difficulties in using an ad hoc wireless sensor network to support multimedia applications that require a high data rate and low delay. In this paper a physical layer link and MAC protocols has been proposed for sensor network. Various techniques for reducing bit error rate and achieving coding gain are also proposed because by achieving the good coding gain the life time of sensor nodes is increased that is the major quality of service parameter for sensor network.

The remaining paper is organized as follows: In Section 2, we describe briefly the various types of proposed transmission methods and MAC Protocols available in the literature that are used in ad hoc sensor networks. Section 3, transmission and link plan with MAC protocols have been proposed for different applications. In Section 4, results have been analyzed in terms of bit error rate. Final concluding remark of the paper is presented in section 5.

2. Related Works

Wang,et.al.(2001)[1],introduced a classical approach to achieve a certain BER at the low signal to noise ratio(SNR) by introducing forward error correction (FEC) in the communication. There are many error correcting codes available for wireless body area networks such as by Petrson,et.al.(2007)[2],In ARQ, the packet should be retransmitted by transmitter again and again if the error has been occurred at receiving side.In Proakis,et.al (2012)[3],for achieving the high SNR it is better to use higher order of modulation whereas the quality of same communication link become poor for lower order of modulation to achieve the same value of BER. In February 2012[4], IEEE80215.6 becomes a standard for WBANs where many ECS for WBANs is introduced. In a survey of wireless body area network by Ullah, et.al.(2012)[5],the power consumed by implant and wearable sensors should be very low because these devices are battery operated and replacement of these devices are very complex medical procedure .Another challenge which are faced in WBANs is the interference cause by the presence of other networks (Barakah& Ammad,2012)[6].A Low duty cycle XT-MAC protocols was introduced by Quan Nguyen-Trung, et.al.(2014)[7], for target tracking in wireless sensor network. This XT-MAC protocol is associated with EMRP routing algorithm to enhance system performance in terms of end-to end delay and energy efficiency. The tracking accuracy should also maintain throughout the cycle. Te-Li Wang,et.al.(2014)[8],introduced a CBA-EVT : a traffic adaptive energy efficient MAC protocol for wireless sensor networks in this both delay and energy consumption are taken into consideration while calculating the sleep schedule.Pei Huang,et.al.(2015)[9],introduced the RC-MAC protocol that integrate duty cycling and receiver centric scheduling.

There are several types of decoding algorithms also exists such as syndrome decoding, maximum likelihood (ML)decoding, viterbi algorithm using trellis and maximum a posteriori (MAP) decoding out of these all decoding methods the best suited method for wireless body area network is MAP decoding with the BCJR algorithm that is used in Turbo codes and belief propagation(BP) by Pearl,et.al.(2014)[10].In 2015. The GMSK modulation schemes is analyzed by Rajoua,et.al.(2015)[11] for coded and uncoded system then it is found that GMSK technique is good for wireless sensor network because of its good BER performance and power efficiency. Praveena,et.al.(2017)[12],calculated the energy efficiency of Raptor code and compare that with BCH code for WBANs in different type of fading channels such as Nakagami, Rayleigh and Rician .In this the authors take the distance between nodes greater than 10m. Ahmed,et.al.(2018)[13] connect two WBANs using UMTS here the medical data uses the adaptive convolutional code and UMTS code and for non-medical data use the UMTS code only for decoding purpose the soft decision decoding and viterbi algorithms is used. In Rajan Kadel,et.al.(2018)[14],the main parameters considered in the performance judgment of any wireless sensor network is Bit error rate(BER),packet error rate(PER),packet acceptance rate(PAR),received signal strength(RSS) and throughput efficiency etc.Ning Ma,et.al.(2019)[15],give a distributed coding scheme of multimedia data compression algorithm for wireless sensor networks. This algorithm is based on gradient-domain ROI which enhance the coding efficiency of the severe motion region and improve the decoded image while reducing the code rate and quality, ultimately reducing the code rate and quality.

3. Materials and Methods for Link Design

Wireless sensor network works in many environments such as military applications where temperature is very low, underwater applications where the water proof arrangement is required and underground applications where the chances of signal loss is very high another aspect of designing is to design a network for a particular work such as wireless multimedia sensor networks which sense audio, video and image. Another example is wireless body area network where wearable and implant sensors inside the human body should be used for monitoring human health issue .so according to the task and environment the selection of transmission strategies and MAC protocols are given in the next section one by one.

3.1. Modeling of biological Sensor network for medical applications

In wireless body area sensor networks basically there are two functions for nodes sensing and transmission. The modulation order also plays an important role in the quality of the physical layer of any communication link. In this type of sensor network the sensors used are basically of digital type and the output of these sensors is in the form of bits so the requirements of DAC and ADC are not here. In wireless body area networks there are three types of nodes (Rajan kadel,2019): implant node, body surface node and external node .The major criteria while selecting an ECC (error correcting codes) and modulation strategies for this network is life time of implanted and wearable sensors because these devices are battery driven and when battery becomes weak then the replacement of this is a complicated surgical procedure in these medical situations. In addition of that the bit error rate and minimum delay in transmission of data also played an important role in these sensor networks. In this section a transmission and reception strategy for this network has been proposed. The proposed network architecture is shown in Figure 1, which shows that data are coming from multiple sensor nodes directly in contact with human body these sensor nodes sensing the data and the data from these sensor node is received in the special modules at discrete time intervals after that various channel coding procedures and modulation processes are performed on data according to the requirement of transmission applications.

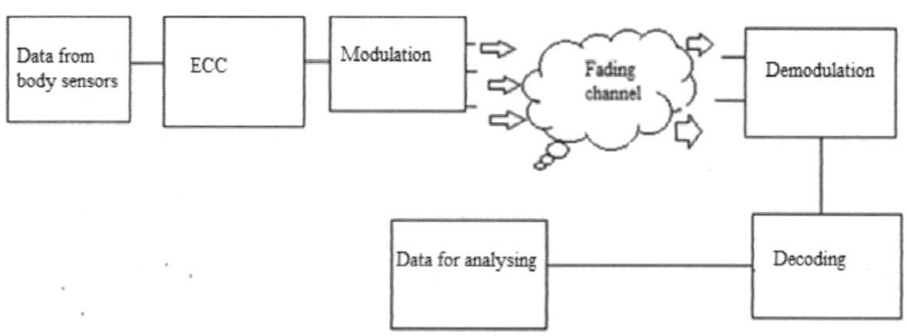

Fig. 1 Data transmission from source to destination in Biological sensor network

These networks are comes in the category of Event driven wireless sensor network and usually operate in light traffic load condition. The Receiver -centric MAC protocol is best suited for that type of biological sensor network. When an event is detected, a large number of packet is generated this protocol easily adapt two situation .To handle bursty traffic triggered by an event this protocol underlying the data gathering tree structure of WSNs and the multichannel technique supported by IEEE802.15.4.Due to the high throughput associated with that protocol the energy efficiency is also improved. This protocol helps us to design multichannel.

3.2. Modeling of Wireless Multimedia sensor network link for multimedia applications

In this type of sensor network the sensors are in the form of cameras and microphones to collect video and audio data. There are a number of limitations associated with wireless sensor network such as battery life time, computational, storage and communication capability of nodes. The quality of service requirement (QoS) of this network is high bandwidth for storage, multimedia information fusion, synchronization between events and device. The requirement of any sensor network is the deployment of a large number of sensor nodes for covering a monitoring area .According to an article on WMSN[15],the network is divided into three levels the function of level one nodes is to collect the environmental information. The function of level two node is to aggregate the information and process that aggregated information and after that the information is transmitted to the level three, the function of level three is to control the overall functioning of the network the high data rate is the major requirement of WMSNs if we focused on level two nodes that are responsible for overall transmission of data to the final destination the by applying OSTBC coding at physical layer level at this stage of transmission then the data transmission rate increase and bit error rate performance also been improved. The figure2,shows the physical layer implementation of that application.

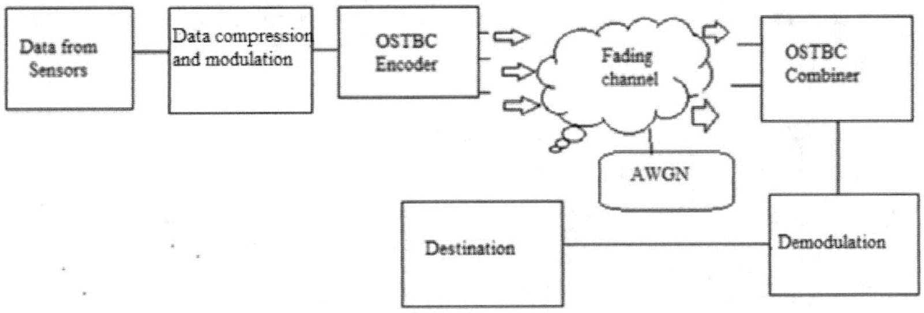

Fig. 2 Data transmission from source to destination in Multimedia sensor network

CBA-EVT protocol at MAC layer is best suited for that type of sensor networks. CBA-EVT uses the cost benefit analysis considering both delay and energy consumption for determining the sleep schedule inside the network. This protocol is designed for a wide range of traffic load condition. This is very effective in case of energy savings and used in wide range of online network condition. This is a receiver oriented approach in which each sensor adjust its time slot when to walk up for packet reception. Each sender node walk up in light of receiver's schedule and goes up to sleep after transmitting all packets to the receiver.

3.3. Modeling of Wireless area sensor network link for security applications

These networks are installed in area for monitoring or tracking unwanted targets. In that data or information is transferred to the destination. That destination is not far away from sender. Long distance transmission is not required in that situation because the aggregated data is transferred to the central hub that are equipped with internet capability so this center node transfer data to the destination from where the controlling action should be taken. The Low duty cycle XT-MAC Protocol is best suited for that application .The designing of target tracking system is need to deal with several factors such as energy consumption, tracking efficiency and processing capability of sensor node. This protocol is used with EMRP routing algorithm to enhance the system performance in terms of end-to-end delay and energy consumption with the management of tracking efficiency.

4. Simulation Results

In this section, the bit error rates of proposed transmission links are discussed. Set of architectures are evaluated in terms of signal to noise ratio and bit error rate. The performance of these transmission network architectures are discussed in the following subsections. A BPSK modulated system with error control coding is considered and performance of system is compared with other transmission methods. For simulation a random binary data is created.

All the parameters used in simulation are shown in Table 1.

Table 1- Simulation Parameters

Type of Channel	Rayleigh Fading Channel
Type of modulation	BPSK
E_{amp} (J/bit/m)	10^{-10}
E_{elec} (nJ/bit)	50
E_b/N_0(dB)	0 to 20 dB

4.1. Bit Error rate performance for biological sensor network transmission

Fig. 3, shows the bit error rate calculation of the transmission link for biological sensor network. When distance is small(less than 40 m) then the difference between energy per information bit for coded and uncoded system is also very less but after that as the distance of transmission increase that difference increases by using Error control codes in the link energy per information bit is reduced so life time of link increase, this reduction is just because of coding gain.

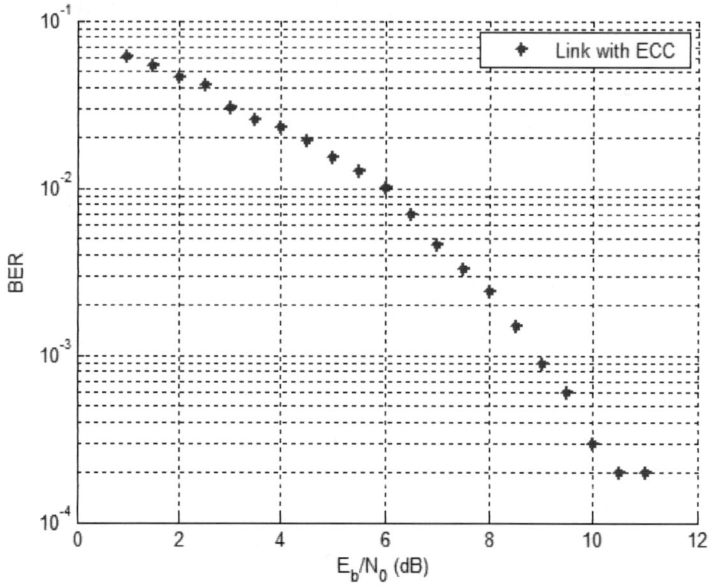

Fig. 3 Bit error rate calculation of link with error control coding

4.2. Bit Error rate performance for multimedia sensor network transmission

Fig. 4, shows the bit error rate calculation of OSTBC transmission that are proposed to use in paper for the transmission of multimedia data .Multimedia data is compressed by using various transformation techniques after that to achieve the high data rate that compressed data is transmitted by OSTBC coding.

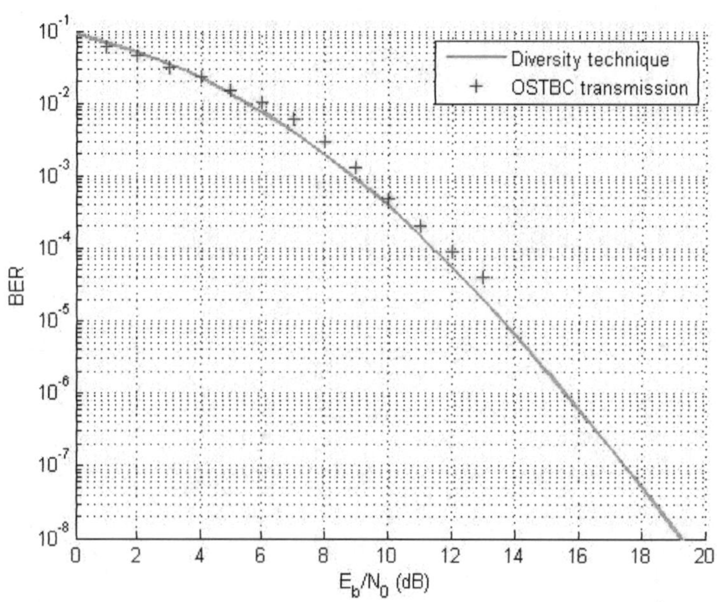

Fig. 4 Bit error rate calculation of link with OSTBC transmission

4.3. Bit Error rate calculation of different transmission methods used in WSNs

For medical field applications, Bit error rate also played an important role. In sensor network for the transmission of signals the transmission strategies are selected according to the requirement of environment and the required accuracy of data. In that section figure5, shows the bit error rate calculation of different transmission methods for WSNs. In this the bit error rate of DSSS(Direct sequence spread spectrum)with ASK modulation, BPSK with diversity order 1 and soft decision decoding with BPSK is shown .Results showed that as the complexity of system increase the bit error rate performance is improved.

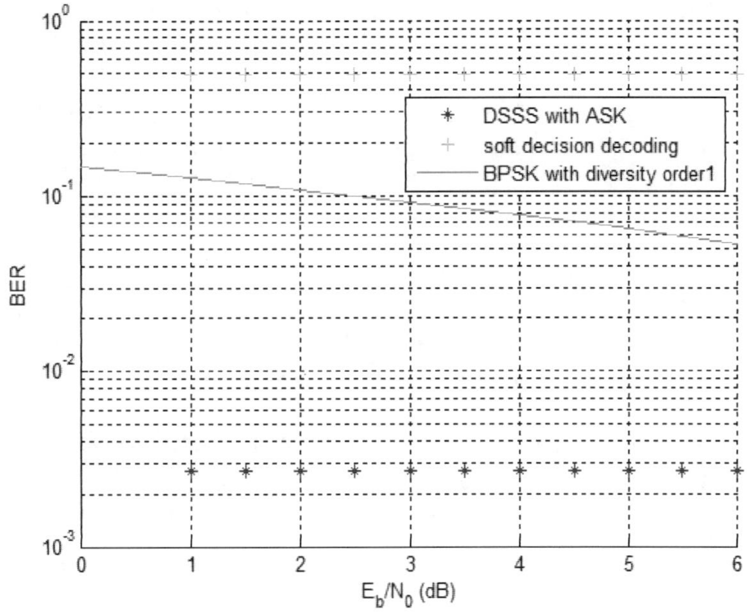

Fig. 5. Different transmission methods for WBASNs

The Performance Comparison of all transmissions methods are collected in Table 2.

Table 2-Comparitive Performance

Mode of transmission	Transmitted bits	Total Error bits	BER	Performance
DSSS with ASK	2000	976	0.4895	Poor BER Performance but complexity is less
Coherent BPSK with diversity order 1	10000	1464	0.1464	Good BER system complexity depends on diversity order
BPSK with soft decision decoding	$3.616x10^4$	101	0.002793	Very Good BER performance but complexity is high

5. Conclusion

In this paper a physical link design and corresponding MAC protocols are proposed for the transmission of data in case of different wireless sensor network applications. In this the bit error rate performance of different schemes are analyzed .every application has their own quality of service requirement. So in this paper we tried to cover maximum possible transmission and MAC protocols combinations. The error control schemes used in biological sensor network provided a wide range of data rate and coding gain. According to requirements of data rate the transmission method should be selected and coding gain helps us to reduce the circuit energy consumption.

References

[1] Wang, C.; Sklar, D.; Johnson, D. Forward error-correction coding. Crosslink 2001, 3, 26-29.

[2] Peterson, L.L.; Davie, B. S., "*Computer Networks: A System Approach*", Elsevier: Amsterdam, The Netherlands, 2007.

[3] Proakis, J.; Salehi, M.; Bauch, G., "*Contemporary Communication Systems using MATLAB*", Nelson Education: Scarborough, ON, Canada, 2012.

[4] IEEE 802.15.6-2012, "*IEEE Standard for local and Metropolitan Area Networks-Part 15.6: Wireless Body Area Networks*", IEEE: New York, NY, USA, 2012.

[5] Ullah, S.; Higgins, H.; Braem, B.; Latre, B.; Blondia, C.; Moerman, I.; Saleem, S.; Rahman, Z.;Kwak, K.S, "A comprehensive survey of wireless body area network", *J. Med. Syst.* 36,(2012):1065-1094. [CrossRef][PubMed]

[6] Barakah, D.M.; Ammad-uddin, M, "A survey of challenges and applications of wireless body area network (WBAN) and role of a virtual doctor server in existing architecture", In Proceeding of the 2012 third International Conference on Intelligent System, Modeling and Simulation (ISMS),Kota Kinabalu, Malaysia,pp.214-219,February 2012.

[7] Quan Nguyen-Trung;Thu Ngo-Quynh;Vinh Tran-Quang, " *A Low Duty –Cycle XT-MAC Protocol for target tracking in Wireless Sensor Networks",*IEEE 978-1-4799-5051-5,2014.

[8] Te-Li Wang;Jung- Chun Kao;Si An Ciou, "*CBA-EVT:A Traffic –Adaptive Energy Efficient MAC Protocol for wireless sensor networks",*IEEE 978-1-4799-1297-1,2014.

[9] Pei Huang;Chen Wang;Li Xiao, "*RC-MAC: A Receiver Centric MAC Protocol for Event-Driven Wireless Sensor Networks",* IEEE TRANSACTIONS ON COMPUTERS, VOL.64, NO.4, APRIL2015.

[10] Pearl, J. *Probabilistic Reasoning in Intelligent systems: Networks of plausible Inference*; Morgan Kaufmann: Burlington, MA, USA, 2014.

[11] Rajoua Anane, Kosai Raoof, Ridha Bouallegue,"On The Evaluation of GMSK Scheme with ECC techniques in Wireless sensor Network.", *International Journal of Wireless & Mobile Networks (IJWMN)* Vol. 7, No.2 April 2015.DOI: 10.5.121/ijwmn.2015.7202.

[12] Praveena, V., Kaythry, P., Kishore, R.,"Fountain code based error control technique in wireless body area network"In Proceedings of the Wireless Communication, Signal Processing and Networking (WiSPNET),Chennai, India, 22-24pp.430-434. March 2017.

[13] Ahmed, E.; Kohno, R. "Error control coding and decoding with medical QoS constraints for WBAN end to end connection via UMTS channel", *ICT Express* , 4, (2018):181-185.[CrossRef]

[14] Rajan Kadel; Nahina Islam; Khandakar Ahmed; Sharly J. Halder, "Opportunities and Challenges for Error Correction Scheme for Wireless Body Area Network- A Survey.", *J. Sens. Actuator Netw.*, 8,1. ;doi:10.3390/jsan8010001.(2019).

[15] Ning Ma, "*Distributed video coding scheme of multimedia data compression algorithm for wireless sensor network",*EURASIP Journal on Wireless Communications and Networking.doi:10.1186/s13638-019-1571-5.(2019)

IOP Conf. Series: Materials Science and Engineering 1020 (2021) 012007 doi:10.1088/1757-899X/1020/1/012007

Feature based analyses of lung nodules from computed tomography (CT) images

Md. Anwar Hussain[a], Lakshipriya Gogoi[a,*]

[a]Department of Electronics and Communication Engineering, North Eastern Regional Institute of Science and Technology, Nirjuli - 791109, India

gogoilakshipriya12@gmail.com

Abstract. Among various lung image modalities, CT (computed tomography) images have been found most suitable and widely used for detection of small lung nodules. Although an expert radiologist can analyse these images quite perfectly, however an efficient CADe (computer-aided detection) system capable of detecting pulmonary nodules automatically may be of great help considering the large number of CT images, a radiologist needs to analyse a day in recent years. A few CADe systems have already been tested within lung cancer screening trial which have enjoyed mixed results. So, the field has enough voids to be filled and research towards development of novel and efficient CADe systems has become an interesting, but challenging field of research. We too have been trying to develop CADe systems that can analyse CT images to detect and identify various lung nodules. In the process, we are reporting herein one of our development for the same. Our CADe system has maintained its 81.25% accuracy in detecting malignant nodules. Hand-crafted features of the selected lung CT images were used in the study. The study is emphasised on developing efficient CADe systems capable of detecting solid nodules (size > 2 mm) at different locations, whether isolated, juxtapleural or juxtavascular nodules. While MATLAB was used to carry out pre-processing, segmentation and nodule detection, testing of datasets was done by a MLP (multilayer perceptron) network by using WEKA (Waikato Environment for Knowledge Analysis) software.

1. Introduction

Lung cancer or lung carcinoma is reported as the deadliest among all the cancers when death toll related to cancer worldwide in last couple of decades are considered [1]. L. Garfinkel et al. in their report in 1991 made a special remark that fatalities in women due to lung cancer had surpassed that caused by breast cancer in the United States in the year 1987 [2]. It is one of the most aggressive kinds of cancers as only 10-16% of the patients are found to be reported to survive for 5-years [3]. The prime reason behind this low value of survival rate is the diagnosis of patients in advance stages of their disease. Reports suggest that the rate could be increased up to 70% by diagnosis of lung cancer in early stages [4,5]. So, a successful clinical treatment of lung cancer relies on procedures and techniques which are capable of accurate detection of lung nodules in early days of the disease. Due to its high structural imaging ability, computed tomography (CT) has become the most suitable and widely used imaging technique for the detection of small lung nodules.

The National Lung Screening Trial (NLST) carried out during the period August 2002 to September 2007 revealed the effectiveness of low-dose computed tomography LDCT images in comparison to chest radiography. Mortality rate in 7 years could be reduced to 20% by screening high risk patients thrice in a year with the former in comparison to the later [6]. This result evolved a genuine hope of changing the scenario of late diagnosis by conducting lung-cancer monitoring programs using LDCT particularly for risk groups such as smokers. U.S. has already implemented such programs while many other countries are likely to follow soon [7]. Indian government too launched national cancer screening program for three types of cancers (cervical, breast & oral) in people more than 30 years of ages in 100 districts [8]. However, we have not found any record of extending the mission to lung cancer till now. The major challenge that needs to be overcome for successful implementation of such programs is the analysis of huge number of CT images recorded. For the best set of radiologists too, there is every chance of human error which may be due to fatigue or any kind of distraction while dealing with such a large number of images [3]. Development of an effective CADe system enabling analyses of CT images to detect lung nodules has been seen as the best possible solution in this scenario and so, researchers throughout the globe have been found actively involved to contribute their part to the cause during the last two decades. Meanwhile, many CADe systems have been evolved and a few of them have been tested within lung cancer screening trial which have enjoyed mixed results [9,10]. However, the answer to the quest of efficient CADe systems for accurate detection of pulmonary nodules till date has been felt to be fairly behind the actual target. CADe systems have not commonly been used in current clinical daily practice due to the drawback of finding high false positive values demanding a radiologist's attention [11].

Realizing the urgency of developing potential CADe systems to detect pulmonary nodules and believing that CADe will progressively be integrated in clinical practice, we have got attracted to this interesting and important field of research. This report describes our developed CADe system capable of detecting solid nodules (size > 2 mm) at different locations.

2. Literature survey

Literature reports suggest that a large number of research groups across the world are engaged in this field of research in the mission of developing efficient CADe systems for pulmonary nodule detection. Although pulmonary nodules are roughly categorized into three wide categories viz. solitary pulmonary nodules (SPNs), juxta-vascular nodules (JVNs) and juxta-pleural nodules (JPNs) based on their spatial locations [1], however large variations in their shapes, scales and densities make it difficult to generalize the nodules into specific groups which in turn increases the difficulties of lung nodule detection [12]. Among the nodules, detection of SPNs is observed to be somewhat easy, while that of other types of nodules demand extra effort due to their complex surroundings. Two different approaches namely feature engineering-based approach and deep learning-based approaches are found to be mostly adopted for the purpose locations [1].

B. Golosio et al. reported a multistage CADe system in 2007 which was found effective for detection of various nodules including juxta-pleural and juxta-vascular nodules [13]. A multi-threshold surface-triangulation approach was adopted in their system for identification of regions of interest (ROIs). They computed five parameters viz. moments of inertia, density, surface area, volume and roundness as functions for thresholding for each ROI using an artificial neural network (ANN) based classifier system. For nodules with diameter ≥3 mm, the system delivered 84% and 71% sensitivity at false positive/scan values of 10 and 4 respectively. Sensitivity increased for nodules with diameter ≥ 4 mm.

W.-J. Choi et al. developed a CADe system which was evaluated using 58 CT scans from LIDC with nodule range from 3-30 mm [14]. Their system was consisted of three steps: a) splitting of CT images into three dimensional block images, b) segmentation of block images and adjustment for nodule candidates' detection and c) classification of candidates into nodule and non-nodule candidates. They carried out entropy analysis of the block images for identification of the informative blocks. Support vector machine (SVM) classifier was used in the final step for classification of the nodule candidates. The developed CADe system was capable of delivering 95.28% sensitivity at 2.27 false positives/scan.

In their report, G. Aresta et. al. displayed their CADe system which was specially designed for detection of juxta-pleural lung nodules only [15]. Lung segmentation was done using region-growing followed by refinement with morphological operations and active contours to include juxta-pleural nodules. Enhancement of sub-solid and non-solid nodules were done with a multiscale Laplacian-of-Gaussian filter. A support vector machine (SVM) with radial basis function was trained and used to reduce the number of false positives. They achieved 57.4% sensitivity with 4 false positives/scan.

M. Javaid et al. reported a CADe method for segmentation and detection of juxtavascular and juxtapleural nodules [16]. They used intensity thresholding for lung segmentation from CT images. They achieved 91.65% sensitivity at 3.19 false positives/scan using their CADe system. Subsolid nodules were found to be less frequent than the solid nodules, although the former show a significantly higher malignancy rate (Silva, Milanese, Seletti, Ariani & Sverzellati, 2018). C. Jacob et al. developed a CADe system for auto-detection of subsolid lung nodules in the year 2014 which was trained, optimized and evaluated using 318 scans collected from two different sites under NELSON trial [17]. In their work, they introduced a novel set of context features along with previously used features viz. intensity, shape and texture features and made special remark that the introduction of the new set features significantly improves the classification performance. Their CADe system reached a sensitivity of 80% for subsolid nodules with diameter varied between 5-34 mm at an average of 1.0 false positive/scan. In recent years, deep learning technique has become very popular as it enjoys certain distinct advantages over traditional method [18]. It provides an alternative way to bypass the troublesome manual feature extraction and thus minimizes both time and effort requirement [13]. In their report, A. A. A. Setio et al. described development of a CADe system in which multi-view convolutional network (ConvNets) was used for false reduction of the CADe system [19]. They trained their CADe system using LIDC-IDRI (Lung Image Database Consortium and Image Database Resource Initiative) data set, while they validate it using both LIDC-IDRI and ANODE09 datasets independently and showed that their proposed multi-view ConvNets is very effective for false positive reduction. Similar to A. A. A. Setio et al., Q. Dou et al. too emphasized on false positive reduction and proposed a CADe system for automated detection of pulmonary nodules from volumetric CT scans [20]. They used 3D convolutional neural networks (CNNs) for false positive reduction. Their proposed scheme was validated in the LUNA 16 challenge [7], where they achieved best competition performance matric (CPM) score for false reduction which showcased the efficiency of their developed methodology.

In India too, a few research groups have been involved in this field of interesting research. G. S. Nair et al. worked on development of segmentation system for lung nodule detection [21]. In their system, the lung lobes and nodules in CT images were segmented using adaptive fissure sweep and adaptive thresholding. In their report, S. Sivkumar et al. demonstrated a CADe system in which nodule segmentation was performed using fuzzy based clustering models while classification of nodule were done using Support Vector Machine (SVM) technique [22]. S. Aswin et al. developed a two stage CADe system in which first stage involved pre-processing while the second step involved ANN architecture [23].

3. Methodology

The steps adopted for successful development of our CADe system has been summarized Figure 1. MATLAB and Weka software were used to carry out the study. Brief description of our approach has been presented here.

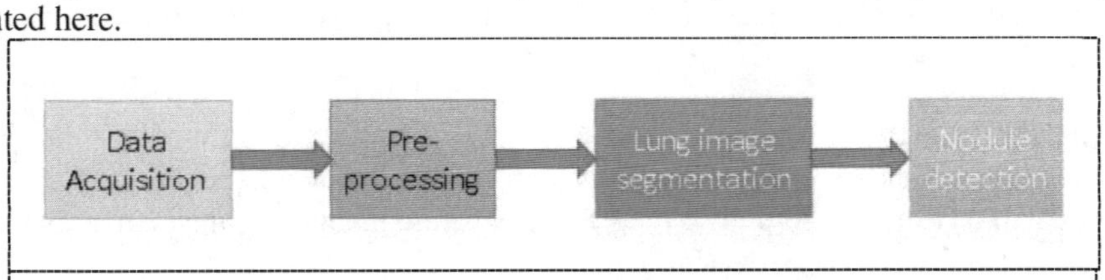

Figure 1. Flowchart of the adopted methodology.

3.1. Data acquisition:

16 lung CT-images were used in our work, 15 of which were collected from LIDC (Lung Image Database Consortium) open source database while one from open web source which was retrieved on 12th June, 2020 using the link https://www.nytimes.com/2007/03/07/health/07lung.html. All the images were in DICOM (Digital Imaging and Communications in Medicine) format.

3.2. Pre-processing:

Collected images were pre-processed using Otsu's global image threshold method (graythreshold) and morphological open operation to improve the quality and interpretability of the input images for better results in pulmonary nodule detection. Otsu's thresholding method was used to achieve spontaneous image thresholding as it can simply separate the image pixels into two main classes (background and foreground) by the help of single intensity threshold. This stage is crucial because lots of fats, vessels and tissues might present in lung parenchyma causing confusion in detecting nodules.

3.3. Lung image segmentation:

Lung parenchyma was extracted from the pre-processed lung images by using morphological operations such as erosion and dilation operators and then digital logical gate operations techniques. Digital gate operations are novel techniques to extract the parenchyma from lung CT images with high accuracy.

3.4. Nodule detection:

As lots of residual parts are generally present around the tumor region and some tumours are not homogenous (there may be some fats, non-solid substances, vessels present in the tumor mass) also, morphological procedure was therefore used on these binary images to refine the margin and content of the tumor images. In this step, the lung CT images were converted into RGB plan during the process of which the region of interests (ROIs), i.e. the nodule parts became red in colour as depicted in Fig. 2. This happens as the nodules' texture and intensities are different from other parts of lungs. In this way, the pulmonary nodules were extracted from the lung parenchyma by removing other lung tissues.

We created a MLP (multilayer perceptron) network using an MLP algorithm with one input layer, one hidden layers and one output layer with learning rate 0.3 and 10- fold cross validation with 66% for training and 34% for testing and validation purpose using WEKA.

WEKA (Waikato Environment for Knowledge Analysis) is a machine learning software based on Java. It was developed at the University of Waikato, New Zealand [24]. It contains a group of algorithms for data analysis and predictive modeling and visualization tools, organized with GUI (graphical user interfaces) for easy access to this functionality.

In MLP, cross-validation was done to evaluate the predictive model by splitting the experimental samples into a training set to train the model, and a test set to evaluate it. MLP is a class of feedforward ANN (Artificial Neural Network) which utilizes a supervised learning method called backpropagation for training. Every nodes in MLP is a neuron except the input nodes and each node uses a nonlinear activation function. The main differences between linear perceptron and MLP are: multiple layers and non-linear activation of MLP. It can differentiate data that is not linearly distinguishable.

4. Result and discussion

All the downloaded images were processed as described in methodology section, the result for one image has been depicted in Figure 2. The lung nodule was segmented properly from lung CT image and its properties were found out using MATLAB.

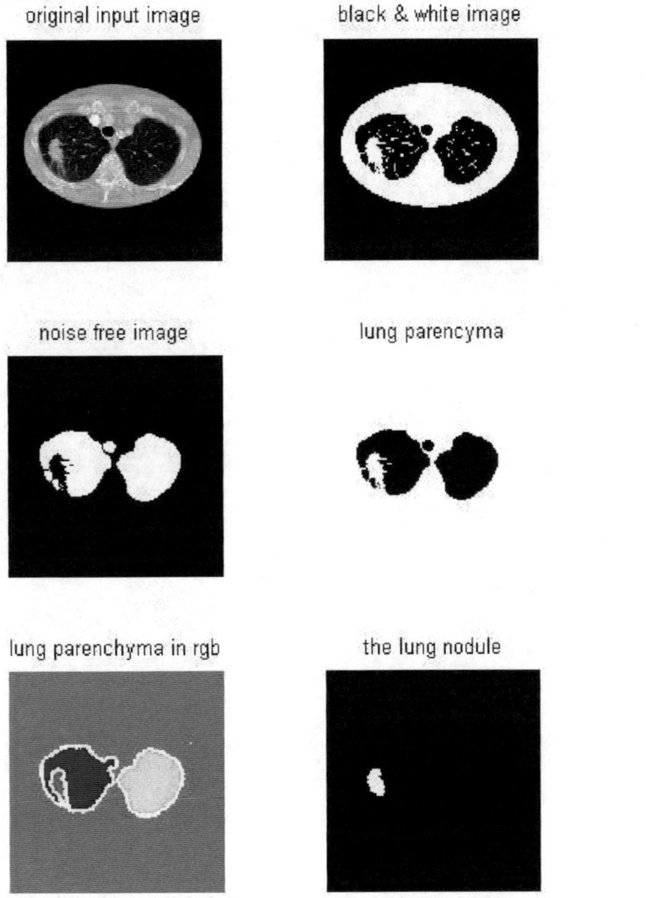

Figure 2. Processed lung CT image.

The region property features were extracted from the ROIs (region of interest) of all the CT images. We extracted seven numbers of features namely area, major-axis, minor-axis, eccentricity, solidity, entropy and perimeter for each CT images and the results are summarized in Table 1. The features under study have been defined briefly.

Area: It is a scalar quantity and refers the total no. of pixels which are present in the ROIs.

MajorAxis Length: It is a scalar quantity specifying the length of the major axis of the ROIs.

MinorAxis Length: It is a scalar quantity that identifies the length of the minor axis of the ROIs.

Eccentricity: It is a real number with non-negative value that can uniquely characterizes the shapes of a ROI. The eccentricity of a circle, ellipse, parabola, and hyperbola are 0, 0 to <1), 1 and >1 respectively.

Solidity: It is a dimensionless value and it measures the complete concavity of a nodule with the value ranges between 0-1. 1 represents very smooth nodule shape, 0 represents an irregular shape.

$$\text{Solidity} = \text{image area}/\text{Convex area}$$

Perimeter: This scalar quantity measures the total distance around the boundary of the ROIs. It actually calculates the distance between each neighbouring pair of pixels nearby the border of the region.

Entropy: The average information or the entropy of an image is a statistical measurement which indicates the degree of randomness in the image. It is used to characterize the texture of the ROIs.

Entropy=-sum.(p.*log2(p))

In the entropy equation p-indicates the normalized histogram counts.

Table 1. Quantitative CT features of the lung nodule under the study.

Image no.	CT features							Nodule (Yes/No)
	Area (mm^2)	Major axis length (mm)	Minor axis length (mm)	Eccentricity	Solidity	Perimeter (mm)	Entropy	
01	1338	58.85	30.49	0.85	0.87	167.96	0.05	yes
02	8	3.65	3.05	0.54	1	7.22	0.0005	yes
03	16	5.03	4.16	0.56	1	11.65	0.0009	Yes
04	8	3.65	3.05	0.54	1	7.22	0.0005	yes
05	8	3.65	3.05	0.54	1	7.22	0.0005	yes
06	5	2.78	2.78	0	1	5.62	0.0003	yes
07	11	5.10	3.07	0.79	1	10.03	0.0006	yes
08	5	2.78	2.78	0	1	5.62	0.0003	yes
09	474	38.19	24.50	0.76	0.52	159.22	0.0191	no
10	291	46.35	20.24	0.89	0.44	127.42	0.0125	no
11	5	2.78	2.78	0	1	5.62	0.0003	yes
12	65	9.63	8.82	0.42	0.97	25.95	0.0033	No
13	65	26.87	3.56	0.99	0.80	49.97	0.0033	no
14	85	26.79	4.26	0.98	0.88	51.93	0.0042	no
15	49	16.06	4.16	0.96	0.85	31.09	0.0026	no
16	5	2.78	2.78	0	1	5.62	0.0003	yes

The developed multilayer neural network for our CADe system has been depicted in Figure 3. Seven features viz. area, major-axis, minor-axis, eccentricity, solidity, entropy and perimeter were fed in the input layer to identify the benign and malignant nodules in the output layer.

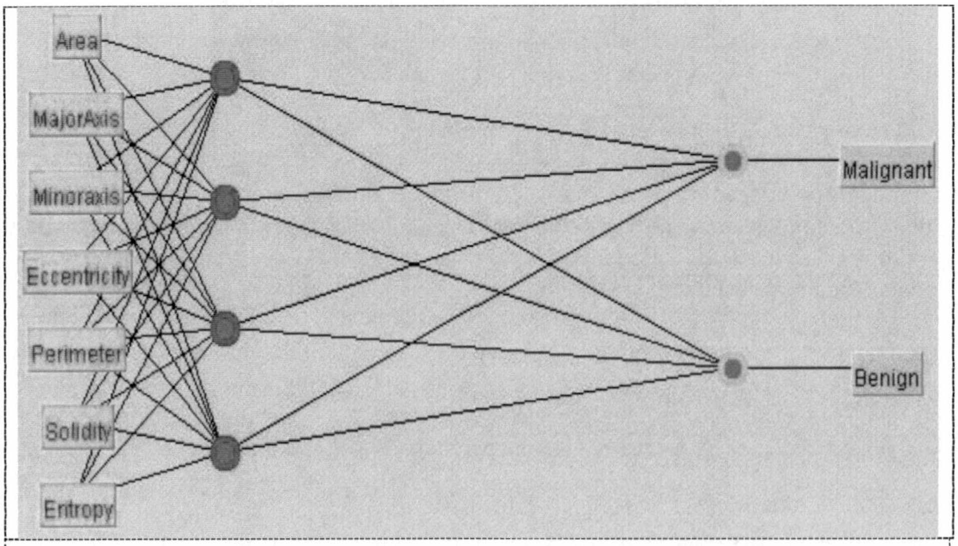

Figure 3. Multilayer neural network developed for the CADe system.

Snapshot of the results of MNP has been depicted in Figure 4. The correctly and incorrectly classified instances display the percentage of test instances that were classified correctly and incorrectly. The total instances can be seen in the confusion matrix representing two class labels: malignant and benign. The percentage of correctly classified instances is called accuracy, thus the accuracy of our system is found to be 81.25%. A positive kappa statistic value (0.5385) was obtained which indicates that the classifier is working satisfactorily.

```
Classifier output

Time taken to build model: 0.06 seconds

=== Stratified cross-validation ===
=== Summary ===

Correctly Classified Instances          13               81.25   %
Incorrectly Classified Instances         3               18.75   %
Kappa statistic                          0.5385
Mean absolute error                      0.2378
Root mean squared error                  0.4341
Relative absolute error                 59.2128 %
Root relative squared error             96.5235 %
Total Number of Instances               16

=== Detailed Accuracy By Class ===

                 TP Rate  FP Rate  Precision  Recall   F-Measure  MCC      ROC Area  PRC Area  Class
                 0.833    0.250    0.909      0.833    0.870      0.545    0.760     0.914     Malignant
                 0.750    0.167    0.600      0.750    0.667      0.545    0.760     0.491     Benign
Weighted Avg.    0.813    0.229    0.832      0.813    0.819      0.545    0.760     0.808

=== Confusion Matrix ===

  a  b   <-- classified as
 10  2 |  a = Malignant
  1  3 |  b = Benign
```

Figure 4. Snapshot of the MLP output..

5. Conclusion

In conclusion, we have developed a CADe for detection of lung nodules of size > 2 mm. Lung nodules of solid, juxtapleural and juxtavascular type could be detected successfully from 16 number of lung CT images collected from LIDC and open web source with an accuracy of 81.25%. While MATLAB was used to carry out pre-processing, segmentation and nodule detection, testing of datasets was done by a MLP network using WEKA software. The suitability of the classifier was indicated by the resulting positive kappa statistic value (0.5385).

Acknowledgements

LG is receiver of DST-INSPIRE fellowship (IF180853) and is grateful to DST for the same.

References

[1] Zhang J, Xia Y, Cui H and Zhang Y 2018 *Biomed. Signal. Process. Control* **43** 138.

[2] Garfinkel L and Silverberg E 1991 *CA Cancer J. Clin.* **41** 137.

[3] Valente I R S, Cortez P C, Neto E C, Soares J M, de Albuquerque V H C and Tavares J M R S 2016 *Comput. Methods Programs Biomed.* **124** 91.

[4] Baldwin D R 2015 *Lung Cancer* **89(1)** 1.

[5] Camarlinghi N 2013 *Eur. Phys. J. Plus* **128** 110.

[6] Aberle D R, Adams A M, Berg C D, Black W C, Clapp J D, Fagerstrom R M, Gareen I F, Gatsonis C, Marcus P M and Sicks J D 2011 *N. Engl. J. Med.* **365(5)** 395.

[7] Setio A A A, Traverso A, de Bel T, Berens M S N, van den Bogaard C, Cerello P, Chen H, Dou Q, Fantacci M E, Geurts B, van den Gugten R, Heng P A, Jansen B, de Kaste M M J, Kotov V, Yu-Hung Lin J, Manders J T M C, Sóñora-Mengana A, García-Naranjo J C, Papavasileiou E, Prokop M, Saletta M, Schaefer-Prokop C M, Scholten E T, Scholten L,

Snoeren M M, Torres E L, Vandemeulebroucke J, Walasek N, Zuidhof G C A, van Ginneken B and Jacobs C 2017 *Med. Image Anal.* **42** 1.
[8] Bagcchi S 2016 *BMJ* **355** i5574.
[9] van Klaveren R J, Oudkerk M, Prokop M, Scholten E T, Nackaerts K, Vernhout R, van Iersel C A, van den Bergh K A M, van't Westeinde S, van der Aslst C, Thunnissen E, Xu D M, Wang Y, Xhao Y, Gietema H A, de Hoop B-J, Groen H J M, de Bock G H, van Ooijen P, Weenink C, Verschakelen J, Lammers J-W J, Timens W, Willebrand D, Vink A, Mali W and de Koning H J 2009 *N. Eng. J. Med.* **361(23)** 2221.
[10] Sverzellati N, Silva M, Calareso G, Galeone C, Marchiano A, Sestini S, Sozzi G and Pastorino U 2016 *Eur. Radiol.* **26(11)** 3821.
[11] Silva M, Milanese G, Seletti V, Ariani A and Sverzellati, N. 2018 *Br. J. Radiol.* **91**, 1083.
[12] Jiang H, Ma H, Qian W, Gao M and Li Y 2018 *IEEE J. Biomed. Health. Inf.* **22(4)** 1227.
[13] Golosio B, Masala G L, Piccioli A, Oliva P and Carpinelli M 2009 *Med. Phys.* **36(8)** 3607.
[14] Choi W-J and Choi T-S 2013 *Entropy* **15** 507.
[15] Aresta G, Cunha A and Campilho A 2017 *SPIE Med. Imag. 101343N.*
[16] Javaid M, Javid M, Rehman M Z U and Shah S I A 2016 *Comput. Methods Programs Biomed.* **135** 125.
[17] Jacobs C, van Rkxoort E M, Twellmann T, Scholten E T, de Jong P A, Kuhnigk J-M, Pudkerk M, de Koning H J, Prokop M, Schaefer-Prokop C and van Ginneken B 2014 *Med. Image anal.* **18** 3745.
[18] Litjens G, Kooi T, Bejnordi B E, Aaa S, Ciompi F, Ghafoorian M, Jawm V D L, Van G B, and Snchez C I 2017 *Med. Image Anal.* **42(9)** 60
[19] Setio A A A, Ciompi F, Litjens G, Gerke P, Jacobs C, van Riel S J, Wille M M W, Naqibullah M, Sanchez C I and van Ginneken B 2016 *IEEE Trans. Med. Imag.* **35 (5)** 1160.
[20] Dou Q, Chen H, Yu L, Qin J and Heng P A 2017 *IEEE Trans. Med. Imag.* **64 (7)** 1558.
[21] Nair G S and Ajij S D 2014 *IOSRJECE* **9(3)** 19.
[22] Sivakumar S and Chandrasekar D C 2013 *Int. J. Eng. Technol.* **5 (1)** 179.
[23] Ashwin S, Ramesh J, Kumar SA and Gunavathi K 2013 International Conference on Emerging Trends in Electrical Engineering and Energy Management Chennai India, 135.
[24] Arora R and Suman 2012 *Int. J. Comp. Appl.* **54** 13.

Prospective Of Artificial Intelligence: Emerging Trends In Modern Biosciences Research

Pradeep Kumar[1]*, Ajit Kumar Singh Yadav[2], Abhishek Singh[3]
[1]Department of Forestry, North Eastern Regional Institute of Science and Technology, Nirjuli-791109 Arunachal Pradesh, India
[2]Department of Computer Engineering, North Eastern Regional Institute of Science and Technology, Nirjuli-791109 Arunachal Pradesh, India
[3]Department of Agricultural Biotechnology, Sardar Vallabhbhai Patel University of Agriculture and Technology, Meerut, India
*Corresponding author: pkbiotech@gmail.com

Abstract

The use of computers and technology to simulate intelligent behaviors and critical thinking comparable to the human beings are termed as Artificial Intelligence (AI). Growing world population causing pressure on researchers to adopt the advanced method to solve the human need and requirements for better healthcare. In the case of human health, research AI playing a vital role in developing various applications for human walefare and society. AI can utilize to solve and interpret complex health care issues using computers for chronic diseases like alzheimer's, diabetes, cardiovascular disease, and various types of cancers. And, this will help in reducing the economic burden and severity of diseases. In the last few years research publication based on AI increases rapidly in various biological fields not only in human health but also in other sectors. Therefore, the current article provides a brief understanding of the emerging treads and use of AI from beginning to 2020 by extracting data from the Pubmed, NCBI using specific words.

Keywords: Artificial intelligence, PubMed, chronic diseases, diagnosis, human health,.

1. Introduction

In recent years AI is one of the most emerging fields of research and incorporated with various displine to get more fruitful results. It was not a new concept and it noticed back to Ramon Llull's theory of a reasoning machine in 1300 CE and even Aristotle's syllogisms in 300 BC (1, 2). During the 1950s a clear definition and application of AI come in to existence (3-5). In the last decayed the increasing interest of AI in various sectors attract researchers and many countries to invest billions of dollars in the AI based programming and innovations (5, 6). The technology is now leading and showing potentials to transform the many aspects of health care like patient administration, clinical decision support, patient monitoring, and healthcare interventions (7). There are several research studies reported that AI can provide better results than human at healthcare task, such as disease diagnosis.

Kelly et al. (8) provide a comprehensive opinion article on various application of AI in healthcare by researchers which includes algorithms for chest radiographs interpretation, cancer mammograms detection, computer tomography scans analysis, identifying brain tumors on magnetic resonance images, and predicting the development of Alzheimer's disease from positron emission tomography (10-14). Along with that, the other important applications have been shown in pathology, indentifying cancers, interpreting retinal imaging, detecting arrhythmias, and hyperkalemia, genomics interpretation, identifying genetic facial condition from facial appearance, and improving *invitro* fertilization (15- 22).

Keeping the broad application of AI in human health care the current research work planned and we perform the survey the current research publications status on AI in healthcare using PubMed search of National Center for Biotechnology (NCBI) database.

2. METHODS

The search engines PubMed, NCBI public domain database was explored for literature search with the suitable keywords to retrieve the qualitative data. Through this search tradition and advanced studies using AI up to November 2020 have been included. Along with this the given additional subjectwise filters are also used to get retrieve the data.

3. Results

A large numbers of research publications available in the public domain of PubMed, NCBI. And during the search when artificial Intelligence (AI) applied for the search a total of 113480 numbers of research paper available since 1951-2020 also besides 25 research paper available for the year 2021. The first paper appears during search is 1951 with the title "Matter with a mind, a neurological research robot" published in Research Journal. Total number of publications appears with ten years interval 1951-60, 1961-70, 1971-80, 1981-90, 1991-2020, 2001-2020 were 7, 36, 57, 1390, 8315, 31512 and 72908 respectively (Table 1, Figure 1). An increasing trend was observed in the use and application of AI based research for human health and welfare research. We excluded the publication of 2021 which was 25 in numbers.

Table 1: Research publication trends in 10 years intervals based on PubMed search using AI.

Year	No. of Publications
1951-60	7
1961-70	36
1971-80	57
1981-90	1390
1991-2000	8315
2001-10	31512
2011-20	72908

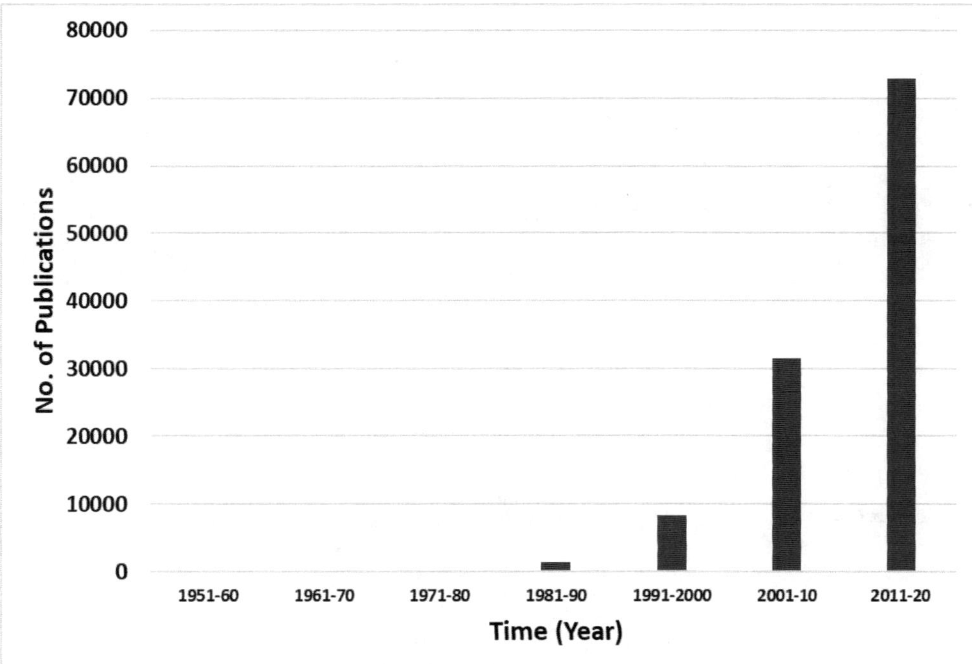

Fig. 1: Research publication trends in 10 years intervals based on PubMed search using AI.

When the applied subjectwise additional filter in PubMed search which shows four categories namely; cancer research, AIDS research, complementary medicine and, systematic review for the different time interval of research and it was observed that a total of 26205 number of research papers available in the domain. The highest number of research papers available for cancer research followed by complementary medicines, AIDS Research and systematic review 20065, 4814, 755, and 571 respectively (Table 2; Figure 2).

Table 2: Research publication trends in the field of Artificial intelligence using subjectwise filters based on PubMed search

Subjectwise filters	No. of Publications
Cancer Research (1961-2020)	20065
AIDS Research (1987-2020)	755
Complementary Medicine (1962-2020)	4814
Systematic Review (2003-2020)	571

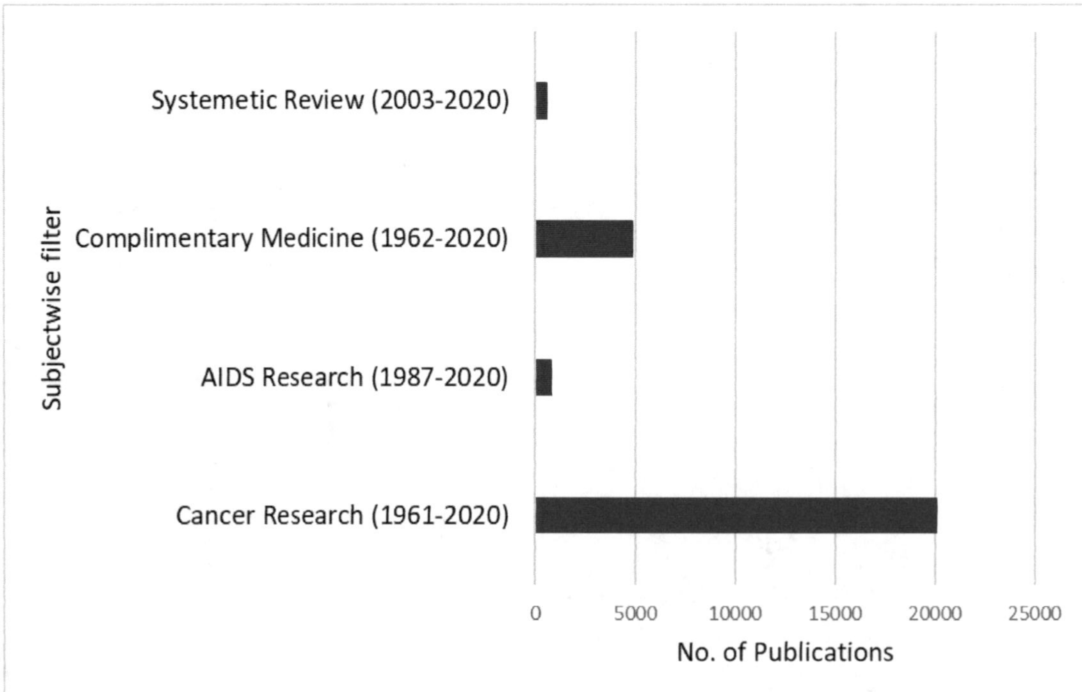

Fig. 2: Research publication trends in the field of Artificial intelligence using subjectwise filters based on PubMed search

4. Conclusions:

Though the search terminology and filter used in the studies is not very specialized and no rigorous screening was done for data search and to obtain the specified result. But still, from these findings we can observe and conclud that an increasing interest of researchers in the field of Artificial intelligence and their applications in human health and welfare research. A specialized and specific machine learning approach and advance engine search are needed to get the more clearcut ideas and accurate results about the increasing trends which will be carried out in near future.

5. Disclosure of interest: There is no conflict of interest among the authors.

6. Acknowledgments: The authors are thankful to their respective department and Institution for the support in doing this research.

7. References

1. Ramesh AN, Kambhampati C, Monson JRT, Drew PJ. Artificial intelligence in medicine. Annals of the Royal College of Surgeons of England. 2004;**86**(5):334-338
2. Warwick K. Artificial Intelligence: The Basics. Abingdon: Routledge; 2012
3. Simmons AB, Chappell SG. Artificial intelligence—Definition and practice. IEEE Journal of Oceanic Engineering. 1988;**13**(2):14-42.
4. Kok JN, Boers EJW, Kosters WA, Van Der Putten P, Poel M. Artificial Intelligence: Definition, Trends, Techniques, and Cases. Oxford, UK: Encyclopedia of Life Support Systems; 2013.
5. Reddy S. Use of Artificial Intelligence in Healthcare Delivery, eHealth - Making Health Care Smarter, Thomas F. Heston, IntechOpen, 2018. DOI: 10.5772/intechopen.74714.
6. Diprose W, Buist N. Artificial intelligence in medicine: Humans need not apply? The New Zealand Medical Journal. 2016;**129**(1434):73-76
7. Reddy S, Fox J, Purohit MP. Artificial intelligence-enabled healthcare delivery. J R Soc Med. 2019 Jan;112(1):22-28.
8. Kelly CJ, Karthikesalingam A, Suleyman M. *et al.* Key challenges for delivering clinical impact with artificial intelligence. *BMC Med.* 2019**; 17,** 195.
9. Nam JG, Park S, Hwang EJ, Lee JH, Jin K-N, Lim KY, et al. Development and validation of deep learning–based automatic detection algorithm for malignant pulmonary nodules on chest radiographs. Radiology. 2019; 290:218–28.
10. Wu N, Phang J, Park J, Shen Y, Huang Z, Zorin M, et al. Deep neural networks improve radiologists' performance in breast cancer screening. arXiv. 2019
11. Yasaka K, Akai H, Abe O, Kiryu S. Deep learning with convolutional neural network for differentiation of liver masses at dynamic contrast-enhanced CT: a preliminary study. Radiology. 2018;286:887–96.

12. Chilamkurthy S, Ghosh R, Tanamala S, Biviji M, Campeau NG, Venugopal VK, et al. Deep learning algorithms for detection of critical findings in head CT scans: a retrospective study. Lancet. 2018;392:2388–96.
13. Kamnitsas K, Ferrante E, Parisot S, Ledig C, Nori AV, Criminisi A, et al. DeepMedic for brain tumor segmentation. In: International Workshop on Brainlesion: Glioma, Multiple Sclerosis, Stroke and Traumatic Brain Injuries; 2016. p. 38–49.
14. Ding Y, Sohn JH, Kawczynski MG, Trivedi H, Harnish R, Jenkins NW, et al. A deep learning model to predict a diagnosis of Alzheimer disease by using F-FDG PET of the brain. Radiology. 2019;290:456–64.
15. Chang HY, Jung CK, Woo JI, Lee S, Cho J, Kim SW, et al. Artificial intelligence in pathology. J Pathol Transl Med. 2019;53:1–12.
16. Esteva A, Kuprel B, Novoa RA, Ko J, Swetter SM, Blau HM, et al. Dermatologist-level classification of skin cancer with deep neural networks. Nature. 2017;542:115–8.
17. Gulshan V, Peng L, Coram M, Stumpe MC, Wu D, Narayanaswamy A, et al. Development and validation of a deep learning algorithm for detection of diabetic retinopathy in retinal fundus photographs. JAMA. 2016;316:2402–10.
18. Hannun AY, Rajpurkar P, Haghpanahi M, Tison GH, Bourn C, Turakhia MP, et al. Cardiologist-level arrhythmia detection and classification in ambulatory electrocardiograms using a deep neural network. Nat Med. 2019;25:65–9.
19. Galloway CD, Valys AV, Shreibati JB, Treiman DL, Petterson FL, Gundotra VP, et al. Development and validation of a deep-learning model to screen for hyperkalemia from the electrocardiogram. JAMA Cardiol. 2019;4(5):428–36.
20. Xu J, Yang P, Xue S, Sharma B, Sanchez-Martin M, Wang F, et al. Translating cancer genomics into precision medicine with artificial intelligence: applications, challenges and future perspectives. Hum Genet. 2019;138:109–24.
21. Gurovich Y, Hanani Y, Bar O, Nadav G, Fleischer N, Gelbman D, et al. Identifying facial phenotypes of genetic disorders using deep learning. Nat Med. 2019;25:60–4.
22. Khosravi P, Kazemi E, Zhan Q, Malmsten JE, Toschi M, Zisimopoulos P, et al. Deep learning enables robust assessment and selection of human blastocysts after in vitro fertilization. NPJ Digit Med. 2019;2:21.
23. Jiang F, Jiang Y, Zhi H, et al. Artificial intelligence in healthcare: past, present and future. Stroke and Vascular Neurology 2017;2: e000101.

Abnormality Detection in Musculoskeletal Radiographs

Ravi Raj choudhary, Somesh Choudhary and Gaurav Meena
Department of Computer Science, Central University of Rajasthan, Rajasthan, India
E-mail: raviraj@curaj.ac.in, gaurav.meena@curaj.ac.in

Abstract. A radiographic study is a typical procedure utilized to find different kinds of abnormalities, where the identification of Musculoskeletal Abnormalities from the norm stands out as a crucial task. in this paper we suggests some deep learning methods to identify such Musculoskeletal Abnormalities using the MURA dataset which is one of the largest collections of upper extremity radiographs. We make use of DenseNet and VGG deep learning models to try to draw some unprecedented findings and results. In addition, we try to visualize all our results and findings using comparative graph study. We compare our model with models proposed during the Stanford ML Group MURA Competition using Cohen Kappa Statistic. The obtained results show that deep convolutional neural networks can achieve results which are very close and even better compared to current state-of-the-art models. We achieve comparable model performance to state-of-the-art performances in three of seven study types.

1. Introduction

Musculoskeletal abnormality detection in radiographic studies is a very important task in the medical field. A variety of diseases or abnormalities can be detected using radiographic studies and this is also the most common way to do it. If such studies are interpreted correctly then the clinicians can rule out abnormalities and rule out the need of further diagnostics. Also, it can help the clinicians to avoid errors while examining a study. In addition, most of the long-term pain, severe pain and disabilities are caused my musculoskeletal abnormalities, which makes the correct identification of such abnormalities a crucial task. Also, some medical centers also have by standing clinicians which work in only emergency scenarios, such clinicians do not have adequate experience and expertise for correct identification of abnormalities on radiographs. In case of busy medical centers, experienced radiologists may not be available. In such a case, clinicians can take hours of valuable time to assess radiographic images and chance of incorrect interpretation of musculoskeletal abnormality is also there which affects the patient care.

There are a number of research articles which discusses and investigates bone abnormality but most of those are limited to bone fracture only. Musculoskeletal abnormalities not only cover bone fracture, but it also covers a wide range of diseases like bone deformities, osteoporosis, congenital diseases, and different kinds of arthritis. There is not much research done on such abnormalities. MURA dataset, which contains thousands of radiographic studies for musculoskeletal abnormality detection, is crafted by Rajpurkar towards this research field and, as far as my knowledge goes, there is no other work for radiographic abnormality detection. The objective of our model is to make use of deep learning methods to find out musculoskeletal abnormalities in radiographic images provided in MURA dataset. We make use of DenseNet169 and VGG16 which are frequently used models for visual tasks. The models are trained and evaluated using MURA dataset crafted by Rajpurkar. The results are compared with Stanford ML Group MURA dataset competition leaderboard (Link) as well as results obtained by Rajpurkar himself. In addition, we also try to visualize the results obtained by our models.

This paper is divided in 5 sections. Section 1 discusses about the basic overview of the problem of Musculoskeletal Abnormality Detection. Available literature decsribe in section II, section III and IV shows methodology and results of proposed work.

2. Literature Review

Detection of abnormality in a radiographic study is a very important task in the medical field: A variety of diseases or abnormalities can be detected using radiographic studies and this is also the most common way to do it. If such studies are interpreted correctly then the clinicians can rule out abnormalities and rule out the need of further diagnostics. Musculoskeletal conditions affect more than 1.7 billion people every year. These conditions can cause of severe pain, long-term pain and even disability. First publication regarding the use of Computer Automated Systems to perform medical task was made by Tian in 2003 in the paper entitled "Computing Neck Shaft Angle of Femur for X-Ray Fracture Detection". In this paper, he made use of feature extraction technique to extract femur contour and classified femur based of neck-shift angle measured by it. This was the first major breakthrough for backing the use of Computer Automated Systems to perform medical task to get a second opinion [1].

Researchers from National University of Singapore made some advancements and proposed a new computer aided system in the paper entitled "Detection of Femur and Radius Fractures in X-ray Images". The paper proposes an approach which uses multiple detection method for detecting fractures in the femur and radius. Different kinds of features are extracted using these methods. For the detection of fracture in femur, neck-shaft angle is extracted. The paper discusses extraction of other general features which may prove vital for detecting fractures. It makes use of SVM and Bayesian classifier for the classification task. [2]. Later in 2005, Lum, Leow and Chen proposed the idea of using a combination of classifiers to deploy a computer aided system to find bone fractures in x-ray images in the paper entitled "Combining Classifiers for Bone Fracture Detection in X-ray Images". They managed to achieve upto 98% accuracy on test set [3]. Machine learning techniques were very popular for the task of image recognition and classification. But in recent times, CNNs gained vast popularity in computer vision related task due to them being more accurate. Deep convolutional neural networks perform great in image classification task due to their operability on large datasets, deeper models and better training algorithms.

In year 2009, Princeton University introduced a huge dataset of around 14 million images divided into around 21,000 categories. They invited people around the world to participate in the image classification task based on this dataset. They also proposed a classification model named ImageNet in the paper entitled "ImageNet: A Large-Scale Hierarchical Image Database" [4] . Over the course of 8 years between 2012-2017 this competition (ILSVRC: ImageNet Large Scale Visual Recognition Challenge) witnessed the introduction of well-known models such as AlexNet, VGG, GoogLeNet, ResNet etc. These models prove to be state-of-the-art in computer vision tasks. First use of CNNs for image classification task on ImageNet dataset was illustrated in 2012 in a paper entitled "ImageNet Classification with Deep Convolutional Neural Networks" [5]. This CNN model outperformed existing state-of-the-art which made CNNs the most favourite technique for image classification task. This model is popularly known as ConvNet. After the success of ConvNet, CNNs became first choice for image classification task. During the ILSVRC-2014, a new model named GoogLeNet was introduced in a paper entitled "Going Deeper with Convolutions" which is a 22 layered CNN[6]. The model was implemented such that it could obtain an architecture with increased depth and width while keeping computational cost at the lower side. In year 2015, Visual Geometry Group of University of Oxford proposed a model named VGG in the paper entitled "Very Deep Convolutional Networks for Large-Scale Image Recognition". This VGG model made use of CNNs which were upto 19 layers deep. VGG model achieved state-of-the-art standards in ILSVRC-2015[7].

The use of deep convolutional neural networks proved to be very successful in case of computer vision tasks. None of the other artificial intelligence techniques could achieve the benchmark set by CNN techniques. As the popularity of CNNs increased, more and more models of CNNs appeared on the surface. People started to use very deep convolutional neural networks which were of the depth two hundred and even more. Further researches in deep CNNs led to the craft of Densely Connected Neural Networks. Dense Convolutional Network or DenseNet was introduced in a paper entitled "Densely Connected Convolutional Networks" in year 2018 by Gao Huang of Cornell University and three other authors. DenseNet has upto 201 layered architecture. They proposed the use of Dense Block in the architecture. These DenseNet models outperformed existing state-of-the-art in nearly all the aspects. Although, these convolutional neural networks models used ImageNet dataset, which consists of non-medical images, for training, they could still work well for medical images also. Many medical researchers made use of theses popular models to conduct researches on medical images also. U-Net is a convolutional neural network model which was introduced in 2015 in a paper entitled "U-Net: Convolutional Networks for Biological Image Segmentation". They used a sliding window convolutional network method for their model and won first prize at ISBI cell tracking challenge 2015[8]. In year 2017, CheXNet model was introduced in a paper entitled "Radiologist-Level Pneumonia Detection on Chest X-ray using Deep Learning"[9] . They developed a convolutional neural network with 121 layers. The model was trained on CheX-ray14 dataset which contained 100,000 images with 14 diseases. The results of model were compared against four radiologists. This model achieved state-of-the-art in case of each 14 diseases. In the same year, researchers from The University of Adelaide published a paper entitled "Detecting hip fractures with radiologist-level performance using deep neural networks". They proposed a 172-layered DenseNet model. The model achieved an accuracy of 0.99 on validation set[10].

All available researcher show that choosing a deep convolution neural network model for musculoskeletal abnormality detection task might prove to be a great selection.In the context of radiographic abnormality detection, only a little literature and research work is available. The major contribution is made by Rajpurkar and Irvin who together developed the model entitled "MURA: Large Dataset for Abnormality Detection in Musculoskeletal Radiographs" in which they used DenseNet architecture and their model outperformed best radiologist performance in wrist abnormality detection[11]. Several other participants deployed their models in the same Stanford ML Group MURA Dataset competition and managed to outperform Rajpurkar's model. However, most of the participants opted for ensemble of models rather than single model. The leaderboard consists of top 70 participants and uses Cohen Kappa statistic to evaluate models. The performance of the models of these 70 participants lies between 0.843 to 0.518[12].

3. Methodology

In this section we explains the theoretical concepts behind the implementation of the model and it also covers the scope of the paper. We define some theoretical background which is essential to understand the theme. Different learning models used in the implementation and evaluation metrics used for performance analysis are described.

3.1. Artificial Neural Networks

Artificial Neural Network (ANN) falls under the category of supervised learning algorithms. It is also called Multi-layer Perceptron. It is designed to replicate the way the human brain works to process information. An ANN comprises multiple layers. It has one input layer, one output layer, and at least one hidden layer. Perceptrons are the fundamental building blocks of the ANN, and each layer might contain several perceptrons. The following figure is an example of ANN.

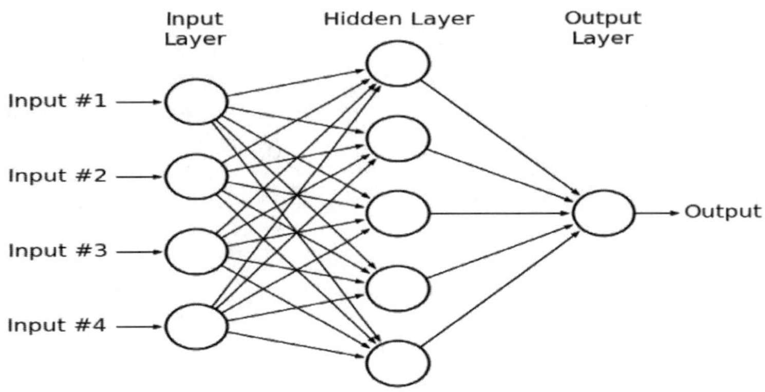

Figure 1. ANN with one hidden layer

3.2. Convolutional Neural Networks

Convolutional Neural Networks belongs to the class of deep neural networks. So basically, it is a deep learning algorithm. It can learn weights and bias for a given input and can assign importance to different objects in an image. Also, it can differentiate various objects of the image. It consists of a Convolutional layer which reduces the number of parameters in ANN which gives CNN an upper hand over ANN. A convolution operation can extract high-level features like edges. For this task, it makes use of a filter or kernel which moves over the complete input image with a stride value. A convolution operation produces feature maps. A CNN also has pooling layers which are used to reduce the dimensionality of feature maps. The following figure shows an example of CNN. CNN are a quite broad subject where there are many

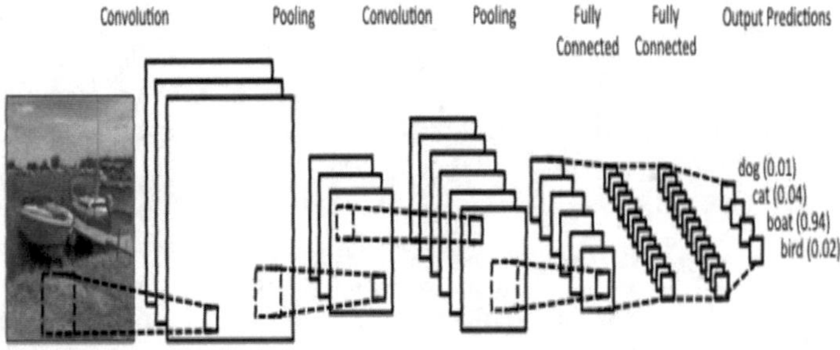

Figure 2. Convolutional Neural Network Architecture

configurations of a single network architecture and we cannot say with full confidence that on which configuration the architecture performs the best for a given task. But there are some models which are well-known for their great performance in image classification task that we are currently dealing with. In general, these models are trained and evaluated on ImageNet dataset. In this section, we are going two describe two models which are used in this paper.

3.2.1. VGG VGG model was introduced by Oxford Visual Geometry Group researchers. VGG is short for Visual Geometry Group model. The salient feature of this model is to add multiple convolutional layers of kernels of size 3*3. This model shows good overall performance on ImageNet dataset. This model generalizes very well for image classification task on other dataset also which makes it a potential solution for our problem. There are many variations of this model based on number of layers used in the architecture. Most popular variants are VGG16 and VGG19. VGG16 has 16 layers whereas VGG19 is a 19 layered architecture. In this paper, we are using VGG16 architecture. Following figure shows overall architecture of VGG. VGG16

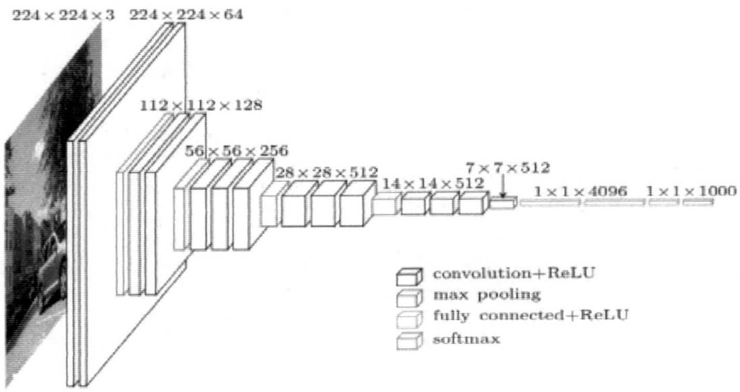

Figure 3. Architecture of VGG [7]

is a simpler model than other deep learning models. It uses kernels of size 3x3 for convolution. It uses max pooling to reduce the volume further. There are two fully-connected layers with 4,096 nodes each. And it makes use of softmax activation function to squeeze the output between 0 and 1. The input to VGG16 is images of size 224*224. The size of images is reducing as they pass through the network.Following figure shows the architecture of VGG16.

3.2.2. DenseNet DenseNet architecture was introduced by Gao Huang of Cornell University and three other authors in a research paper entitled "Densely Connected Convolutional Networks". DenseNet is built on research that CNNs are more efficient and accurate for training task if the links between layers close to inputs and layers close to outputs are shorter. To implement this research, authors introduced a Dense Block, where every layer treats the feature maps produced by its previous layers as an input for itself. And, the feature maps produced by each layer are used as inputs for all the subsequent layers. This reduces the training time because all the layers have direct access to gradient of loss function. The following figure shows the example of Dense Blocks. The traditional CNNs have only one link between each layer and it's following layer that means L number of links for L number of layers. Whereas, DenseNet has $\frac{L(L+1)}{2}$ direct links, L being number of layers. These Dense Blocks are linked together using transition layers. Transition layers perform convolution and pooling operations. The following figure shows overall architecture of DenseNet. Just like VGG, DenseNet also has

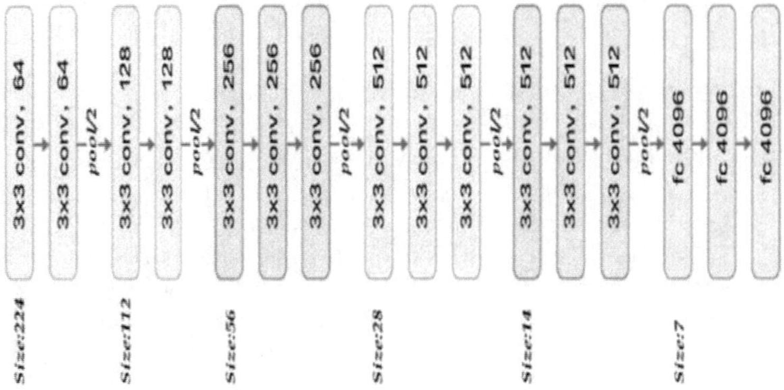

Figure 4. Architecture of VGG16[13]

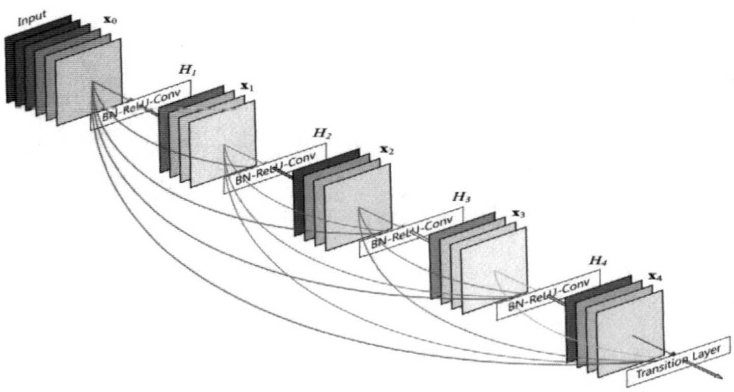

Figure 5. Example of five-layer dense block

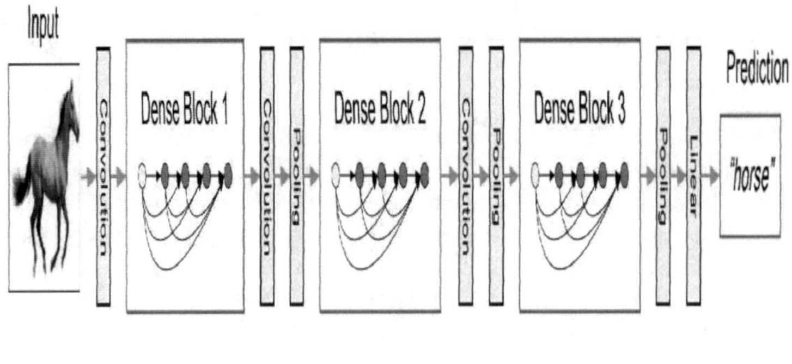

Figure 6. Architecture of three-layer deep DenseNet

IOP Conf. Series: Materials Science and Engineering 1020 (2021) 012009 doi:10.1088/1757-899X/1020/1/012009

different models depending upon the layers used in the model architecture. Some of those are DenseNet121, DenseNet169, DenseNet201 and DenseNet264.

4. Proposed Method

In this paper, we implementd two deep learning models, VGG16 and DenseNet169. Our proposed models are somewhat different from existing models. Following section provides details about both models.

4.1. VGG16

The model is forged using 5 convolution blocks. After each convolutional block, pooling operation is performed. The type of pooling performed is Max-pooling with stride value 2. First convolutional block consists of 64 filters which is the depth of the block. Each filter is of size 3x3. Number of convolutional layers in the block is 2.

(i) Activation function used at the end of block is ReLu activation. Second convolutional block consists of 128 filters which is the depth of the block. Each filter is of size 3x3. Number of convolutional layers in the block is Activation function used at the end of block is ReLu activation.
(ii) Third convolutional block consists of 256 filters which is the depth of the block. Each filter is of size 3x3. Number of convolutional layers in the block is 3. Activation function used at the end of block is ReLu activation.
(iii) Fourth convolutional block consists of 512 filters which is the depth of the block. Each filter is of size 3x3. Number of convolutional layers in the block is Activation function used at the end of block is ReLu activation.
(iv) Last convolutional block consists of 512 filters which is the depth of the block. Each filter is of size 3x3. Number of convolutional layers in the block is 3. Activation function used at the end of block is ReLu activation.

At the output layers, sigmoid activation function is used which gives an output value between 0 and 1.

4.2. DenseNet169

A total of 5 dense blocks are used in the architecture. A $1x1$ convolution is applied at each transition layer. Sigmoid function is used at the output layer to obtain a output value between 0 and 1. The following table shows the architecture of used DenseNet169 model.

4.3. Experimental Setup

We trained our VGG16 and DenseNet169 models on MURA dataset. The MURA dataset needs not to be partitioned in training and validation set since it is already partitioned. Training set about 90% of all the images and validation set has remaining 10% images. The model has to be trained on study basis and not on per image basis because every study has multiple images of same extremity of same patient taken from different view angles. A single image of a study does not guarantee the detection of abnormality. A different view angle might show an abnormality which is not visible in one view as per Figure 8.The following figure shows basic pipeline used in this paper. We resize the variable size images to a fixed size of 224*224 pixels according to ImageNet standards and then we augment the images as mentioned in section 5.1.1. For the evaluation of our models, we used the quantitative evaluations metrics described in section 5.2, which then are compared with our baseline model given by Rajpurkar et al. initial work.

Layer	Output Size	Layer Description
Convolution	112×112	7×7 conv, stride 2
Pooling	56×56	3x3 max pool, stride 2
Dense Block(1)	56×56	$\begin{bmatrix} 1 \times 1conv \\ 3 \times 3conv \end{bmatrix}$ X6
Transition Layer(1)	56×56 28×28	1×1 Conv 2x2 avg pool, stride 2
Dense Block (2)	28×28	$\begin{bmatrix} 1 \times 1conv \\ 3 \times 3conv \end{bmatrix}$ X12
Transition Layer(2)	28×28 14×14	1x1 conv 2x2 avg pool, stride 2
Dense Block (3)	28×28	$\begin{bmatrix} 1 \times 1conv \\ 3 \times 3conv \end{bmatrix}$ X 32
Transition Layer(3)	14×14 7×7	1x1 conv 2x2 avg pool, stride 2
Dense Block (4)	7×27	$\begin{bmatrix} 1 \times 1conv \\ 3 \times 3conv \end{bmatrix}$ X 32
Classification Layer	14×14	7× 7 global avg pool, 1000D fully-connected, sigmoid

Table 1. Architecture of DenseNet169

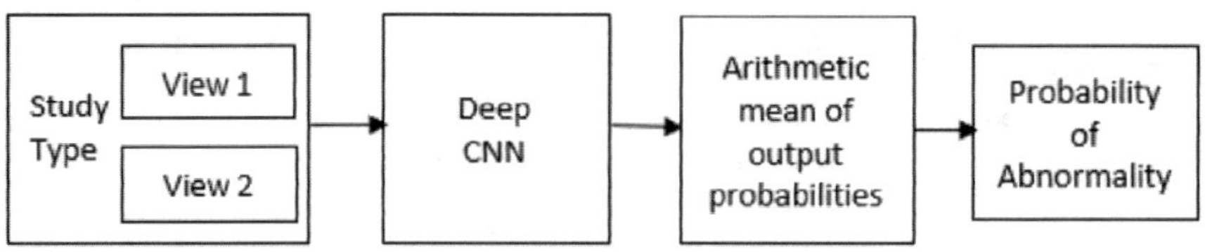

Figure 7. Block Diagram of Model

4.4. Dependencies

- Keras 2.1.6 is used with TensorFlow as backend
- TensorFlow 1.15.0 GPU version is used
- NVIDIA 940MX GPU is used and in order to run TensorFlow on this GPU, following applications are prerequisite:
  - Compute Unified Device Architecture (CUDA) 10.1
  - NVIDIA CUDA Deep Neural Network Library (cuDNN) 7.6.5
- Python version 3.6 is used
- Scipy 1.1.0
- Numpy
- Pandas
- OpenCV-python
- Keras-vis
- Matplotlib
- Tqdm

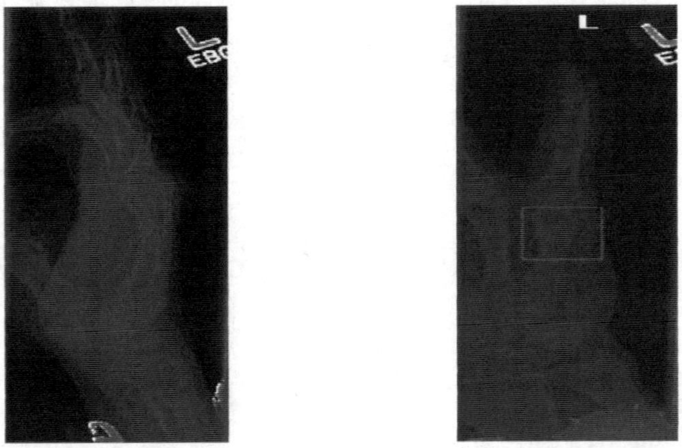

Figure 8. Comparison of Two Radiographic Images(fracture and non-fracture) belonging to the same study

It is better to setup a separate virtual environment for this task so that these particular versions of dependencies do not interfere with existing configurations of the machine. We used Anaconda (64-bit version) to setup a separate virtual environment for deploying our models. Further we require internet connectivity while running the models so it can download ImageNet weights. If the machine is not GPU-enabled, then the user can change "tensorflow-gpu==1.15" to "tensorflow==1.15" written in 'prerequisite.txt' file to make use of CPU only. These dependencies can be installed by simply running the command 'pip install -r requirements.txt' and 'pip install -r prerequisite.txt' in the command prompt of corresponding software installed on the machine.

5. Results

5.1. Data Set

We are using MURA dataset in this paper which is a huge collection of over 40 thousand radiographic images. The objective here is introducing a classification model to classify whether the study has any musculoskeletal abnormality or not. By musculoskeletal abnormality we mean fractures, degenerative bone diseases, inserted hardware and other kind of abnormalities. The dataset contains only upper extremity radiographic studies where each study contains one or more images showing different radiographic views. The input from the dataset is a study, which is basically a set of images, and the output by the model is a binary prediction $y\epsilon\{0,1\}$, where 0 indicates the absence of abnormality and 1 indicates presence of abnormality. Each study belongs to one of the seven standard upper extremity radiographic study types: Elbow, forearm, finger, hand, humerus, shoulder and wrist. Each study was labeled as abnormal or normal manually by board certified radiologists of Stanford Hospital. Table 2 summarizes the distribution of abnormal and normal studies. The dataset is already splitted in training and validation sets. Dataset contains a total of 14,863 studies conducted on a total of 12,173 patients. Total number of images in the dataset is 40,561.

5.1.1. Data Augmentation As we are dealing with upper extremity studies, it is a possibility that the dataset contains unbalanced studies on training set because it might take only one side of the extremity in account. For example, it might have studies of only right hand. So, augmentation of dataset images is crucial so that high-level features are learned by our models. These features are unvaried of transformations like horizontal flips and small degree rotations. Such transformations can occur in real-life scenarios also at the time of capturing radiographs. We make two such transformations which are as follows:

(i) Horizontal Flip: We randomly make horizontal flips on studies to make our data symmetric so that the model can learn both sides of the extremity.

(ii) Small Rotations: Random rotations upto 30 degrees are applied on studies. It enforces the models to deal with small variations that might occur in a normal radiographic study.

5.1.2. Loss Function The process of learning from data or to find the solution to the problem, ideally, we have labels making learning Supervised Learning; in this process training data is used to generate a new predictor function . This function attempts to map input data to labels. We expect this function to generalize for unseen data also. For this, we produce some output label, desired output, for every data point in our dataset. Obviously, the predicted labels are not identical to actual labels, so we try to calculate error value by comparing predicted labels with actual labels. To move close to a good solution, we always try to minimize this error value. The set of operations that we propose to find the error value is known as Loss function. Loss functions might be different for different models depending on the data. So basically, loss functions are generally generated depending upon properties of any given dataset.

5.1.3. Binary Cross Entropy Cross-entropy loss is a metric to evaluate the performance of a classification model having probability value as output. As the predicted label moves away from actual label, cross entropy loss increases. So, lower loss value indicates a better model. For a perfect model loss value is 0. Cross entropy for a binary classification model is given by:

$$-ylog(P) + (1 - y)log(1 - P) \tag{1}$$

5.2. Quantitative Results
We compare results obtained by VGG16 and DenseNet169 models with the baseline proposed by Rajpurkar. We also compare overall performance of each of our models with leader board of Stanford ML Group MURA Dataset Competition.

5.2.1. Baseline The baseline model which we are using to compare our results against is themodel proposed by Rajpurkar. This model obtained a Kappa value of 0.705, sensitivity of 0.815 and specificity of 0.887. The table below shows the results obtained by baseline model:

Study	Elbow	Finger	Forearm	Hand	Humerus	Shoulder	Wrist
Kappa	0.710	0.389	0.737	0.851	0.600	0.729	0.931

Table 2. Performance of Baseline model

5.2.2. VGG16 We trained VGG16 model over 70 epochs with a learning rate of 0.00001 and default parameters. The results obtained are tabulated below: VGG16 achieved overall accuracy of 0.767, sensitivity of 0.701, specificity of 0.827 and a Kappa score of 0.532. Whereas, the baseline model achieved considerably higher scores of sensitivity, specificity and Kappa compared to our model. The following table shows the comparison between results obtained by VGG16 and baseline model. Table 4 clearly shows that baseline model clearly outperforms VGG16 model in all the metrics. Nevertheless, in case of finger study, baseline model obtains a Kappa of 0.389

Study	Accuracy	Sensitivity	Specificity	Precision	Kappa
Elbow	0.800	0.760	0.838	0.821	0.599
Finger	0.750	0.732	0.771	0.786	0.501
Forearm	0.750	0.668	0.833	0.801	0.501
Hand	0.747	0.539	0.892	0.778	0.453
Humerus	0.753	0.771	0.736	0.734	0.507
Shoulder	0.714	0.669	0.757	0.729	0.427
Wrist	0.830	0.749	0.895	0.853	0.652
Overall	0.767	0.701	0.827	0.789	0.532

Table 3. Result analysis of VGG16

Model	Accuracy	Sensitivity	Specificity	Precision	Kappa
Baseline	-	0.815	0.887	-	0.705
VGG16	0.767	0.701	0.827	0.789	0.532

Table 4. Comparison of results between VGG16 and Baseline

whereas VGG16 obtains a Kappa of 0.501. So VGG16 was able to outperform baseline model in only one case. Still VGG16 showed a respectable performance. We visualised the performance of VGG16 graphically using number of epochs versus metrics graph. The following figure shows that graph. We also visualised the loss functions of training and validation set over the epochs.

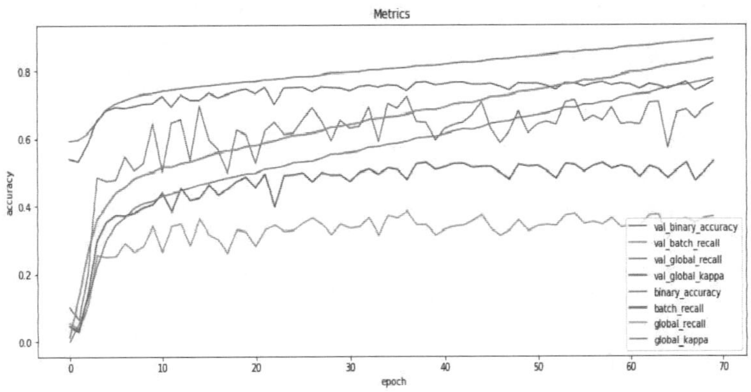

Figure 9. Visual analysis of VGG16 performance

Following Figure 10 shows that graph. In conclusion, we can say that VGG16 could not perform excellently on medical radiographic studies, but it still showed respectable performance.

5.2.3. DenseNet-169 We trained DenseNet model over 40 epochs with initial learning rate of 0.0001, ImageNet weights and default parameters. Every time the validation loss hits a plateau, learning rate is decayed by a factor of 10 in subsequent epoch. The results obtained by DenseNet169 are tabulated below. DenseNet169 obtained an accuracy score of 0.809, sensitivity of 0.721, specificity of 0.890, precision of 0.858 and a Kappa score of 0.616. It achieved specificity more than the baseline model and a very considerable accuracy. The following table shows the comparison between the performance of baseline model and DenseNet169. Our DenseNet169

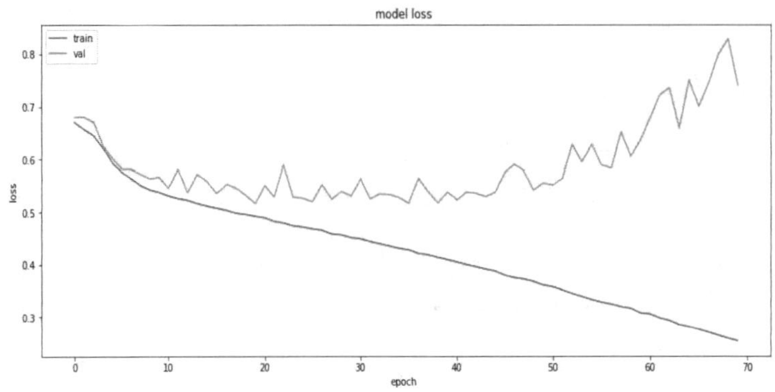

Figure 10. Loss function of VGG16

Study	Accuracy	Sensitivity	Specificity	Precision	Kappa
Elbow	0.855	0.769	0.940	0.926	0.711
Finger	0.776	0.684	0.883	0.871	0.558
Forearm	0.837	0.735	0.940	0.925	0.674
Hand	0.780	0.571	0.926	0.843	0.523
Humerus	0.833	0.817	0.817	0.815	0.666
Shoulder	0.756	0.817	0.817	0.787	0.512
Wrist	0.843	0.769	0.903	0.866	0.680
Overall	0.809	0.721	0.890	0.858	0.616

Table 5. Result analysis of DenseNet169

Study	Kappa (DenseNet169)	Kappa (Baseline)
Elbow	0.711	0.710
Finger	0.558	0.389
Forearm	0.674	0.737
Hand	0.523	0.851
Humerus	0.666	0.600
Shoulder	0.512	0.729
Wrist	0.680	0.930
Overall	0.616	0.705

Table 6. Comparison of results between DenseNet169 and baseline

model outperforms baseline model in the study of elbow, finger and humerus. It also outperforms the best radiologist's performance in finger study. So, DenseNet169 shows a good overall performance. We visualised the performance of DenseNet169 graphically using number of epochs versus metrics graph. The following figure shows that graph. We also visualised the loss functions of training and validation set over the epochs. Following Figure 12 shows that graph. As we mentioned previously, learning rate changes whenever the loss function of validation set hits a plateau. Following figure shows the visualization of decay of learning rate overtime.

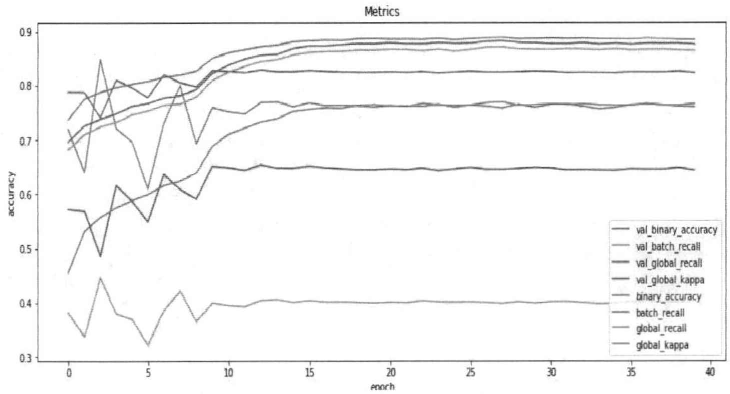

Figure 11. Visual analysis of DenseNet169

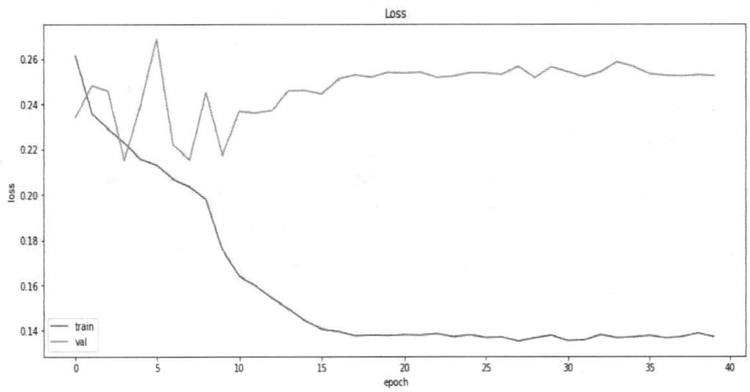

Figure 12. Loss function of DenseNet169

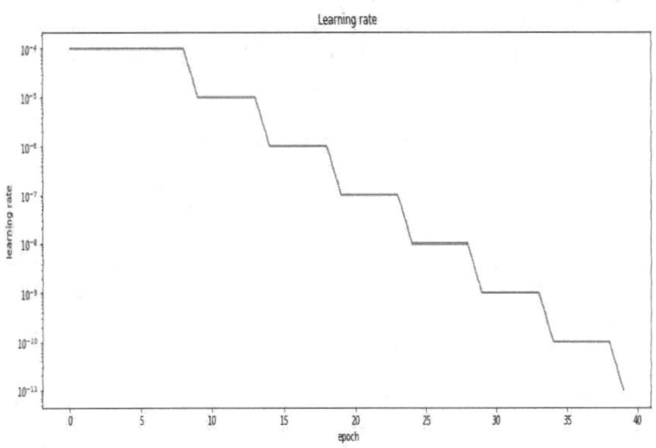

Figure 13. Visualization of learning rate decay

6. Conclusion

In this paper, we looked at the effect of using well-known models which are trained on ImageNet dataset, a dataset consisting non-medical image, to detect musculoskeletal abnormalities using the MURA dataset. We used VGG16 and DenseNet169 models for the cause. Although, our models outperformed baseline models in some aspects, but none of these could beat the overall performance of baseline model. But, on the bright side, our models achieved a very good accuracy score. DenseNet169 performed better than our baseline in many aspects. It even gained higher specificity score than baseline model. Our models secured respectable positions on Stanford University ML Group MURA Dataset competition leaderboard. In conclusion, although these models and even current state-of-the-art models have not achieved the performance level of clinicians, but they may prove to be of vital help for providing second opinion and even for prioritising work in emergency scenarios.

References

[1] Tian T P, Chen Y, Leow W K, Hsu W, Howe T S and Png M A 2003 *Computer Analysis of Images and Patterns* ed Petkov N and Westenberg M A (Berlin, Heidelberg: Springer Berlin Heidelberg) pp 82–89 ISBN 978-3-540-45179-2

[2] Lim S E, Xing Y, Chen Y, Leow W K, Howe T S and Png M A 2004

[3] Vineta Lai Fun Lum, Wee Kheng Leow, Ying Chen, Tet Sen Howe and Meng Ai Png 2005 *IEEE International Conference on Image Processing 2005* vol 1 pp I–1149

[4] Deng J, Dong W, Socher R, Li L, Kai Li and Li Fei-Fei 2009 *2009 IEEE Conference on Computer Vision and Pattern Recognition* pp 248–255

[5] Krizhevsky A, Sutskever I and Hinton G E 2012 *Advances in Neural Information Processing Systems 25* ed Pereira F, Burges C J C, Bottou L and Weinberger K Q (Curran Associates, Inc.) pp 1097–1105 URL `http://papers.nips.cc/paper/4824-imagenet-classification-with-deep.pdf`

[6] Szegedy C, Liu W, Jia Y, Sermanet P, Reed S, Anguelov D, Erhan D, Vanhoucke V and Rabinovich A 2015 *Proceedings of the IEEE conference on computer vision and pattern recognition* pp 1–9

[7] Simonyan K and Zisserman A 2014 Very deep convolutional networks for large-scale image recognition (*Preprint* `1409.1556`)

[8] Ronneberger O, Fischer P and Brox T 2015 *Medical Image Computing and Computer-Assisted Intervention – MICCAI 2015* ed Navab N, Hornegger J, Wells W M and Frangi A F (Cham: Springer International Publishing) pp 234–241 ISBN 978-3-319-24574-4

[9] Chandra T B and Verma K 2020 *Proceedings of 3rd International Conference on Computer Vision and Image Processing* (Springer) pp 21–33

[10] Gale W, Oakden-Rayner L, Carneiro G, Bradley A P and Palmer L J 2017 *CoRR* **abs/1711.06504** (*Preprint* `1711.06504`) URL `http://arxiv.org/abs/1711.06504`

[11] Rajpurkar P, Irvin J, Bagul A, Ding D, Duan T, Mehta H, Yang B, Zhu K, Laird D, Ball R L, Langlotz C, Shpanskaya K, Lungren M P and Ng A Y 2017 Mura: Large dataset for abnormality detection in musculoskeletal radiographs (*Preprint* `1712.06957`)

[12] `https://stanfordmlgroup.github.io/competitions/mura/`

[13] 2009 `https://mc.ai/the-architecture-behind-cnn-and-vggnet-16/`

Analysis of EC over Gamma Shadowed α-η-μ Fading Channel

Poonam Yadav*, Sandeep Kumar+, Rajesh Kumar*

*Dept. of EC, NERIST, Nirjuli, AP, India

+CRL, BEL, Ghaziabad, UP, India

*poonam.yadav.25@gmail.com, +sann.kaushik@gmail.com, *itsrk2006@gmail.com

Abstract-Composite fading channels are used to characterize the real time wireless communication systems, as it can account for multipath fading as well as shadowing effect on the channel. To analyse the performance of the wireless communication systems supporting real time applications effective capacity is a crucial performance measure. In this paper, analysis of effective capacity for the α-η-μ fading channel under the shadowing effect is performed. To model shadowing gamma distribution is used. Moreover, asymptotic expressions are derived in the pursuance of obtaining insight behaviour of the channel.

Keywords-Delay constraint, effective capacity, generalized α-η-μ fading channel, QoS provision

1. Introduction

Effective Capacity (EC) proposed by the authors in [1] is drawing a heavy attention as a key performance criteria to quantify the channel performance including Quality of Service (QoS) metric.The fading effect in the channel is one of the major reason for variation of the channel capacity. There is extensive literature available on the analysis of effective rate on to the various fading channels. EC is analysed over generalized and multiple input single output (MISO) system in [2] and [3] respectively. Authors have derived tractable expression in terms of well known mathematical function such as meijer G and fox-H . To analyse EC closed-form expressions for MISO system over the κ-μ , η-μ and α- μ fading channel is presented in [4], [5], [6]. In [7], EC of the generalized inverse gamma fading channel is studied by deriving a closed-form expression.

Enormous work has been devoted to the system EC analysis over several composite fading channels [8], [9], [10]. A generalized fading distribution α-η-μ was introduced by the authors in [11]. The versatility of this distribution to account for the real world propagation aspects such as irregularity of the medium and number of multipath clusters have drawn keen attention to analyse the performance of communication system. Spectrum sensing study based energy detection for the gamma shadowed α-η-μ and α-κ-μ is presented [12] . In [13] simplified closed-form expression using fox H function over the lognormal(LN) shadowed α-η-μ and α-κ-μ fading channel is presented by the authors whereas ED performance is studied in [14] . On the extensive literature survey it has been found that no work has been presented on the study of the EC of composite α-η-μ channel experiencing gamma shadowing. Motivated by this, authors here present a simple closed-form expression for the EC analysis of this composite fading channel.

This paper is systemized as follows. In section 2 channel model comprising of α-η-μ/gamma distribution is discussed. Generalized expression for EC of the mentioned composite fading channel is derived in section 3, followed section 4 comprising of asymptotic analysis in section 4. Section 5 puts light on the numerical and simulation results following with conclusion in section 6.

2.Channel Model

2.1. α-η-μ/gamma Distribution:

This distribution elucidates the non-homogeneity as well as non-linearity of the propagation channel. The composite PDF for received SNR γ for the channel is given in [12] as

$$f_\gamma(\gamma) \approx C\gamma^{m-1} \sum_{k=1}^{K} w_k\, a_k\, e^{-b_k\gamma} \quad (1)$$

where a_k , w_k and K are the abscissas ,weight factors, and number of Gaussian –Laguerre summation terms..So the composite PDF expression for gamma shadowed α-η-μ channel can be given by (1) and other parameters of (1) are given as

$$C=\frac{\sqrt{\pi}\, 2^{\left(\frac{2m}{\alpha}-\mu+1/2\right)}\mu^{2m/\alpha} h^{\left(\frac{2m}{\alpha}-1/2\right)}}{\Gamma(\mu)\Gamma(m)H^{(\mu-1/2)}\bar{\gamma}^{m}}\ , \quad a_k = x_k^{\left(\mu-\frac{2m}{\alpha}-\frac{1}{2}\right)} I_{(\mu-1/2)}\left(\frac{Hx_k}{h}\right)\ , \quad b_k = \frac{1}{\bar{\gamma}}\left(\frac{2\mu h}{x_k}\right)^{\frac{2}{\alpha}} \quad (2)$$

The parameters h and H are defined in [12], the values of H and h for both the formats are given in table 1.

Table1. Value of h and H

Format	h	H
1 (0< $\eta < \infty$)	$(1+\eta)^2/4\eta$	$(1-\eta)^2/4\eta$
2 (-1< $\eta < 1$)	$1/(1-\eta^2)$	$\eta/(1-\eta^2)$

3. Effective Capacity

The analytical expression for normalized EC for a block fading channel is given by [2]

$$R = -\frac{1}{A} log_2(\mathbb{E}[(1+\gamma)^{-A}]) \quad (3)$$

Where,

$$\text{Delay constraint metric, } A = \theta TB/ln2 \quad (4)$$

θ= delay exponent ;T = frame duration; B =bandwidth of system; $\mathbb{E}[.]$ being expectation operator.

$$R = -\frac{1}{A} log_2\left(\int_0^\infty (1+\gamma)^{-A} f_\gamma(\gamma)d\gamma\right) \quad (5)$$

3.1 EC analysis over α-η-μ/gamma channel

Substituting (1) in (5) we can deduce the expression for the EC of gamma shadowed α-η-μ fading channel, which can be written as

$$R^{\alpha-\eta-\mu/\text{gamma}} = -\frac{1}{A} log_2\left(C \sum_{k=1}^{K} w_k\, a_k \int_0^\infty (1+\gamma)^{-A} \gamma^{m-1} e^{-b_k\gamma} d\gamma\right) \quad (6)$$

The integral can be obtained by using equation 8.4.2.5 given in [15]

$$R^{\alpha-\eta-\mu/\text{gamma}} = -\frac{1}{A} log_2\left(C \sum_{k=1}^{K} w_k\, a_k \int_0^\infty \frac{1}{\Gamma A}\gamma^{m-1}. e^{-b_k\gamma} G_{1,1}^{1,1}\left[\gamma \middle| \begin{matrix} 1-A \\ 0 \end{matrix}\right] d\gamma\right) \quad (7)$$

Transforming the exponential term by using equation (07.34.03.0228.01) of [16] , (7) can be given as

$$R^{\alpha-\eta-\mu/\text{gamma}} = -\frac{1}{A} log_2\left(C \sum_{k=1}^{K} w_k\, a_k \int_0^\infty \frac{1}{\Gamma A}\gamma^{m-1} G_{0,1}^{1,0}[b_k\gamma|0] G_{1,1}^{1,1}\left[\gamma \middle| \begin{matrix} 1-A \\ 0 \end{matrix}\right] d\gamma\right) \quad (8)$$

Solving (8) using equation 07.34.21.0011.01 of [16] and rearranging the terms, the closed form expression for the EC of gamma shadowed α-η-μ fading channel is given as

$$R^{\alpha-\eta-\mu/\text{gamma}} = -\frac{1}{A} log_2\left(C \sum_{k=1}^{K} w_k\, a_k \frac{1}{\Gamma A} b_k^{-m} G_{2,1}^{1,2}\left[\frac{1}{b_k} \middle| \begin{matrix} 1-A, 1-m \\ 0 \end{matrix}\right] d\gamma\right) \quad (9)$$

4. Asymptotic Analysis

4.1. High SNR Analysis

The closed-form expression explains limited physical significance of the quantitative effects of different parameters. So, high SNR analytical study is done to get the deeper understanding of the influence of the various parameters on the channel. Under high SNR regime where, $\gamma \to \infty$, (2) can be reduced to

$$R = -\frac{1}{A} log_2(E[(\gamma)^{-A}]) \tag{10}$$

At $\gamma \to \infty$, EC for α-η-μ/gamma is given by substituting (9) in (13) and Using equation 3.326.2 given in [17] we get

$$R^{\infty} == -\frac{1}{A} log_2\left(C \sum_{k=1}^{K} w_k\, a_k\, \frac{\Gamma(m-A)}{b_k(m-A)}\right) \tag{11}$$

4.2. Low SNR Analysis

In many wireless communication applications, power levels of the transmitted signals are considered to be very low so for the efficient design of such systems low SNR approximation provides a better insight of the system. In the low SNR sovereignty EC is approximated in the terms of transmitted normalized energy per information bit, $\frac{E_b}{N_0}$ [18] which is mathematically given as

$$R(\frac{E_b}{N_0}) \approx S_0 log_2(\frac{E_b/N_0}{E_b/N_{0min}}) \tag{12}$$

where S_0 is the rate against the SNR slope and is expressed asE_b/N_{0min} is the minimum transmit energy, needed to reliably convey non-zero transmission rate. Mathematically expressed as

$$S_0 = \frac{-2\left(R'(0,\theta)\right)^2 ln2}{R''(0,\theta)} \quad ; \quad E_b/N_{0min} = \lim_{\rho \to 0} \frac{\rho}{R(\rho,\theta)} = \frac{1}{R'(0,\theta)} \tag{13}$$

Calculating the Ist order and IInd order partial derivatives of the EC function we get

$$R'(0,\theta) = \frac{E\{\gamma\}}{ln2}\ , R''(0,\theta) = \frac{-(A+1)E\{\gamma^2\}}{ln2} + \frac{A(E\{\gamma\})^2}{ln2} \tag{14}$$

using equation 3.326.2 of [19] $E\{.\}$ is evaluated as

$$E\{\gamma^{-A}\} = C \sum_{k=1}^{K} w_k\, a_k\, \Gamma(m-A)/b_k(m-A) \tag{15}$$

By keeping *A=-1* and *A=-2*, first and second moment of the received SNR can be obtained respectively.

$$E\{\gamma\} = C \sum_{k=1}^{K} w_k\, a_k\, \Gamma(m+1)/b_k(m+1) \tag{16}$$

$$E\{\gamma^2\} = C \sum_{k=1}^{K} w_k\, a_k\, \Gamma(m+2)/b_k(m+2) \tag{17}$$

Substituting (16) and (17) in (14) and then keeping the result in (13), S_0 can be calculated as

$$S_0 = 2\Big/\left(\frac{(A+1)C\sum_{k=1}^{K} w_k a_k\, \Gamma(m+2)/b_k(m+2)}{\left(C\sum_{k=1}^{K} w_k a_k\, \Gamma(m+1)/b_k(m+1)\right)^2} - A\right) \tag{18}$$

By substituting (18) in (12), the generalized expression for EC over α-η-μ/gamma composite fading channels for low SNR can be given as

$$R(\frac{E_b}{N_0}) \approx \frac{2log_2\left(\frac{(C\sum_{k=1}^{K} w_k a_k\, \Gamma(m+1)/b_k(m+1))E_b/N_0}{ln2}\right)}{\left(\frac{(A+1)C\sum_{k=1}^{K} w_k a_k\, \Gamma(m+2)/b_k(m+2)}{\left(C\sum_{k=1}^{K} w_k a_k\, \Gamma(m+1)/b_k(m+1)\right)^2} - A\right)} \tag{19}$$

5. Results and Discussion

This section includes the comprehensive explanation of results obtained. The parameters *h* and *H* are evaluated using format 1 as mentioned in the analysis. The results are validated with Monte Carlo simulations using 10^6 samples. The analytical expressions obtained have been simulated using MATLAB 2015a. Fig. 1 depicts impact of fading parameter on the presented channel. It is clear from the graph that increment in the value of the fading parameters, enhances EC of the channel. This is followed by the fact

IOP Conf. Series: Materials Science and Engineering 1020 (2021) 012010 doi:10.1088/1757-899X/1020/1/012010

that, larger value α leads to larger fading gains, larger value of η means higher power of in-phase component and μ represents the number of multipath cluster. Hence, as these parameters increases performance improves. The asymptotic curves tightly follow the analytical curves at high SNR. Fig 2 reveals profile of EC curve under the impact of the delay constraint A. Quantitatively, At A=10, the EC degrades by 1 bit/Hz/s as $\bar{\gamma}$ falls to 5 dB from 20 dB. At $\bar{\gamma}$ =20dB as A increases from 4 to 8 there is a decrement in the EC by 0.5 bits/Hz/s. On analysing the behaviour it can be clearly stated that as the delay increases the performance of the system degrades, however at a fixed delay value performance improves with increase in $\bar{\gamma}$.

Figure 3 explains how the shadowing parameter affects the channel performance. It can be observed that for the higher values of m the EC high which indicates that shadowing is becoming less prominent. Also it is observed that as the $\bar{\gamma}$ is increasing the EC is also improving. Figure 4 reveals the EC behaviour in the under the low SNR condition. From the plots it can be stated that average transmit energy decreases as the shadowing index is increased. For example, as shadowing index increases to 1.5 from 1, average transmit SNR decreases by 1.3 dB also change in A does not affects the EC performance in low SNR regime.

6. Conclusion

In this paper a closed-form analytical expression for the EC over the α-η-μ/gamma composite fading environment is derived. The profile of the EC for the shadowing index as well as fading parameters is comprehensively studied. The impact of the delay constraint parameter A is shown in the results, which concludes that on increasing delay the EC performance of the channel ameliorates. Additionally, asymptotic analysis reveals the insight behaviour of the of the channel under the fading environment. Furthermore, results presented here can be elaborated to study over the techniques to maximize the capacity of the channel.

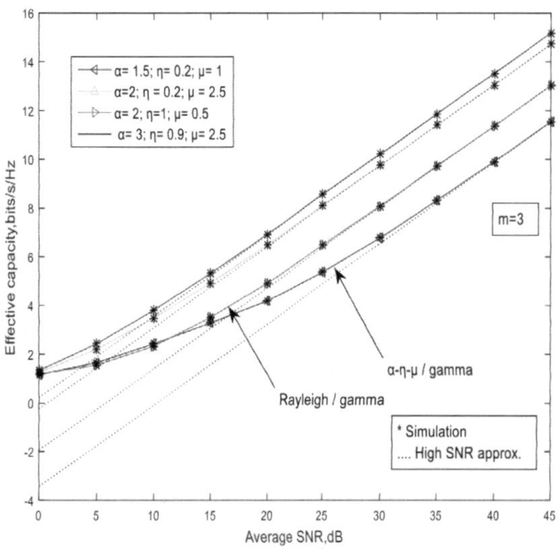

Figure 1. EC versus Average SNR for the for distinct values of α, η and μ at A=2

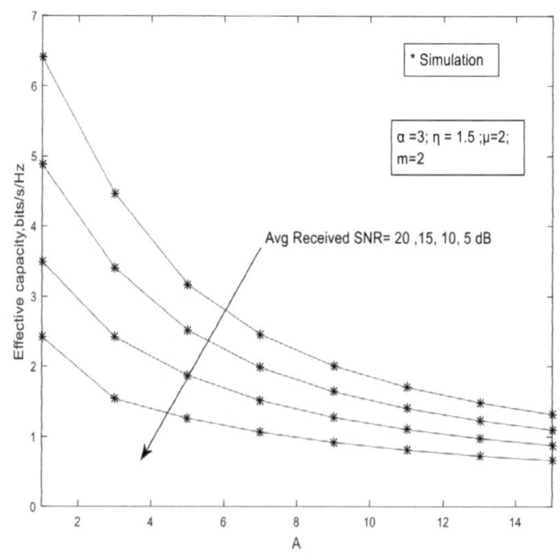

Figure 2. EC versus A for distinct values of $\bar{\gamma}$ at B=1Hz and T=1 sec

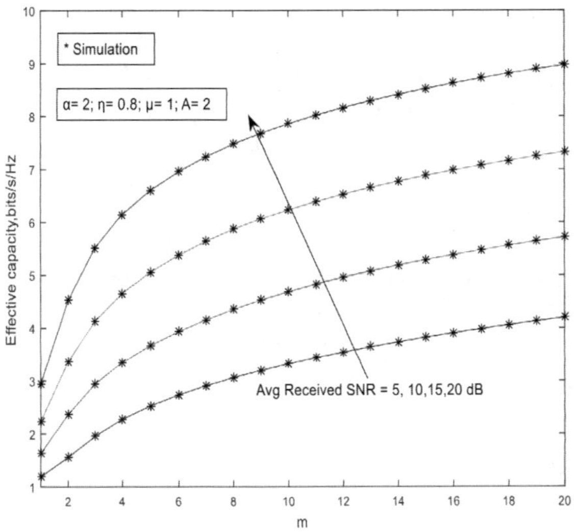

Figure 3.EC versus m for distinct values of $\bar{\gamma}$

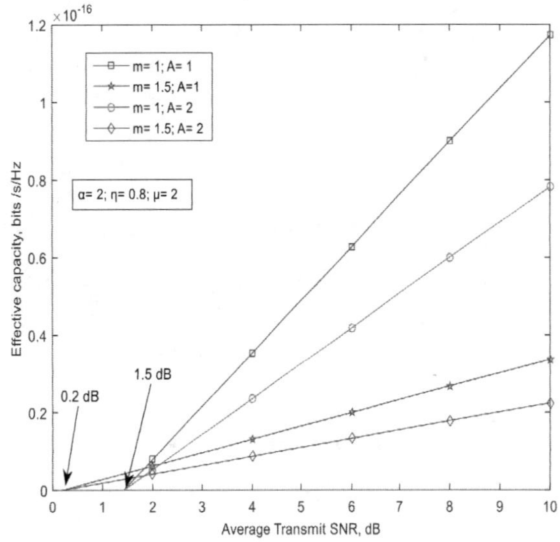

Figure 4. Low SNR approximation versus average transmit SNR

References

[1] Wung D and Negi R 2003 Effective capacity: A wireless link model for supportof quality of service *IEEE Trans. Wireless Commun.* **2** *630-643.*

[2] Ji Z, Dong C, Wang Y and Lu J 2014 On the analysis of effective capacity over generalized fading channels *IEEE International Conference on Communications (ICC), Sydney, NSW, Australia.*

[3] You M,Sun H,Jiang J and Zhiang J 2016 Effective rate analysis in Weibull fading channels *IEEE Wireless Commun. Lett* **5**, *340-343.*

[4] Zhang H, Tang Z, Wang H, Huang Q and Hanzo L 2014 Throughput of MISO systems over κ-μ fading channels, *IEEE Trans Veh. Technol.,* **63**, *943–947*

[5] Zhang J,Matthaiou M, Tan Z and Wang H 2013 Effective rate analysis of MISO η-μ fading channels , *IEEE Int. Conf. Commun. (ICC),*, Budapest, Hungary.

[6] J. Zhang, L. Dai, Z. Wang, D. W. K. Ng and W. H. Gerstacker, 2015 Effective Rate Analysis of MISO Systems over α-μ Fading Channels, *IEEE Global Communications Conference (GLOBECOM)*, San Diego, CA, USA, 6-10 .

[7] You SK,Cotton SL,Sofotasios PC,Muhadait S and Karagiannidis GK, 2020 Effective Capacity Analysis Over Generalized Composite Fading Channels *IEEE Access,* **8**, *123756-123764.*

[8] Almehmadi FS, Badarneh OS. 2018 On the effective capacity of fisher–Snedecor F fading channels. *Electronics Letters.* **54** *1068-1070.*

[9] Zhang J and Gerstacker WH , "Effective capacity of communication systems over κ–μ shadowed fading channels," *Electronic Lett.,* vol. (19).51, p. 1540–1542, 2015.

[10] Al-Hmood H, Al-Raweshidy HS. 2019 Effective throughput of MISO systems over *κ-μ* shadowed fading channels: MGF based analysis. 2018.https://arxiv.org/abs/1804.10991.

[11] Fraidenraich G, Yacoub MD. The *α-ν-μ* and *α-κ-μ* fading distributions. Paper presented at: IEEE Ninth International Symposium on Spread Spectrum Techniques and Applications; August 28–31, 2006; Manaus-Amazon, Brazil: 16–20.

[12] Kumar S, Kaur M, Singh NK, Singh K, Chauhan PS. 2018 Energy detection based spectrum sensing for gamma shadowed α-η-μ and α-κ-μ fading channels. *AEU-Int J Electron C.***93** *26-31.*

[13] Yadav P, Kumar R, Kumar S. 2020 Effective capacity analysis over generalized lognormal shadowed composite fading channels. *Internet Technology Letters.* e171.https://doi.org/10.1002/itl2.171

[14] Bhatt M and Soni SK 2018 A unified performance analysis of energy detector over α−η−μ/lognormal and α−κ−μ/lognormal composite fading channels with diversity and cooperative spectrum sensing *Int. J. Electron. Commun. (AEÜ), 367-376.*

[15] Prudnikov AP, Brychkov YA, Marichev OI. 1990 *Integrals and series: more special functions*, **3**. New York, NY: Gordon & Breach Sci. Publ.

[16] WOLFRAM.Wolfram function site. http://functions.wolfram.com/PDF/MeijerG.pdf. Accessed August 30, 2020.

[17] I. S. Gradshteyn and I. M. Ryzhik, Table of Integrals, Series, and Products, 7th ed., Elsevier: Academic Press, 2007.

[18] Tulino AM, Verdú S, Lozano A.2003 Multiple-antenna capacity in the low-power regime. *IEEE Trans Inf Theory.* **49** *2527-2544.*

Single Dual-X Current Conveyor based BASK/BPSK Modulators

Deva Nand
Department of Electronics and Communication Engineering
Delhi Technological University, Delhi, India.
devkamboj07@gmail.com

Abstract. Two voltage mode digital modulators have been proposed i.e. Binary Amplitude Shift Keying (BASK) modulator and Binary Phase Shift Keying (BPSK) modulator. Both circuits use Dual-X Second Generation Current Conveyor (DXCCII) as the active building block and have the advantages of proper input impedances and use of grounded resistor. Detailed mathematical analysis of both circuits are given. The circuits have been simulated with SPICE using 0.35μm CMOS technology parameters. The supply voltage of ±1.8 V is used. Much satisfactory results have been obtained with the proposed circuits.

Keywords: BASK; BPSK; DXCCII; modulators.

1. Introduction

Modulation is a process where one of the parameters of the carrier signal, which can be amplitude, phase or frequency, is varied linearly according to the variations in the message signal or modulating signal. With the modulation, the information contained in the message signal is transferred to the carrier signal. When modulation is done for the digital signals then it is known as the digital modulation. The device that performs this modulation is known as the digital modulator. The digital signal is a basically a binary signal in the form of 0s and 1s and it is a low pass signal and can't be transmitted through the wireless medium, so in order to transmit the digital signal through wireless medium, modulation is required. In Binary Amplitude Shift Keying (BASK), the binary 1 is represented by the presence of carrier signal while binary 0 is represented by the absence of carrier signal whereas in Binary Phase Shift Keying (BPSK), the binary 1 is represented by the presence of in phase carrier signal while binary 0 is represented by the presence of 180° out of phase carrier signal. Infrared (IR) remote controls, transmitter and receiver in optical fiber communication employ BASK, and broadband modems, satellite communication, mobile phones employ BPSK.

Many new active building blocks have been introduced in recent past to explore above mentioned applications and one such trendy block is second generation current conveyor (CCII) [1] and inverting CCII [2]. Literature survey shows that one another most trending block that can be explored to implement modulation process is dual X second generation current conveyor (DXCCII) [3,4]. As per literature survey, two EXCCII based voltage mode digital modulator circuits are reported in [5] and [6] which provide simultaneous ASK and BPSK modulation processes. A current mode BASK modulator circuit has been reported in [7] making use of two MO-CCCCTA. A voltage mode BASK modulator using only one CCIII+ has been reported in [8]. Voltage mode memristor based BASK and BPSK modulators are presented in [9]. DPCCII based BASK modulator is reported in [10]. Further voltage mode MO-CCCC, digitally programmable CDTA and MO-OTA based BASK and BPSK modulator circuits are reported in [11], [12] and [13] respectively. A DXCCII based modulator topology has also been presented in [6] which performs BPSK modulations but unexplored and unanalyzed.

The newly proposed BASK modulator circuit employs single DXCCII, one NMOS transistor and one grounded resistor while BPSK modulator circuit uses single DXCCII, one NMOS and one PMOS transistor and one grounded resistor. This article includes total of six sections including introduction in Section-1. DXCCII block description is given in Section-2 followed by proposed BASK and BPSK modulators in Section-3 & 4 respectively. Proposals simulation verification is reported in Section-5 and concluded in Section-6.

2. DXCCII Description

The DXCCII, conceptually, is a combination of CCII [1] and ICCII [2]. It is therefore more versatile building block as it has the features of both the current conveyors i.e. CCII and ICCII. The block diagram of DXCCII [3] is shown below in Fig. 1. It has 5 terminals named as Y, Xp, Xn, Zp and Zn. Xp is called as the non-inverting X terminal while Xn is called as the inverting X terminal. Y is an input terminal with high impedance. Xp and Xn are terminals with low impedance while Zp and Zn are terminals with high impedance.

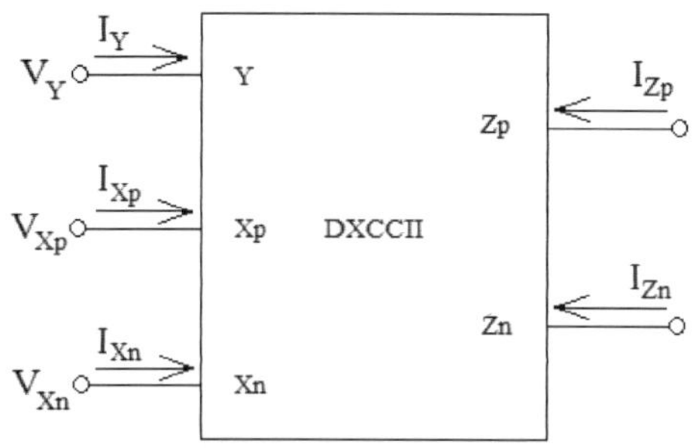

Figure 1. Block diagram of DXCCII [3].

Table 1. (W/L) Ratio of CMOS for DXCCII [3].

Transistor	W(µm)	L(µm)
M_1, M_2	1.4	0.7
M_3, M_7, M_8	2.8	0.7
M_4, M_5	2.4	0.7
M_6, M_9, M_{10}	4.8	0.7
M_{11} to M_{20}	9.6	0.7

Ideally, the voltages and currents relationships of various terminals can be described by the following matrix [3] mentioned below in (1).

$$\begin{bmatrix} I_Y \\ V_{Xp} \\ V_{Xn} \\ I_{Zp} \\ I_{Zn} \end{bmatrix} = \begin{bmatrix} 0 & 0 & 0 \\ 1 & 0 & 0 \\ -1 & 0 & 0 \\ 0 & 1 & 0 \\ 0 & 0 & 1 \end{bmatrix} \begin{bmatrix} V_Y \\ I_{Xp} \\ I_{Xn} \end{bmatrix} \quad (1)$$

The CMOS implementation of the DXCCII used in the proposed circuits is shown in Fig. 2. The aspect ratio of transistors used in the implementation of DXCCII are given in Table 1 [3].

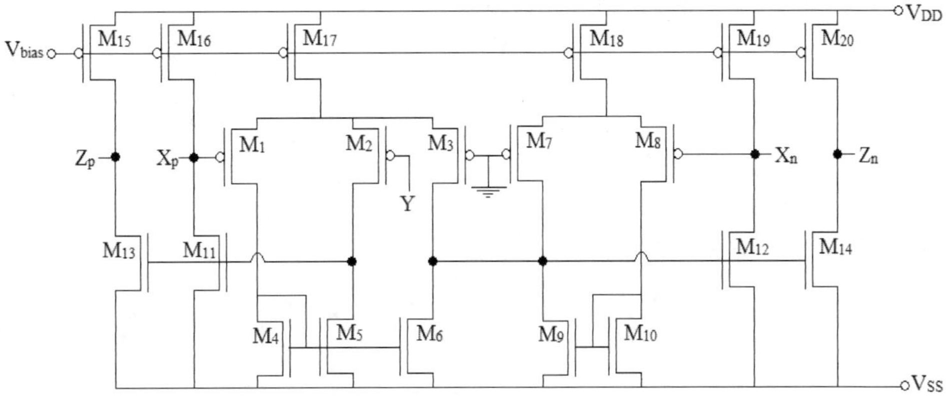

Figure 2. CMOS implementation of DXCCII [3].

3. Proposed BASK Modulator

The proposed BASK modulator is shown in Fig. 3 below.

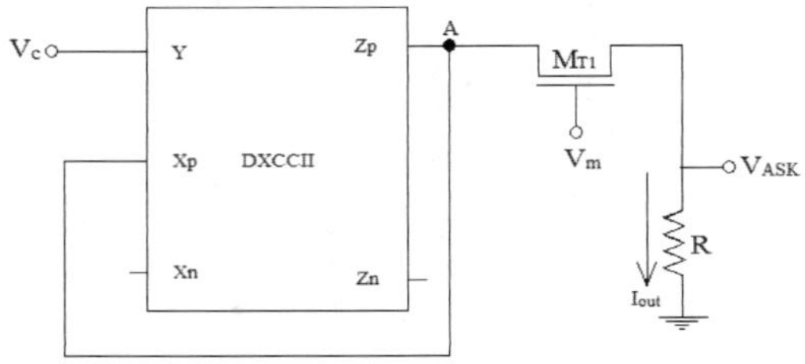

Figure 3. Proposed BASK modulator

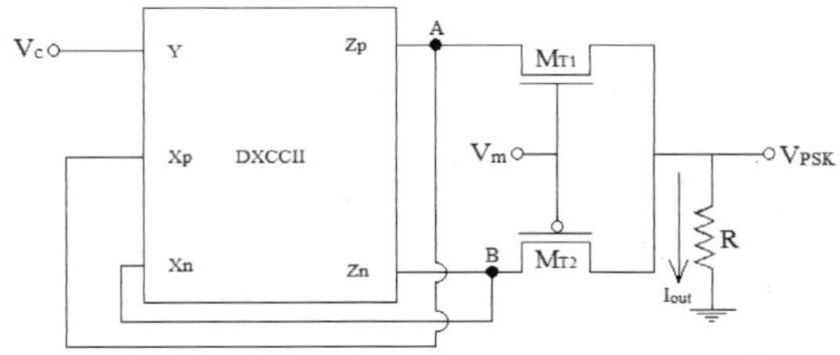

Figure 4. Proposed BPSK modulator

3.1 Operation of the proposed BASK modulator

V_c is the sinusoidal carrier signal. It carries the modulating signal V_m. Voltage V_c is applied at the high input impedance terminal Y of the DXCCII. Voltage V_c is given as

$$V_c = A_c \mathrm{Sin}(2\pi f_c t) \tag{2}$$

Where

A_c is the amplitude of the carrier signal voltage V_c,
t is the time period and f_c is the frequency in hertz (Hz) or cycles/second.

This carrier signal voltage V_c is copied at the Xp terminal of the DXCCII, that is

$$V_{Zp} = V_{Xp} = V_c \tag{3}$$

voltage V_m is the bipolar modulating signal.
voltage V_{ASK} is the BASK modulated output voltage signal.

The M_{T1} is used as a switch. The modulating signal V_m is applied at the high impedance gate terminal of this transistor. When voltage V_m is positive, the M_{T1} turns on and works as a closed switch, and when voltage V_m is negative, the M_{T1} turns off and works as an open switch. The gate voltage of M_{T1} depends on the amplitude of the modulating signal V_m. And the drain voltage of M_{T1} depends on the amplitude of the carrier signal V_c, which is very small as voltage V_c is much smaller in magnitude than voltage V_m. The source voltage of M_{T1} depends on the voltage V_{ASK} developed across the resistor R.

Therefore, whenever M_{T1} is on, it always works in the non-saturation region or the ohmic region as it always holds the condition of working in ohmic region which is given as

$$V_{DSn} < V_{GSn} - V_{THn} \tag{4}$$

where
V_{DSn} is the drain to source voltage of M_{T1}, V_{GSn} is the gate to source voltage of M_{T1},
and V_{THn} is the threshold voltage of M_{T1}, V_{THn} for NMOS is positive.

The on resistance of this M_{T1} used as a switch is denoted by $R_{on,n}$. From [6], for the ohmic region, the resistance $R_{on,n}$ is therefore expressed as

$$R_{on,n} = \frac{1}{(W/L)_n \mu_n C_{ox,n}(V_{GSn} - V_{THn} - V_{DSn})} \tag{5}$$

Where
$(W/L)_n$ is the aspect ratio of M_{T1}, μ_n is the mobility of electron, and $C_{ox,n}$ is the gate oxide capacitance per unit area, of M_{T1}.
It can be seen from the equation (5) that the $R_{on,n}$ varies with voltage V_{GSn} and voltage V_{DSn}. But, since the on resistance $R_{on,n}$ is much smaller in magnitude than the magnitude of the external resistor R, therefore the effect of variation of $R_{on,n}$ on the BASK modulated output voltage V_{ASK} is negligible. Mathematically, it can be expressed by the following equations written below

$$R_{on,n} \ll R \tag{6}$$

The BASK modulated output voltage is given as

$$V_{ASK} = I_{out} \times R \tag{7}$$

The voltage V_A at node A is equal to the voltage V_{Xp}, that is

$$V_A = V_{Xp} \tag{8}$$

Also, when the M_{T1} is on, then the total resistance $R_{total,n}$ between node A and ground is equal to the sum of the on resistance $R_{on,n}$ of the M_{T1} and the external resistance R, i.e.

$$R_{total,n} = R_{on,n} + R \tag{9}$$

The output current I_{out} is given as

$$I_{out} = \frac{V_A}{R_{total,n}} \tag{10}$$

From equations (8), (9) and (10)

$$I_{out} = \frac{V_{Xp}}{R_{on,n} + R} \tag{11}$$

From equations (6) and (9)

$$R_{total,n} = R_{on,n} + R \cong R \tag{12}$$

Therefore from equations (11) and (12), I_{out} is given as

$$I_{out} \cong \frac{V_{Xp}}{R} \tag{13}$$

Thus from the equations (7) and (13), it can be concluded that the variation in the output current I_{out} because of $R_{on,n}$ is negligible and therefore, the effect on the output voltage V_{ASK} because of the variations in $R_{on,n}$ is also negligible. When the M_{T1} working as a switch is in off condition than the associated resistance is denoted by $R_{off,n}$. $R_{off,n}$ is ideally infinite as the M_{T1} is in non-conducting state. Now when the voltage of the modulating signal V_m is positive (logic high), M_{T1} turns on and the equal amount of the currents developed at Xp and Zp terminals of the DXCCII flow towards the node A, which then get added up at the node A as no current goes to the gate terminal of M_{T1} and then the net current flows through the transistor M_{T1} and then flows through the resistor R and then goes to the ground. The two currents developed at Xp and Zp terminals of the DXCCII linearly depend on the voltage V_{Xp} as can be observed from the equation (13). As the voltage V_{Xp} is AC in nature therefore the currents developed because of it are also AC in nature and since these two currents add up at the node A and therefore the resultant current is also AC in nature which flows through the resistor R developing an AC voltage V_{ASK} in return across it which is in phase with the carrier signal voltage V_c. When the voltage of the modulating signal V_m is negative (logic low), M_{T1} turns off and therefore no current flows through the resistor R and the voltage V_{ASK} developed across it is zero. This way, the ASK modulated output voltage V_{ASK} is thus obtained across the resistor R.

The overall operation of the proposed BASK modulator can therefore be summarized by these two equations as mentioned below

$V_{ASK} = I_{out} \times R \cong V_c = A_c\text{Sin}(2\pi f_c t)$; when V_m is positive or logic high (NMOS M_{T1} is on)

$V_{ASK} = 0$; when V_m is negative or logic low (NMOS M_{T1} is off)

4. Proposed BPSK Modulator

The proposed BPSK modulator is shown in Fig. 4.

4.1 Operation of the proposed BPSK modulator

V_c is the sinusoidal carrier signal. It carries the modulating signal V_m. V_c is applied at the high input impedance terminal Y of the DXCCII. V_c is given as

$$V_c = A_c\text{Sin}(2\pi f_c t) \quad (14)$$

Where

A_c is the amplitude of the carrier signal voltage V_c, t is the time period and f_c is the frequency in hertz (Hz) or cycles/second. This carrier voltage V_c is copied at the Xp terminal of the DXCCII, that is

$$V_{Zp} = V_{Xp} = V_c = A_c\text{Sin}(2\pi f_c t) \quad (15)$$

And the voltage developed at the Xn terminal of DXCCII is the inverted carrier voltage V_c, that is

$$V_{Zn} = V_{Xn} = -V_c = A_c\text{Sin}(2\pi f_c t) \quad (16)$$

V_m is the bipolar modulating signal. V_{PSK} is the BPSK modulated output voltage signal.

The M_{T1} and M_{T2} are used as the two switches. The modulating signal V_m is applied at the high impedance gate terminals of these transistors, M_{T1} and M_{T2}. When voltage V_m is positive M_{T1} turns on and works as a closed switch while M_{T2} turns off and works as an open switch. And when voltage V_m is negative M_{T1} turns off and works as an open switch while M_{T2} turns on and works as a closed switch. The gate voltages of M_{T1} and M_{T2} depend on the amplitude of the modulating signal V_m. And the drain voltage of M_{T1} and the source voltage of M_{T2} depend on the amplitude of the carrier signal V_c, which are very small as voltage V_c is much smaller in magnitude than voltage V_m. The source voltage of M_{T1} and the drain voltage of M_{T2} depend on the voltage developed across the resistor R. Therefore, whenever M_{T1} and M_{T2} are on, they always work in the non-saturation region or the ohmic region as they always hold the condition of working in the ohmic region. The conditions for operation in the ohmic region for M_{T1} and M_{T2} are given below in equations (17) and (18) respectively.

For M_{T1}

$$V_{DSn} < V_{GSn} - V_{THn} \quad (17)$$

For M_{T2}

$$V_{DSp} > V_{GSp} - |V_{THp}| \tag{18}$$

where
V_{DSn} is the drain to source voltage of M_{T1}, V_{GSn} is the gate to source voltage of M_{T1},
V_{THn} is the threshold voltage of the M_{T1}. V_{THn} for NMOS is positive,
V_{DSp} is the drain to source voltage of M_{T2}, V_{GSp} is the gate to source voltage of M_{T2}, and
V_{THp} is the threshold voltage of the M_{T2}. V_{THp} for PMOS is negative.

The on resistances of the transistors M_{T1} and M_{T2} used as switches are denoted by $R_{on,n}$ and $R_{on,p}$ respectively. From [6], for the ohmic region, the resistances $R_{on,n}$ and $R_{on,p}$ can be expressed as

$$R_{on,n} = \frac{1}{(W/L)_n \mu_n C_{ox,n}(V_{GSn} - V_{THn} - V_{DSn})} \tag{19}$$

and

$$R_{on,p} = \frac{1}{(W/L)_p \mu_p C_{ox,p}(V_{GSp} - V_{THp} - V_{DSp})} \tag{20}$$

where
$(W/L)_n$ is the aspect ratio of the M_{T1}, μ_n is the mobility of electron, $C_{ox,n}$ is the gate oxide capacitance per unit area, of M_{T1}, $(W/L)_p$ is the aspect ratio of M_{T2}, μ_p is the mobility of hole, $C_{ox,p}$ is the gate oxide capacitance per unit area, of M_{T2}.

It can be seen from the above equations (19) and (20), that $R_{on,n}$ varies with the V_{GSn} and V_{DSn}, while $R_{on,p}$ varies with the V_{GSp} and V_{DSp}. But, since both of these on resistances i.e. $R_{on,n}$ and $R_{on,p}$ are much smaller in magnitude than the magnitude of the external resistor R therefore the effect of variations of $R_{on,n}$ and $R_{on,p}$ on the BPSK modulated output voltage V_{PSK} is negligible. Mathematically, it can be expressed by the following two equations written below

$$R_{on,n} \ll R \tag{21}$$

$$R_{on,p} \ll R \tag{22}$$

The BPSK modulated output voltage is given as

$$V_{PSK} = I_{out} \times R \tag{23}$$

The voltage V_A at node A is equal to the voltage V_{Xp}, that is

$$V_A = V_{Xp} \tag{24}$$

The voltage V_B at node B is equal to the voltage V_{Xn}, that is

$$V_B = V_{Xn} \tag{25}$$

Also, only of the transistors is on at a time i.e. either M_{T1} or M_{T2}. When M_{T1} is on, then the total resistance $R_{total,n}$ between node A and ground is equal to the sum of the on resistance $R_{on,n}$ of the M_{T1} and the external resistance R. And when the M_{T2} is on, then the total resistance $R_{total,p}$ between node B and ground is equal to the sum of the on resistance $R_{on,p}$ of M_{T2} and the external resistance R. Mathematically $R_{total,n}$ and $R_{total,p}$ are given as

$$R_{total,n} = R_{on,n} + R \tag{26}$$

$$R_{total,p} = R_{on,p} + R \tag{27}$$

When only M_{T1} is on, then the output current I_{out} is

$$I_{out} = \frac{V_A}{R_{total,n}} = \frac{V_{Xp}}{R_{on,n} + R} \tag{28}$$

From equations (21) and (26)

$$R_{total,n} = R_{on,n} + R \cong R \tag{29}$$

IOP Conf. Series: Materials Science and Engineering 1020 (2021) 012011 doi:10.1088/1757-899X/1020/1/012011

Therefore, from equations (28) and (29)

$$I_{out} \cong \frac{V_{Xp}}{R} \quad (30)$$

Now when only M_{T2} is on, then the output current I_{out} is

$$I_{out} = \frac{V_B}{R_{total,p}} = \frac{V_{Xn}}{R_{on,p} + R} \quad (31)$$

From equations (22) and (27)

$$R_{total,p} = R_{on,p} + R \cong R \quad (32)$$

Therefore from equations (31) and (32), I_{out} is given as

$$I_{out} \cong \frac{V_{Xn}}{R} \quad (33)$$

Thus from the equations (30) and (33), it can be concluded that the variation in the output current I_{out} due of $R_{on,n}$ and $R_{on,p}$ is negligible and therefore, the effect on the output voltage V_{PSK} because of the variations in $R_{on,n}$ and $R_{on,p}$ is also negligible.

When M_{T1} and M_{T2} working as the switches are in off condition than the associated resistances are denoted by $R_{off,n}$ and $R_{off,p}$ respectively. $R_{off,n}$ and $R_{off,p}$ are infinite ideally as the M_{T1} and M_{T2} transistors are in non-conducting state or off state. Now when the voltage of the modulating signal V_m is positive (logic high), M_{T1} turns on and M_{T2} turns off and the currents developed at Xp and Zp terminals of DXCCII flow towards node A, which then get added up at node A as no current goes in to the gate terminal of M_{T1} and then the net current flows through the transistor M_{T1} and then through the resistor R and then goes to the ground. The two currents developed at Xp and Zp terminals of DXCCII linearly depend on the voltage V_{Xp} as can be seen in equation (30). As the voltage V_{Xp} is AC in nature therefore the currents developed because of it are also AC in nature and since these two currents add up at the node A and therefore the resultant current is also AC in nature which flows through the resistor R developing an AC voltage V_{PSK} in return across it which is in phase with the carrier signal voltage V_c. And when the voltage of the modulating signal V_m is negative (logic low), M_{T1} turns off and M_{T2} turns on and the currents developed at Xn and Zn terminals of DXCCII flow towards node B, which then get added up at node B as no current goes in to the gate terminal of M_{T2} and then the net current flows through the M_{T2} and then through the resistor R and then goes to the ground. The two currents developed at Xn and Zn terminals of DXCCII linearly depend on the voltage V_{Xn} and it can be seen in equation (33). As the voltage V_{Xn} is AC in nature therefore the currents developed because of it are also AC in nature and since these two currents add up at the node A and therefore the resultant current is also AC in nature which flows through the resistor R developing an AC voltage V_{PSK} in return across it which is 180° out of phase with the carrier signal voltage V_c.

The operation can be summarized by these two equations as mentioned below

$V_{out} = I_{out} \times R \cong V_c = A_c Sin(2\pi f_c t);$ when V_m is positive or logic high (NMOS M_{T1} is on and PMOS M_{T2} is off)

$V_{out} = I_{out} \times R \cong -V_c = -A_c Sin(2\pi f_c t);$ when V_m is negative or logic low (NMOS M_{T1} is off and PMOS M_{T2} is on)

In conclusion, the output voltage V_{ASK} of the proposed BASK modulator and the output voltage V_{PSK} of the proposed BPSK modulator can mathematically be written in the tabulated form as shown below in the Table 2.

Table 2. Summary of outputs of the proposed BASK and BPSK modulators

Modulating signal (V_m)	**BASK output (V_{ASK})**	**BPSK output (V_{PSK})**
Logic HIGH (1V)	$A_c Sin(2\pi f_c t)$	$A_c Sin(2\pi f_c t)$
Logic LOW (−1V)	0	$-A_c Sin(2\pi f_c t)$

5. Simulation results of the proposed BASK and BPSK modulator

The BASK & BPSK modulators are simulated with SPICE using CMOS implementation of the active block DXCCII [35] and 0.35μm CMOS technology parameters. The supply voltage for the active block DXCCII is taken as $\pm 1.8V$ and the voltage V_{bias} is set to 1.05V. The aspect ratios (W/L) of the transistors M_{T1} and M_{T2} used as the switches are same i.e. $30\mu m/0.7\mu m$. The value of external resistor R is selected as 5KΩ. The waveform of the input carrier voltage signal V_c is shown in Fig. 5. The carrier signal V_c is having a peak amplitude A_c of 100mV and frequency f_c of 20MHz. The waveform for the modulating signal V_m is shown in Fig. 6. The waveforms for the ASK modulated output voltage V_{ASK} and PSK modulated output voltage V_{PSK} are shown above in Fig. 7 and Fig. 8 respectively.

The logic high and logic low levels of the modulating signal V_m are defined as follows

$V_m = 1V$; for logic high or logic 1 or bit 1

$V_m = -1V$; for logic low or logic 0 or bit 0

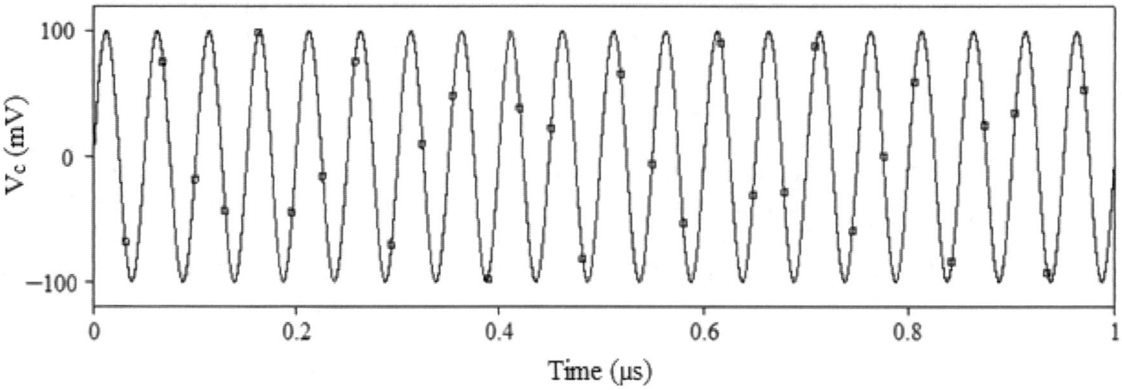

Figure 5. Waveform of the carrier signal V_c

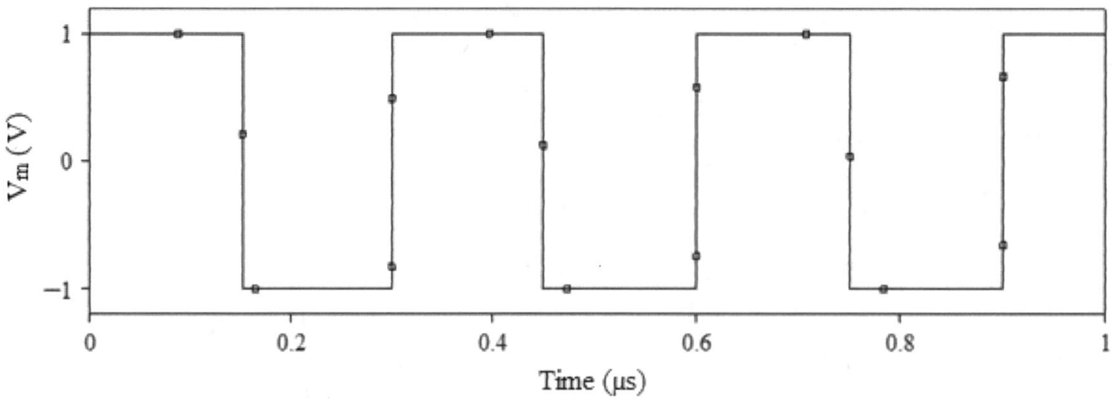

Figure 6. Waveform of the modulating signal V_m

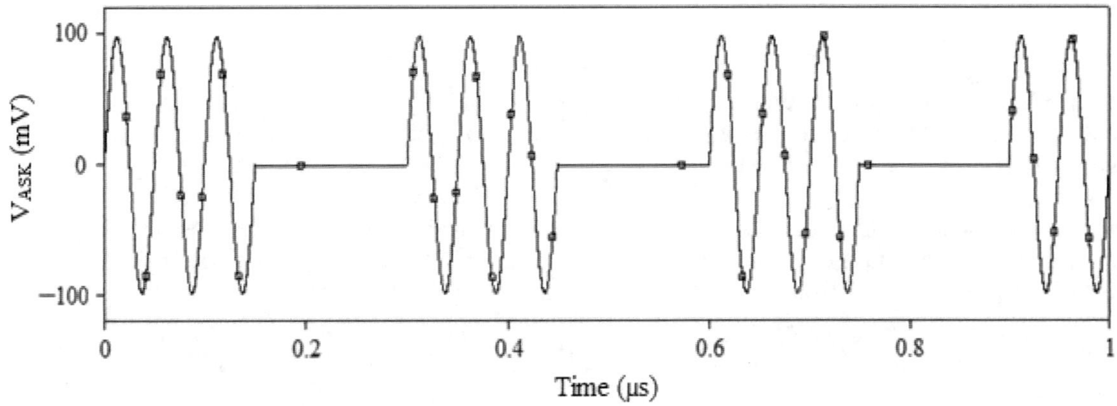

Figure 7. ASK modulated output voltage V_{ASK}

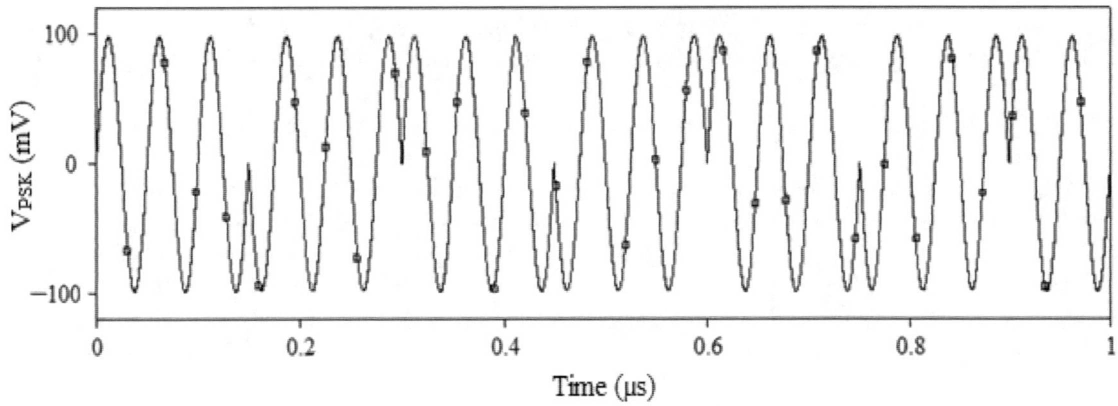

Figure 8. PSK modulated output voltage V_{PSK}

Table 3. Comparison of the various Modulators

Ref. no.	Active elements	No. of floating/ grounded resistors	No. of capacitors	No. of MOS	BASK/ BPSK	Max. carrier frequency (Hz)	Supply voltage
[5]	1 EXCCII 1 MOS	1/0	0	23+1	Both	20M	±1.25
[6]	1 EXCCII 2 MOS	0/3	0	21+2	Both	20M	±1.25
[7]	2 MO-CCCCTA	0/0	3	2×26	ASK	2.11M	±0.9
[8]	1 CCII+	0/1	2	NA	ASK	20.78M	±1.5
[9]	1 VCCS Memristors	0/1	1	NA	Both	1K	NA
[10]	1 DPCCII	0/2	0	18	ASK	125K	±0.75
[11]	1 MO-CCCC	1/2	3	NA	Both	1.5M	NA
[12]	1 Digitally programmable CDTA	0/2	2	67	Both	34.39M	±0.8
[13]	2 MO-OTA	0/1	3	NA	Both	NA	NA
Proposed	1 DXCCII, 1 MOS	0/1	0	20+1	ASK	20M	±1.8
Proposed	1 DXCCII, 2 MOS	0/1	0	20+2	PSK	20M	±1.8

6. Conclusion

Two voltage mode digital modulators have been proposed namely BASK and BPSK. BASK employs one DXCCII, one grounded resistor and one NMOS transistor while BPSK employs one DXCCII, one grounded resistor, one NMOS and one PMOS transistor. Both the above proposed circuits provide the advantage of having proper input impedances and use of grounded resistor. A comparative study table of various digital modulators available in the open literature is given in table 3. All the proposed circuits have been simulated with SPICE using 0.35μm CMOS technology parameters. The supply voltage of $\pm$1.8V is used. Convincing results have been obtained with all the proposed circuits.

References

[1] Sedra, A. S., Smith, K. C., "A second generation current conveyor and its applications", IEEE Transactions on Circuit Theory, 17, pp. 132-134, 1970.
[2] Awad, I. A., Soliman, A. M., "Inverting second generation current conveyor: the missing building blocks, CMOS realizations and applications", Int. J. Electron. (AEÜ), 86, pp. 413-432, 1999.
[3] Zeki, A., Toker, A., "The dual-X current conveyor (DXCCII): a new active device for tunable continuous-time filters", Int. J. Electron. (AEÜ), 89(12), pp. 913-923, 2002.
[4] Mohan, J., Chaturvedi, B., Maheshwari, S., "Novel current-mode all-pass filter with minimum component count", I.J. Image, Graphics and Signal Processing, 12, pp. 32-37, 2013.
[5] Maheshwari, S., "Voltage-mode full-wave precision rectifier and an extended application as ASK/BPSK circuit using a single EXCCII", Int. J. Electron. (AEÜ), 84, pp. 234-241, 2018.
[6] Maheshwari, S., "Realization of simple electronic functions using EXCCII", 26(11), Journal of Circuits, Systems, and Computers, 2017.
[7] Chen, H., Hwang, Y., Ku, Y., "Voltage-mode and current-mode resistorless third-order quadrature oscillator", Appl. Sci., 6, 2016.
[8] Khan A. M., "Generation of ASK and FSK from digitally controlled CCCII+ oscillator", The International Journal of Engineering and Science (IJES), 2(10), pp. 19-22, 2013.
[9] Goknar, I. C., Oncul, F., Minayi, E., "New memristor applications: AM, ASK, FSK and BPSK modulators", IEEE Antennas Wave Propag. Mag., 55(2), pp. 304-313, 2013.
[10] Khan, I. A., Simsim, M. T., "An ASK modulator for RFID applications using low voltage digitally controlled programmable CMOS-CCII", Saudi Int. Electronics, Communication, and Photonics Conf., pp. 1-4, 2013.
[11] Abuelma'Atti M.T., "New ASK/FSK/PSK/QAM wave generator using a single current-controlled multiple output current conveyor", Int. J. Electron. (AEÜ), 89(1), pp. 35-43, 2002.
[12] Xia, Z., Wang, C., Jin, J., Du, S., Lin, H., Yang, H., "Novel AM/FM/ASK/FSK/PSK/QAM signal generator based on a digitally programmable CDTA", Circuits Syst Signal Process, 34(5), pp. 1635-1653, 2014.
[13] Abuelma'Atti, M.T., "New ASK/FSK/PSK/QAM wave generator using multiple-output operational transconductance amplifiers", IEEE Trans. Circuits Syst. I, Fundam. Theory Appl., 48(4), pp. 487-490, 2001.

Deva Nand did his B.Tech in Electronics and Communication Engineering, M.Tech in Microelectronics and VLSI Design from Kurukshetra University, Kurukshetra, India and Ph.D. from Delhi Technological University, Delhi, India. At present he is Assistant Professor in Department of Electronics and Communication Engineering, Delhi Technological University, Delhi, India. A life member of ISTE, member of IEEE, USA, life member of IAENG and member of IAASSE. His research interests include analog mixed signal VLSI design and digital system design.

IOP Conf. Series: Materials Science and Engineering 1020 (2021) 012012 doi:10.1088/1757-899X/1020/1/012012

Fraud Detection in Financial Statements using Text Mining Methods: A Review

Ajit Kr. Singh Yadav[#], *Marpe Sora*[*]

Asst. Prof. Dept. of CSE, NERIST, Itanagar, Arunachal Pradesh &
Research Scholar, Dept. of CSE, RGU, Itanagar, Arunachal Pradesh
Email: ajityadav101@rediffmail.com

* Asst. Prof. Dept. of CSE, RGU, Itanagar, Arunachal Pradesh
Email: marpe.sora@rgu.ac.in

Abstract: In the financial industry, financial fraud is one of the ever-growing hazards with far concerns. Financial statements are the fundamental papers which replicate economic position of a corporation. Users of the financial information like public, creditors etc. are the major foundations of a decision-making process for financing stakeholders. Financial fraud has extremely damaged the sustainable growth of financial markets and enterprises. The amount of financial reporting fraud cases keeps on developing. Each incident is a thick hit to partners, banks, and financial specialists and it costs human progress significantly. One of the serious issues is to recognize the financial reporting fraud by utilizing formation of an active model. The aim of this paper is to identifying frauds using various text mining techniques and guard, the public's investments. This investigation will benefit auditors and financial governors.

Keywords: *Text Mining, Fraud Detection, Financial statements, Adaptive Crime, Computational Fraud detection model.*

1. Introduction

In the modern world, the text is one of the common means for exchanging data. Text mining is an interdisciplinary field which includes ordinary linguistic processing, information retrieval, web mining, computational linguistics, data extraction and data mining. From formless and semi-structured papers, automatically structured information were extracted in text mining. It has a very high commercial value. For evaluating huge assortments of formless documents, it is an advance innovation technique for the purposes of extracting and knowledge or non-trivial patterns [1]. Document files have several formats like text files or flat files and PDF files. These files are gathered from multiple sources such as message boards, newsgroups, emails, online chat, SMS, web pages and so on. Humans can easily handle the problems and have the ability to identify and apply linguistic patterns to text. But computers cannot easily handle such as spelling, contextual meaning, slang and variations [2]. Nevertheless, to know unstructured data, the language abilities permit us and computer capability to process the text at high speeds or in large volumes. For analysis task on unstructured data, text mining helps to the computer. Fraud Detection has great prominence to Financial Organizations. In worldwide, financial fraud is a critical commercial issue. Numerous kinds of fraud include in financial fraud such as health-care fraud, mortgage fraud, corporate fraud, safeties and commodities fraud, financial organisation fraud, and others. Fraud detection takes into action when preclusion of fraud fails. Fraud may be unaware so that fraud preclusion has failed. Therefore, detection of fraud must be used repeatedly [3].

Criminals often change their place when they get information about the usage of the fraud detection scheme in a certain process. Plenty of criminals don't aware of this kind of protection scheme which causes ease of fraud identification [4]. Application of early detection tools is very helpful but it is difficult to process to exchange ideas in fraud detection schemes. Public description of fraud detection scheme is not good. Because criminals will easily access all details so that it is important to make more secure schemes [5]. Conventional detection methods are not good in terms of economy, security, time consumption, law, and computational complexity. Assurance claims, credit card businesses, tax reappearance claims and cell phones are some possible ways of crime. These kinds of criminalities make more loss to governments and businesses due to a liberal manner of security [6].

Fraud is additionally called an adaptive crime. Therefore, it needs some superior techniques of intelligent data investigation to prevent and detect it. In several areas of fraudulent crimes, they are delivering applicable and successful solutions. Due to data investigation, these methods belongs to data mining, text mining, machine learning, statistics and knowledge discovery in databases (KDD) [7]. Financial organizations and banks create huge volumes of detailed transaction data. Data mining includes some primary applications such as trend analysis, direct marketing, fraud detection and risk assessment of potential customers at banks. For forecasting, most of the requirements dominate in the financial field. With an extraordinary level of estimation, commodity prices and predicting of stock prices can great incomes [8]. In predicting, options and bond trading, portfolio management, the neural network techniques are used and also applied in acquisitions and mergers. A whole sympathetic of nature, consequence and fraudulent financial reporting activities needs a suitable explanation of financial statement fraud. Association of certified fraud examiners (ACFE) describes financial statement fraud as the intentional, misstatement or accounting information or omission of material facts. That is deceptive when considered with all the information made obtainable; it would cause the reader to modify her or his decision [9].

Financial statement fraud is known as when the managers of a company deliver false financial information. Financial statement audits guarantee that a company's financial reports are free from material misstatement and fraud. In this scheme, financial statement fraud is typically committed [10]. To fight fraudulent activities, organizations want to be prepared in a recent challenging economy. By a business administration, reporting fraud includes the modification of financial statement data to attain a fraudulent result [11]. By the top management of publicly traded companies, financial statement fraud is a deliberate and illegal act devoted. Three features of frauds are available such as rationalization or attitude, opportunity and pressure or motive. In every year, financial statement fraud costs millions of dollars to the world economy because of the failure of many companies [12]. It is a type of the management fraud which operates the financial information. In many organizations, prevention of financial statement fraud has one of the major concerns [13]. Financial reporting fraud contains activities such as reporting sales which did not happen. In the recent year, reporting income into the current year essentially fits. In the current year, reporting an expense or capitalizing expenses in the next year that should be reported [14].

For discovery and identification of fraud, fraudulent observations are generally applied. The observations are the result of error or circumvent procedures in the prevention of fraud. A measure to break fraud in the first place from happening is known as fraud prevention [15]. In an extra structured spreadsheet layout, feature selection and feature generation events are accomplished to characterize the amorphous text documents based on these recovered and pre-processed documents. Feature selection procedures need an extensive exploration of all subdivisions of features to recognize the significant features [16]. For the applications of the knowledge discovery process, Information Extraction, Information Retrieval, the Text Mining methods are integrated after the suitable collection of features. Mostly classification is used to classify each item from a set of data into one of the predefined set of classes [17]. Classification procedure contracts with the software that can acquire to organize the data substances into groups. This scheme is also known as a process of identifying a motivating, inferred and innovative data in KDD.

In developing countries, financial statement fraud is more severe and it is a serious economic and social problem [18]. In yearly and quarterly, a company needs stock exchange to distribute its financial statements such as revenue, statements of recollected earnings, profit and loss account and cash flow statements. By analysing, financial report of a company replicates its genuine financial health. In the company, shareholders can form a wise decision about financing [19]. Financial fraud accomplishments of money laundering and credit card fraud gradually increase in recent years. It causes the loss of individual and/or enterprises' assets. Due to the profit from fraud, they threaten the safety of a country which may go to terrorism. Financial fraud detection and sketching fraud is critical and needed. Fraud detection is a difficult challenge for the reason of transactions and intricate trading systems. Many fraud detection models work with attribute-value data points which are obtained from transactions data [20]. Majority of fraud detection studies are applying data mining procedures which focussed on organised information using numerical approaches.

2. A BRIEF REVIEW OF FRAUD DETECTION IN FINANCIAL STATEMENTS

Financial fraud detection was talked about by Huang *et al* [21] with inconsistency feature detection. This article gives the CoDetect innovation to play out the fraud detection on highlight and chart based likeness matrix promptly. The innovation exhibits another strategy to uncover the idea of commercial exercises from fraud examples to disbelieving assets. In addition, to recognize the fraud on a sparse matrix, the structure gives an increasingly interpretable way. Exploratory results are tried on manufactured and true datasets which demonstrates the recommended structure can successfully identify guarded features and the fraud decorations.

Hybrid data mining plan based financial statement fraud detection was portrayed by Yao *et al* [22]. The primary motivation behind this investigation is to enhanced financial fraud detection conspire to consolidate feature selection and machine learning classification. Re-enactment considers obviously demonstrated that irregular forest gives better execution thought about than different frameworks. XGBOOST gives the better execution in the two feature selection strategies. In light of the exploration work, two or five factors are progressively adequate in this investigation.

Utilizing machine learning, credit card fraud detection was proposed by Yee *et al* [23] in the data mining process. Based on anomalies and standardized data, data mining strategies were utilized to examine the patterns and features of suspicious and non-suspicious exchanges. Something else, the machine learning techniques were utilized to consequently anticipate the non-suspicious and suspicious transactions by utilizing classifiers. By learning the patterns of the data, the blend of data mining and machine learning approaches could recognize the genuine and non-genuine exchanges.

ATM card fraud detection framework was presented by Prakash *et al* [24] utilizing machine learning strategies. This plan lessening the measure of false alert and can demonstrates accurate in deducting fraudulent transactions. Machine learning is another technique as far as application field. The likelihood of fraud transactions can be anticipated after ATM card transactions when this technique is utilized into bank ATM card fraud detection framework. To keep banks safe from incredible losses and decrease risks, the arrangement of anti-fraud strategies can be accepted. The motivation behind the investigation was taken a few issues contrasted than we had an alternate misclassification cost.

An effective financial statement fraud detection model was introduced by Jan [25] for manageable progress of financial souks confirmation obtained from Taiwan. This model actualizes the various data mining strategies. To screen out critical factors, an artificial neural network (ANN) and a support vector machine (SVM) schemes are masterminded in the main stage. For classification purposes, the four sorts of decision trees, for example, chi-square automatic interaction detector (CHAID), classification and regression tree (CART), quick unbiased efficient statistical tree (QUEST) and C5.0 are created in the second stage. To build an exceedingly accurate model, both non-financial and financial factors are chosen to recognize fraudulent financial detailing.

Financial fraud detection was produced by Macro *et al* [26] by thinking about human conduct. This article presents detecting fraud, a hypothetical structure that permits to order and synopsis a gathering of individuals inside a financial organization who submit extortion, upheld by the fraud triangle hypothesis. Misrepresentation discovers process in the techniques for a constant audit that will be in dependable of social occasion data of specialists introduced in user apparatus. In view of semantic methods, the data is connected by means of the accumulation of expressions by the clients behind the investigation for later being transferred to a storehouse for later examination. In the reduction of instances of financial fraud, this examination contributes to the cybersecurity field.

Credit card fraud detection was expected by Kuldeep *et al* [27] by utilizing lion's share voting and AdaBoost techniques. To detect credit card fraud, the machine learning method is utilized in this article. At the outset, the standard models are utilized. After that, the hybrid strategies are connected which incorporates the majority voting and AdaBoost procedures. A freely accessible credit card data set is utilized to survey the model adequacy. A genuine credit card data set is investigated which are gotten from the financial institution. Besides, the noise is added to the data samples and also evaluates the toughness of the procedures. In distinguishing fraud cases, the reproduction results demonstrate that the dominant part voting framework gives the better accuracy contrasted with others in the credit card field.

A mining corporate yearly report was depicted by Hajek [28] for detection of financial fraud in relative investigation of machine learning structures. By deriving the consolidating explicit features, this examination recognizes the upgraded fraud detection from financial data and business annual reports. This work needs to finish up a low frequency of harmful words in yearly files it might show non-fraudulent businesses. From freely accessible financial proclamations and predictions of revenues and earnings, it is basic to utilize separated data to recognize the fraudulent firms.

Anticipating fraudulent financial explanations was examined by Stamatis *et al* [29] by utilizing active learning techniques. This work analyses the effectiveness of AL hypothesis over detection of FFS issue contrasted and managed learning techniques. In the relative field, the analog upgrade of the anti-fraud frameworks ought to be utilized to empowering the machine learning instruments. Because of the powerlessness, the plan to gather dependable datasets that assign the equivalent proportions of a firm. It has gone with fraud activities; approaches misuse the presence of few marked events for finding valuable designs from a pool of unlabelled data could be checked extremely capable.

On big data based misrepresentation detection plot was exhibited by Chen *et al* [30] for financial explanations of business gatherings. In finance and economics, this investigation deliberates the several features. Moreover, estimate the big data to upgrade the fraud detection conspire to precisely anticipate the financial statement misrepresentation for financial reports of commercial gatherings. Moreover, improving the investment decision making and diminishing speculation hazards and losses help creditors and investors.

A main impact of social media for financial fraud disclosure was depicted by Dong *et al* [31] in text mining based investigation. In Internet-based life, this examination is used to leverage huge measure of user-produced substance for fraud detection. A logical text structure is framed to detect amorphous internet-based life substances onto words weights, topics, and emotion features. Most of the web-based social networking highlights, proceeding to the fraud which covers the fake and matched identical firms.

For 10-fold cross validation, it gives the average accuracy of 81.43%. For financial fraud revelation, the scheme is prominent aftereffect of social media substance.

Financial fraud detection was presented by Gupta and Nasib [32] by utilizing text mining. For both non-fraud and fraud associations, fraud detection shows a given collection of financial proclamations. In financial proclamations, these two associations are trailed by pre-processing which incorporates a lexical examination of present content. From that point onward, a sack of words plot has been chosen for expelling data hidden in the text which results in vector spaces for both non-fraudulent and fraudulent associations.

Review of the uses of text mining was produced by Kumar and Ravi [33] in the financial domain. For various uses of text mining, this article gives the cutting edge overview. The text ideas are classified as client relationship management (CRM), stock market and forex rate prediction. Some issues are arising which are dominant in optional and consumer advance features. This investigation can be extremely valuable to researchers in this field; the same number of open issues is featured.

By utilizing phonetics, an investigating financial fraud cases based text mining framework was proposed by Zaki and Babis [34]. In various financial settings, this idea bolsters the market observing surveillance frameworks to allow reprocessing of these learning assets, comprehend and share financial fraud logic activities and update the awareness of the procedures of financial fraud. From SEC process, a semantic-based text mining strategy is built up to capture the data and comprehend the unique market manipulation assortments. It helps the examination of fraud during the time spent examination.

An innovative scheme was displayed by Glancy and Yadav [35] for financial statement fraud detection. On textual data, CFDM (computational fraud detection model) utilizes quantifiable technique. For fraud detection, it incorporates a few techniques that utilizes all of the data contained in the textual data in a general sense. In low and high computer-mediated communication (CMC), existing work conveys establishment for perceiving deception. CFDM offers a systematic procedure which prospective for robotics. Imitation annoyed from 10-K filings based on the administration negotiations and investigation. From non-fraudulent scheme, it could discriminate the deceitful filings. CFDM act as a selection device when misleading is distrusted.

Reproduction tried on the MD&A (Management's Discussion and Analysis) from 10-K filings and it could distinguish fraudulent filings from a non-fraudulent scheme. CFDM can help as a screening tool where misleading is suspected.

3. Problem Statement

To provide decision support to stakeholder, different data mining strategies have been utilized on the grounds. The financial statement fraud is a troublesome issue for government and community regulator. Public companies make fraudulent financial data just not for the purpose of critical financial loss to the wide investors but also it demotivates the financial market. During the general auditing, the ordinary auditing rehearses that may principally concentrate on factual examination. In financial statements, most financial fraud detection looks into limit their examination just too numerical data. From authentic data, fraudulent financial data could scarcely be recognized because of deliberate covering as well as accounting shenanigans. The common absence of field for financial fraud statement can be hard to diagnose, the rarity in which it happens, and the way that it is typically dedicated by proficient individuals inside the industry who are equipped for marketing their deceit.

4. Methodology: A Text Mining Approach

Most of the companies release the financial statement documents that explain all details about their loans, profits, expenses, and income. From management, it can additionally include some comments based on the business performance. In the future, the predictable problems may arise. The different financial statements provide the pictorial explanation of each company status which is obtained from the company. The stock prices of the successful company are indicated, by various financial statements from which they are identifying whether it is applicable for loans or not.

In this paper, we propose financial statement fraud detection by using effective feature selection and classification technique. This method consists of four stages: 1) pre-processing, 2) sampling, 3) feature selection, and 4) clustering and classification.

In the pre-processing process, data correlation analysis and data cleaning process which cleans the noise data are performed. Also, data transformation, integration, and reduction are included in this process. The following process is the sampling process which evaluates dataset with various ratios for verification through a hybrid sampling method. Feature selection is an effective scheme which used for solving the problems of data mining and machine learning methods. Three kinds of feature selection schemes are available such as filter method, wrapper method, and an embedded method. The wrapper system to assess the selected features based on the analytical presentation of a predefined learning procedure. The filtering scheme to estimate the significance of characteristics depends on the exact features of data and this method is independent in any learning algorithms. With the model learning, an embedded method is to embed the feature selection and it is a mutual technique between the wrapper and filter methods. By using filter-based learning scheme, a feature selection procedure has accomplished. Filter is a method of eliminating the characteristic of low score after score of characteristic based on the score given by evaluation condition. The probability of occurrence of over fitting is relatively low and it is good for large-scale data. Evaluate the importance of characteristics based on reliable data features that are independent of learning algorithms.

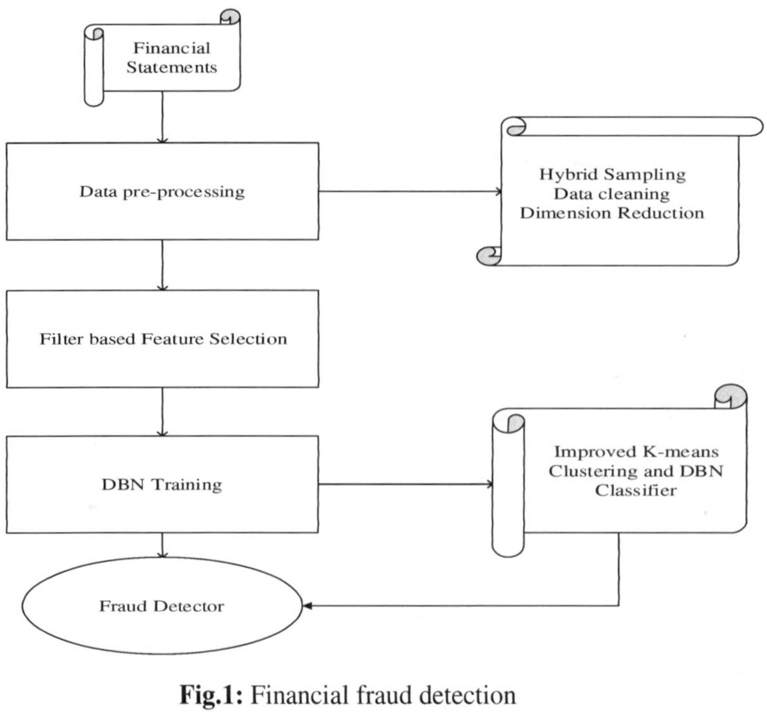

Fig.1: Financial fraud detection

Clustering process performs using improved K-means after a process of feature selection and this result will be used as training and validation set of the classification process (Fig.1). The optimization process minimizes the objective function (Error Rate) inside the deep belief network. So the optimized deep learning can enhance the fraud detection accuracy.

Financial statement fraud detection can be tested on the datasets such as COMPUSTAT, PagSeguro, Athens stock exchange and 1610 firm-year samples from U.S. listed companies and on any other datasets available in near future. In these datasets, 70% for training and 30% testing can be used.

The performance of the proposed method can be analysed in the form of statistical measures such as recall, sensitivity, F-measure, specificity, and classification accuracy. The simulation outcomes will be evaluated and compared with the earlier classification technique.

5. Conclusion

The classification process uses optimized deep belief network (DBN) to get the better accuracy. Moreover, it leads the lower computational cost. A DBN generally have three layers of RBMs (Restricted Boltzmann Machines), the first layer of RBMs takes the input as the visible layer, and the second layer takes the hidden layer as the visible layer. The hidden layer of the last RBM takes the output of DBN as visible layer which is considered as one unit. The above classification processes include the methods:

I. To analyse the several state-of-art techniques related to financial fraud detection.
II. To solve the conflicts among the different financial statement fraud.
III. To analyse the various unsupervised learning algorithms for clustering process of financial statements.
IV. To get the better classification accuracy for financial statements fraud by extracting the several optimal features under the feature selection process.

The performance of the proposed method can be analysed in the form of statistical measures such as recall, sensitivity, F-measure, specificity, and classification accuracy. The simulation outcome may be evaluated and compared with the earlier classification techniques.

6. References

[1] Sumathy, K. L., and M. Chidambaram. "Text mining: concepts, applications, tools, and issues-an overview." International Journal of Computer Applications 80, no. 4 (2013).

[2] Aggarwal, Charu C., and Haixun Wang. "Text mining in social networks." In Social network data analytics, pp. 353-378. Springer, Boston, MA, 2011.

[3] Mostafa, Mohamed M. "More than words: Social networks' text mining for consumer brand sentiments." Expert Systems with Applications 40, no. 10 (2013): 4241-4251.

[4] Netzer, Oded, Ronen Feldman, Jacob Goldenberg, and Moshe Fresko. "Mine your own business: Market-structure surveillance through text mining." Marketing Science 31, no. 3 (2012): 521-543.

[5] Fuller, Christie M., David P. Biros, and Dursun Delen. "An investigation of data and text mining methods for real-world deception detection." Expert Systems with Applications 38, no. 7 (2011): 8392-8398.

[6] Othman, Rohana, Nooraslinda Abdul Aris, Ainun Mardziyah, Norhasliza Zainan, and Noralina Md Amin. "Fraud detection and prevention methods in the Malaysian public sector: Accountants' and internal auditors' perceptions." Procedia Economics and Finance 28 (2015): 59-67.

[7] Dong, Wei, Shaoyi Liao, and Liang Liang. "Financial Statement Fraud Detection using Text Mining: A Systemic Functional Linguistics Theory Perspective." In PACIS, p. 188. 2016.

[8] Fu, Kang, Dawei Cheng, Yi Tu, and Liqing Zhang. "Credit card fraud detection using convolutional neural networks." In International Conference on Neural Information Processing, pp. 483-490. Springer, Cham, 2016.

[9] Rawte, Vipula, and G. Anuradha. "Fraud detection in health insurance using data mining techniques." In Communication, Information & Computing Technology (ICCICT), 2015 International Conference on, pp. 1-5. IEEE, 2015.

[10] Dilla, William N., and Robyn L. Raschke. "Data visualization for fraud detection: Practice implications and a call for future research." International Journal of Accounting Information Systems 16 (2015): 1-22.

[11] Kanapickienė, Rasa, and Živilė Grundienė. "The model of fraud detection in financial statements by means of financial ratios." Procedia-Social and Behavioral Sciences 213 (2015): 321-327.

[12] West, Jarrod, and Maumita Bhattacharya. "Some Experimental Issues in Financial Fraud Mining." In ICCS, pp. 1734-1744. 2016.

[13] Kim, Yeonkook J., Bok Baik, and Sungzoon Cho. "Detecting financial misstatements with fraud intention using multi-class cost-sensitive learning." Expert systems with applications 62 (2016): 32-43.

[14] Olszewski, Dominik. "Fraud detection using self-organizing map visualizing the user profiles." Knowledge-Based Systems 70 (2014): 324-334.

[15] Albrecht, Chad, Daniel Holland, Ricardo Malagueño, Simon Dolan, and Shay Tzafrir. "The role of power in financial statement fraud schemes." Journal of Business Ethics 131, no. 4 (2015): 803-813.

[16] West, Jarrod, Maumita Bhattacharya, and Rafiqul Islam. "Intelligent financial fraud detection practices: an investigation." In International Conference on Security and Privacy in Communication Systems, pp. 186-203. Springer, Cham, 2014.

[17] Lin, Chi-Chen, An-An Chiu, Shaio Yan Huang, and David C. Yen. "Detecting the financial statement fraud: The analysis of the differences between data mining techniques and experts' judgments." Knowledge-Based Systems 89 (2015): 459-470.

[18] McMahon, Richard, Diana Pence, Linda Bressler, and Martin S. Bressler. "New Tactics in Fighting Financial Crimes: Moving Beyond the Fraud Triangle." Journal of Legal, Ethical & Regulatory Issues 19, no. 1 (2016).

[19] Wong, Shirley, and Sitalakshmi Venkatraman. "Financial accounting fraud detection using business intelligence." Asian Economic and Financial Review 5, no. 11 (2015): 1187.

[20] Throckmorton, Chandra S., William J. Mayew, Mohan Venkatachalam, and Leslie M. Collins. "Financial fraud detection using vocal, linguistic and financial cues." Decision Support Systems 74 (2015): 78-87.

[21] Huang, Dongxu, Dejun Mu, Libin Yang, and Xiaoyan Cai. "CoDetect: Financial Fraud Detection with Anomaly Feature Detection." IEEE Access 6 (2018): 19161-19174.

[22] Yao, Jianrong, Jie Zhang, and Lu Wang. "A financial statement fraud detection model based on hybrid data mining methods." In 2018 International Conference on Artificial Intelligence and Big Data (ICAIBD), pp. 57-61. IEEE, 2018.

[23] Yee, Ong Shu, Saravanan Sagadevan, and Nurul Hashimah Ahamed Hassain Malim. "Credit Card Fraud Detection Using Machine Learning as Data Mining Technique." Journal of Telecommunication, Electronic and Computer Engineering (JTEC) 10, no. 1-4 (2018): 23-27.

[24] Prakash, B., G. Venu Madhava Murthy, P. Ashok, B. Pavan Prithvi, and S. Sai Harsha Kira. "ATM Card Fraud Detection System Using Machine Learning Techniques." (2018).

[25] Jan, Chyan-long. "An Effective Financial Statements Fraud Detection Model for the Sustainable Development of Financial Markets: Evidence from Taiwan." Sustainability 10, no. 2 (2018): 513.

[26] Sánchez, Marco, Jenny Torres, Patricio Zambrano, and Pamela Flores. "FraudFind: Financial fraud detection by analyzing human behavior." In Computing and Communication Workshop and Conference (CCWC), 2018 IEEE 8th Annual, pp. 281-286. IEEE, 2018.

[27] Randhawa, Kuldeep, Chu Kiong Loo, Manjeevan Seera, CHEE PENG Lim, and Asoke K. Nandi. "Credit card fraud detection using AdaBoost and majority voting." IEEE ACCESS 6 (2018): 14277-14284.

[28] Hajek, Petr, and Roberto Henriques. "Mining corporate annual reports for intelligent detection of financial statement fraud–a comparative study of machine learning methods." Knowledge-Based Systems 128 (2017): 139-152.

[29] Karlos, Stamatis, Georgios Kostopoulos, Sotiris Kotsiantis, and Vassilis Tampakas. "Using Active Learning Methods for Predicting Fraudulent Financial Statements." In International Conference on Engineering Applications of Neural Networks, pp. 351-362. Springer, Cham, 2017.

[30] Chen, Yuh-Jen, and Chun-Han Wu. "On Big Data-Based Fraud Detection Method for Financial Statements of Business Groups." In Advanced Applied Informatics (IIAI-AAI), 2017 6th IIAI International Congress on, pp. 986-987. IEEE, 2017.

[31] Dong, Wei, Shaoyi Liao, Yujing Xu, and Xiaoqian Feng. "Leading effect of social media for financial fraud disclosure: A text mining based analytics." (2016).

[32] Gupta, Rajan, and Nasib Singh Gill. "Financial statement fraud detection using text mining." Editorial Preface 3, no. 12 (2012): 189-191.

[33] Kumar, B. Shravan, and Vadlamani Ravi. "A survey of the applications of text mining in financial domain." Knowledge-Based Systems 114 (2016): 128-147.

[34] Zaki, Mohamed, and Babis Theodoulidis. "Analyzing financial fraud cases using a linguistics-based text mining approach." (2013).

[35] Glancy, Fletcher H., and Surya B. Yadav. "A computational model for financial reporting fraud detection." Decision Support Systems 50, no. 3 (2011): 595-601.

Hydrological significance of Himalayan surface water and its management considering anthropogenic and climate change aspects

Pema Tshering Lepcha[1], Pankaj Kumar Pandey[1] and Prem Ranjan[1]

[1]Department of Agricultural Engineering, North Eastern Regional Institute of Science and Technology, Nirjuli, Arunachal Pradesh, 791109, India

E-mail: thingtempema.2904lepcha@gmail.com

Abstract. Water plays a key role in maintaining different ecosystem services. The anthropogenic and climate change impacts studies on Himalayan water resources has been trending in recent years. The Indian Himalayan region cover several states like Jammu & Kashmir, Ladakh (Union territory), Himachal Pradesh, Uttarakhand, Sikkim, Arunachal Pradesh and the hilly regions of Assam and West Bengal. Water resources in this region are the life support for humankind and rich biodiversity. Researchers has reported that the warming in the Himalayas is more than the global average, thereby climate change can affect significantly in the mountainous water resources so as the snow and glacial melt is the main lifeline of the perennial rivers' flow. Almost every year, the heavy monsoon rainfall brings floods on the upstream highlands and then the downstream plains causing enormous human suffering, loss of property, loss of life, agricultural crops damage as well as increased economic losses. The main challenges being associated to the frequency and magnitude of extreme weather events like monsoon rainfall which leads to flash floods, landslides and debris flow; and then the drought during dry season also, has had huge impacts on regional water security. Human actions like economic development and population growth have also brought changes in land cover/ use to Himalayan watersheds recently. As a result of farming, grazing, tree planting, urbanization, etc., the changes in land cover have altered the water balance and transformed the water-flow paths in the global hydrological cycle. In contrast to freshwater ecosystems, demand is high, but their supply is limited, also depend on the specific flow pattern of rivers. Although, with interference of various infrastructure development upstream results in changes in flow pattern downstream causing non-availability of water to support aquatic life and communities that depend on for livelihoods. Initiation of monitoring and observation system for water resources and use of appropriate hydrological model are preferable to better understand the hydrological processes of the Himalayas.
Keywords: Climate change, Ecosystems, Freshwater, Himalayas, Land Cover, Water Balance.

1. Introduction

The Himalayan regions, are generally characterized by hilly terrain with complex hydrology. Almost every river on our planet are originated from mountainous regions. The volume of water that accumulated in the mountains joins the river systems to the downstream areas in which higher fractions of it is utilized for agricultural purposes like crop production. Considering mountain regions as an area lying higher than 1000 meters above mean sea level summed up to only 27 per cent of the earth's mainland areas [1], it provides enormous volume of water to the world's population which largely exceeds this value. Because of this state, Himalayas are frequently stated as natural "water towers". Mountainous region with their vast storage of water compared to low lying or plain areas, are of highly hydrological significance. It is the massive suppliers of fresh water to the world and are the part of joined ocean-atmosphere and terrestrial systems [2]. It also contributes about 30 to 60 per cent of the fresh water accessible at the low-lying humid areas and 70 to 95 per cent of the water that is accessible in arid and semi-arid regions [3].

Mountainous region with widespread snow and glacial coverage indicates climate change warnings [4]. The Himalayas most vital resource springs are also under threat due to climate and land use changes [5]. Both climate as well as land cover changes affect the water resource and hydrological dynamics in river basins [6; 7]. For researchers and policymakers, defining the negative impact of land use/cover (LULC) and climate changes on hydrology is always challenging work, due to the complication of hydro-metrological phenomena in mountainous regions. It has also direct influences on streamflow's, natural hazards, ecosystem, etc. and on zonal climate, specifically, the timing and strength of the monsoon in Asia [8]. Climate change has been known as one of the extensive ongoing environmental issues that have impact on agriculture, ecology, hydrology and socio-economic frameworks [9]. Of late, the climate change impact on the hydrology has been broadly explored [8, 10-12]. The mountainous regions are also considered as most sensitive indicators for climate change [13]. As a result, water resource pressure has become a serious problem that ultimately affects the socio-economic growth of the area [14]. Hence, evaluating the comparative consequence of anthropogenic activities and climate change is crucial not just for understanding the cycle of hydrological reaction but also for management of this vital resource to avoid the droughts and floods [15]. Climate change effects the hydrologic cycle by changing the precipitation pattern, inconsistencies of evapotranspiration and streamflow [16]. The linear trend of global warming over the 100 years is 0.74°C. It is believed that the recent deviations in temperature have made known noticeable influences on physical and biological systems [17]. Many studies have showed that water resources have altered within the river basins due to climate change [18, 19].

The recent economic development and population growth in mountainous region have brought land cover changes in the Himalayan watersheds. Mountain settings are highly critical to actions of humankind, predominantly to changes in land cover [20]. Land cover changes is one of the chief issues manipulating the hydrological pattern of the catchments [21]. Many studies have been carried out on hydrological responses to land cover change with assuming climatic settings as constant [21, 22]. Water resources management and Land-use pattern are inter-reliant. Due to urbanization, grazing, farming, tree plantation, etc., the changes in land cover have altered the water balance and modify the flow paths of water in the global water cycle [23]. A decrease in areal coverage of cropland, wetland and grassland and an increase in build-up area has resulted in increased water yield [24]. Dynamic development of urban land use is extremely complex and is greatly affected by the cultural, economic, natural, political aspects, etc. At the present time, urban land use transformed further under the joint effect of physical characteristics and anthropological environmental system, disturbing all features of sustainable urban growth.

The Cellular Automata (CA), a separate grid dynamic model, in which the state, time and space are considered as discrete, whose spatial collaboration and connection on time are defined based on the local relation. CA is appropriate to study the complex spatiotemporal physical system, especially for urban land use, and is widely adopted in urban LULC modelling [25]. However, the model has been widely accepted in future LULC simulation and forecasting purposes, but maximum studies remained in the simulation results for the warning, analysis, application and cellular sensitivity analysis purposes. The CA-Markov model is appropriate for three-dimensional land use modelling and land-cover reconstructions. The simulation of future LULC maps by the CA-Markov model can assist in early warning system of the future aspects of LULC change [26]. LULC prediction by means of model is vital to highlight the future potential changes that are likely to happen to the land in the near future. Especially, these changes happen due to rapidly growing population, the soil type, environmental issues, etc. Predicting future LULC changes can bring advantages in many applications like urban planning by means of rural and urban growth modelling [27]; choosing significance areas for conservation and setting alternate solutions [28]; reviewing the shifting cultivation dynamics [29]; and modelling the land use dynamics under different watershed [30].

The exploration on LULC and climate changes impacts on hydrology and water-related ecological services, and who will (or will not) get benefited from it on different approaches, is not sufficient for the preparation of useful policies [31]. To access the impacts of climate and LULC changes, various hydrological models are employed like Soil and Water Assessment Tool [32], Topography based hydrological model [33], and the Variable Infiltration Model [34].

2. Challenges and issues in surface water management in Himalayas

In Himalayan region, the fieldwork is very difficult due to rough terrain, inaccessibility due to high cliffs, rocks, snow cover, etc. Along with this, a long-term series of good quality hydro-meteorological data to characterize the dynamics of Himalayan rivers hydrology is limited. Due to inadequate quality and quantity of data for model, add up more challenges in employing hydrological modelling due to parametric uncertainty [35]. As a result, the hydrologists might not quantitively defined the relationships between runoff and various hydrological parameters with respect to change in land use/cover and climate [36]. Changed environments are probable since hydrological systems are frequently affected by modification of land use/ cover (afforestation, deforestation, changes in configuration of tree species, terracing, tillage, drainage, urbanization, road building) and climate changes. Hydrologic modelling incorporates framing the scientific representations to characterize the hydrologic processes (precipitation, snowmelt, evapotranspiration, interception, infiltration, surface and sub-surface flow) and the interaction amongst them. Hydrologic modelling encompasses non-linear operations, complex relations and more spatial inconsistencies within the basin [37].

Due to the social, economic and physical environment of Himalayan region, they are highly vulnerable to disaster related to water such as earthfall, floods, and droughts. Almost every monsoon, the heavy rainfall brings floods on the upstream highlands and the downstream plains causing human suffering, property damage, financial expenses and loss of life [38]. The recurrent existence of droughts and floods also has negative consequences on regional food and energy security. In the current scenarios of increasing population, large demand and utilization of water in agriculture for irrigation and industrial purposes are the main causes of increasing water withdrawals from the basin.

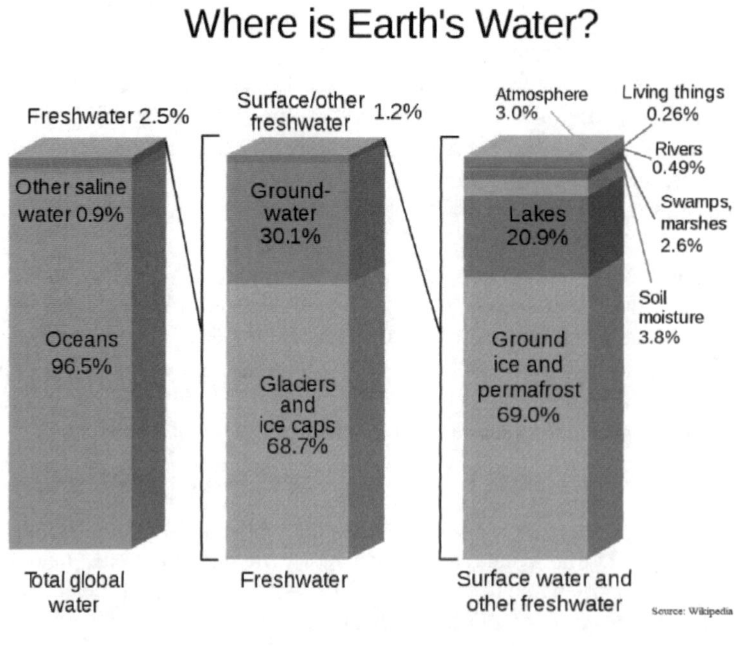

Figure 1 Distribution of planet's water. Total freshwater is less than 3% of the planet's water. Out of which, freshwater at surface accounts for 1.2%, ice caps and glaciers (68.7%) and groundwater (30.1%). Surface freshwater (lakes, rivers and swamps) is 0.3% of the planet's freshwater reserves.

Freshwater accounts for less than 3 per cent of our planet's water (figure 1), and less than 50 per cent of it is in the form of liquid; the remaining is stored in polar ice caps and glaciers. Due to this limitations, freshwater ecologies are a valuable resource to living beings. Now, the freshwater availability has been a worldwide issue, and the reliable estimation and quantification of freshwater existing within the basin is essential to brace the sustainable water management [39]. In recent years, the deterioration in quality of water has also observed all around due to increased urban expansion and industrial development. In addition to these, it is well known that the increased use of agro-fertilizers, domestic sewage discharge and deprived sanitation services have degraded the quality of water downstream and worsen the rives ecosystems downstream.

Water plays a significant part in sustaining various eco-system services. In contrast to significance of freshwater (lakes, ponds, rivers, streams, springs, etc.) ecosystems, demand is high, but their supply is limited, also depend on the specific flow pattern of rivers. Although, with interference of various infrastructure development upstream results in changes in flow pattern downstream causing non-availability of water resources to support aquatic life and communities that depend on for livelihoods. However, availing a minimum flow to sustain freshwater supply and backing ecosystems that depend on it is a big concern.

The India, Indus river provide irrigation for about 144,900 ha, the Ganges provides 156,300 ha, and the Brahmaputra of about 6000 ha of agricultural land [40]. Any change in access to water resources especially for agricultural food production and the sustainable management of water will raise concern beginning from the local to national level. Most of the rivulets and small rivers in eastern part of Indian Himalayas originates from Bhutan, Tibet, and China joined the tributaries of Brahmaputra river. These small networks of water channels are the main and only water sources use for drinking, irrigation, fishing and the basis for livelihoods for the local community. The number of consecutive rainy days has decreased, and the pattern of rainfall has also changed to high-intensity lower duration, leading in sudden floods. However, during the dry season, channels now run with very low discharge and some even dried up. Many of the region is now facing the acute shortage of water during the dry season. As most of the areas is rainfed, without irrigation, most of the farmers are unable to cultivate their land. Women and children have to travel long distance to collect water for domestic use. Further, lack of mechanism to channel water during rainy season, the canals and embankments fail to withstand the impact of turbulent water. The rapid expansion of urban land, deforestation and climate change have made the scenarios of water scarcity and led local communities subjected to condition of hunger and poverty.

3. Scope of anthropogenic and climate change studies on Himalayan surface water

The climatic conditions of Indian Himalayas are non-homogeneous and is aggressively monitored by the south Asian monsoon in the eastern and the mid-latitude western part. Studies suggested that climate change is likely to have greater influence on water resources, particularly in Himalayas [41], and even worse in the high-altitude region where the increase in temperature, being more than the average at global level, has a substantial consequence on the cryosphere environment [42, 43]. Due to ever increasing global warming, concerning superiority in Himalayan regions [44], the glaciers and snowpack have been receding continuously. The proportion and timing of glacial liquification, the quantity of snow melt that decrease glacier coverage, fluctuate temporally and the reasons are very difficult to explain. In most occasions, there may be many combinations of factors that interact in complicated ways. Further, every Himalayan river receive discharge from snow, glacial melt and rain. The continuous decline in snow and glacial melting due to global warming in Himalayas may increase the availability of surface water including freshwater in streams, springs, lakes, ponds, etc. and streamflow temporally to some extent provided no change in precipitation quantity, while after sometime, they may decline which alarms about the availability of water supplies in future and its associated consequences. It is reported that, the erection of various dams and reservoirs linked with many domestic and industrial stresses give rise to decreased stream flows downstream [45]. This may result in imbalanced water cycle in the Himalayas that might deteriorate the balanced ecosystem of the regions. In most of the low-lying regions, the quantity and timing of streamflow resulting from melting of glacial and snow are very important for water supply, agricultural production, hydropower generation, and the operational and well-being of ecosystems.

Amid accelerated ecological, social and economic changes, the increasing population will need more food and water, and the unbiased access to key resources is a big concern. An appropriate hydrological modelling concerning to quantification of snowmelt and the resulting runoff in snow coverage Himalayas should be implemented for surface water management by adopting different possible climate scenarios keeping in mind the various policies for sustainable watershed management and planning. Also, a thorough analysis on anthropogenic activities (implications of various water resource development projects, agricultural practices like jhum cultivation commonly practiced in hilly regions of India, rapid and unplanned infrastructure development, intentional/ unintentional forest fires, road construction, etc.) happening within the basin, comparing their historical- and present changes in quantity, pattern and trends of surface water is necessary. Since, the glaciers is one of the major water resources, their probable consequences of change in snow coverage area on the inhabitants living within the hilly watersheds, its significance on regional water balance and river ecosystem is important in terms of surface water managements in the Himalayan regions.

Apart from above discussions, the following may address in regards to sustainable water management:

- Extent of climate change impacts on water availability in hilly as well as in low-lying areas.
- Vulnerability of floods and droughts conditions due to anthropogenic and climate change.
- What measures/ policies needed to be adopted in local to national, national to international level to monitor and control the negative aspects of climate change and to what extent?
- Agricultural is one of the major contributors to Green House Gases (GHGs). What measures/ policies need to be adopted in agricultural sector so as to feed the ever-increasing population, with main objective being monitoring and controlling the release of GHGs to lesser proportion.
- Almost, every individual living in this earth might have the knowledge on global warming issues. However, the main challenges arise is how to implements various policy from local to international level, how individual plays a key role in controlling global warming?
- Sensitiveness to climate and other ecological change on glaciology.
- Potential impressions of climate changes on glaciers, the timing and volume of flow in the regions and the probable implications for water supplies and extreme climatic events such as floods.
- An alternate water management system to adapt changes in regional hydrologic systems.
- Vulnerabilities to changes in water supplies of downstream populations.

4. Mitigation measures to anthropogenic and climate change

Climate change results in occurrence of drought, floods, melting of glaciers, rise of sea level, change in rainfall pattern, etc. with severe consequences. It has a direct consequence on water security and associated conflicts. Some of the mitigation measures of climate change to surface water are discussed below.

4.1 Towards water conservation

- *Surface water storage structures*: Construction of structure for storing water for various purposes including irrigation, domestic purposes, flood control and so on. The sizes of structures may vary from storing rainwater or diverted streams, rivers, etc.
- *Efficient use of water*: Enhancing the water use efficiency is one of the best measures one can adopt in agricultural sector, while maintaining best productivity. Modern irrigation technologies have the capability to monitor and control the supply of water to crops at optimal amount with precise volumetric water applications to an extent of soil and crop water requirements. Hence, more crop per drop can be achieved.
- *Water related awareness programmes*: Dissemination of environment and socio-economic benefits of water conservation is one of the key aspects that need to be addressed to every

beneficiary through the awareness campaigns. The communication medium may be through social media, workshops, etc.

- *Harvesting of rainwater*: Rainwater harvesting is one of the best ex-situ water harvesting techniques that can be employed to conserve water. It involves collection of rainwater from the roof top and stored it in tank, for various purposes like drinking, sanitation, agriculture, and so on.

4.2 Controlling GHGs emissions to limit the climate change consequences:

The aim is to limit the rate of warming either by reducing sources of greenhouse gases or availing the "sinks" that store these gases (such as the oceans, forests and soil) and its associated effects. This usually involves reducing human's contribution to GHGs emissions. Some of the mitigation measures are discussed here:

- *Afforestation — GHGs sink and reduced emissions*: Forests, when sustainably managed, can have key role in climate change mitigation and adaptation. They regulate ecosystems, protect biodiversity, and play an integral part in the carbon sequestration.
- *Reducing emissions from deforestation and degradation: protecting climate, forests and livelihoods*: Cutting down of forest for urban expansion, road construction, timber, paper production and for various purposes, is now a global issue and have negative consequences in diverse ecosystems. Deforestation, apart from its undesirable effects on biodiversity and the livings of individuals that rely on forest, is the main factor of ecological transformation and contribute in greenhouse gas emissions worldwide.
- *Irrigation*: It has significant effects on total energy consumption as required to pump and treat water especially in arid and semi-arid regions. Energy requirements for accessing groundwater sources are high, which are frequently achieved by burning fossil fuels which led to emissions of greenhouse gases. For example, China alone contributes 16 - 25 Mt of carbon dioxide emissions, and India accounts for about 4 - 6 per cent of total of national emissions [46].

  With use of non-conventional sources of energy like solar energy in irrigation is a suitable alternative for reducing GHGs emissions linked to agriculture. Solar energy is reliable, clean energy and comparatively low cost, option for agricultural water management. However, being associated with their low cost of operation, there is a high potential of over extraction of water resources.
- *Development of alternative from the existing scenarios*: Adopt the resource conservation technologies, which are innovative and affordable to reduces the GHGs emissions. There are many examples of adoption of alternatives of conventional energy sources. Few of them are implementation of wind energy and bio-diesel in India, micro-hydropower and biogas in Nepal, etc. The government of India has also taken various initiative to tackle the adverse consequences of climate change. For instance, National Action Plan for Climate Change focused on reducing the carbon emission by the promoting Solar Mission, Enhanced Energy Efficiency, Sustainable Habitat, Sustaining Himalayan Ecosystem, Water Mission, Green India, Sustainable Agriculture, and Strategic Knowledge for Climate Change; State Action Plan on Climate Change emphasis on climate change mitigation, energy efficiency, and resource conservation to climate adaptation; Pradhan Mantri Ujjwal Yojana schemes provides LPG connections (below poverty line) to reduce the use of fossil fuel and conventional fuel.

5. Conclusion

The study on anthropogenic and climate change is crucial to quantify its influence on various surface water resources of the Himalayan region. As the hilly regions receives high rainfall (the high intensity rainfall events deliver too much water at once, causing water to runoff instead of infiltrate into the ground) during the monsoon season and the frequent occurrence of floods has resulted to various losses of human life residing in low land areas, losses of property, resources, etc. Most of the hilly regions in India has suffered drastic changes in terms of areal coverage of vegetation due to Jhum cultivation, overgrazing, forest fires, urbanization, unplanned development of various infrastructure, the rapid growth rate of population, over exploitation of natural resources etc. which adversely affect the ecological and hydrological balance of the region.

To characterize the dynamics of Himalayan rivers hydrology, the hydrologists might not quantitively represent the relationships between runoff and the associated hydrological parameters with respect to change in land use/cover and climate change. This is due to the fact that, the topography of Himalayas is very complex and to carry out fieldwork is a big challenge, which often leads to limited data availability. In addition to these, a good long-term hydro-meteorological data of Himalayan rivers is limited which add up additional challenges in employing hydrological modelling due to parametric uncertainty.

Even though, in spite of the eminent importance and necessity of such studies, there is not much of attempts have been made to evaluate the hydrological behaviour of Himalayan surface water resources. The study of Himalayan hydrology will assist policy makers to cope-up with many water-related issues like hydrological extremes, efficient water resource planning and management, conservation of soil and water, and thereby improve the livelihood of locality, balance ecosystem and biodiversity.

Acknowledgements
Authors are grateful to the Department of Agricultural Engineering, North Eastern Regional Institute of Science and Technology, Nirjuli, Arunachal Pradesh for their valuable support and suggestions for bringing this paper.

References

[1] Ives J D, Messerli B and Spiess E 1997 Introduction in B. Messerli, J. D. Ives(eds) *Mountains of the world: a global priority*, Parthenon, New York and London 1-15.
[2] Viviroli D and Weingartner R 2002 The significance of mountains as sources of the world's fresh water. *GAIA-Ecological Perspectives for Science and Society* **11** 3 182-186.
[3] Messerli B, Viviroli D and Weingartner R 2004 Mountains of the world: vulnerable water towers for the 21st century. AMBIO: *A Journal of the Human Environment* **33** 13 29-34.
[4] Xu J, Shrestha A, Vaidya R, Eriksson M and Hewitt K 2007 The melting Himalayas: regional challenges and local impacts of climate change on mountain ecosystems and livelihoods. *The melting Himalayas: regional challenges and local impacts of climate change on mountain ecosystems and livelihoods.*
[5] Ranjan P and Pandey P K 2020 Reviving, Development and Protection of Springs to Increase Water Security in the Himalayan Region. *Proceedings of the 5th International Conference on Computers & Management Skills* 25-29.
[6] Vorosmarty C J, Green P, Salisbury J and Lammers R B 2000 Global water resources: vulnerability from climate change and population growth. *science* **289** 5477 284-288.
[7] Wang G, Liu J, Kubota J and Chen L 2007 Effects of land-use changes on hydrological processes in the middle basin of the Heihe River, northwest China. *Hydrological Processes: An International Journal* **21** 10 1370-1382.
[8] Xu Y P, Zhang X, Ran Q and Tian Y 2013 Impact of climate change on hydrology of upper reaches of Qiantang River Basin, East China. *Journal of Hydrology* **483** 51-60.
[9] Tanzeeba S and Gan T Y 2012 Potential impact of climate change on the water availability of South Saskatchewan River Basin. Climatic Change **112** 2 355-386.
[10] Ashraf V, Mousavi S, Abbaspour S, Srinivasan R K and Yang H 2014 Analyses of the impact of climate change on water resources components, drought and wheat yield in semiarid regions: Karkheh river basin in Iran. *Hydrol Process* **28** 2018–2032.
[11] Perazzoli M, Pinheiro A and Kaufmann V 2013 Assessing the impact of climate change scenarios on water resources in southern Brazil. *Hydrolog Science Journal* **58** 77–87.
[12] Feng X, Zhang G and Yin X 2011 Hydrological responses to climate change in Nenjiang river basin, northeastern China. *Water resources management* **25** 2 677-689.

[13] Shrestha A B and Aryal R 2011 Climate change in Nepal and its impact on Himalayan glaciers. *Regional Environmental Change* **11** 1 65-77.

[14] Yang D, Gao B, Jiao Y, Lei H, Zhang Y, Yang H and Cong Z 2014 A distributed scheme developed for eco-hydrological modeling in the upper Heihe River. *Science China Earth Sciences* **58** 1 36-45.

[15] Ye X, Zhang Q, Liu J, Li X and Xu C Y 2013 Distinguishing the relative impacts of climate change and human activities on variation of streamflow in the Poyang Lake catchment, China. *Journal of Hydrology* **494** 83-95.

[16] Ponpang-Nga P and Techamahasaranont J 2016 Effects of climate and land use changes on water balance in upstream in the Chao Phraya River basin, Thailand. *Agriculture and Natural Resources* **50** 4 310-320.

[17] IPCC 2007: Climate Change 2007: Synthesis Report. Contribution of Working Groups I, II and III to the Fourth Assessment Report of the Intergovernmental Panel on Climate Change [Core Writing Team, Pachauri, R.K and Reisinger, A.(eds.)]. IPCC, Geneva, Switzerland, 104.

[18] Rees H G and Collins D N 2006 Regional differences in response of flow in glacier-fed Himalayan rivers to climatic warming. *Hydrological Processes: An International Journal* **20** 10 2157-2169.

[19] Chen Y, Takeuchi K, Xu C, Chen Y and Xu Z 2006 Regional climate change and its effects on river runoff in the Tarim Basin, China. *Hydrological Processes: An International Journal* **20** 10 2207-2216.

[20] Bosch J M and Hewlett J D 1982 A review of catchment experiments to determine the effect of vegetation changes on water yield and evapotranspiration. *Journal of hydrology* **55** 1-4 3-23.

[21] Viola M R, Mello C R, Beskow S and Norton L D 2014 Impacts of land-use changes on the hydrology of the Grande river basin headwaters, Southeastern Brazil. *Water resources management* **28** 13 4537-4550.

[22] Zhou F, Xu Y, Chen Y, Xu C Y, Gao Y and Du J 2013 Hydrological response to urbanization at different spatio-temporal scales simulated by coupling of CLUE-S and the SWAT model in the Yangtze River Delta region. *Journal of Hydrology* **485** 113-125.

[23] Chhabra A, Geist H, Houghton R A, Haberl H, Braimoh A K, Vlek P L, Patz P, Xu J, Ramankutty N, Coomes O and Lambin E F 2006 Multiple impacts of land-use/cover change. In *Land-use and land-cover change* (71-116). Springer, Berlin, Heidelberg.

[24] Li S, Yang H, Lacayo M, Liu J and Lei G 2018 Impacts of land-use and land-cover changes on water yield: A case study in Jing-Jin-Ji, China. *Sustainability* **10** 4 960.

[25] Wang S Q, Zheng X Q and Zang X B 2012 Accuracy assessments of land use change simulation based on Markov-cellular automata model. *Procedia Environmental Sciences* **13** 1238-1245.

[26] Kamusoko C, Aniya M, Adi B and Manjoro M 2009 Rural sustainability under threat in Zimbabwe–simulation of future land use/cover changes in the Bindura district based on the Markov-cellular automata model. *Applied Geography* **29** 3 435-447.

[27] Arsanjani J J, Helbich M, Kainz W and Boloorani A D 2013 Integration of logistic regression, Markov chain and cellular automata models to simulate urban expansion. *International Journal of Applied Earth Observation and Geoinformation* **21** 265-275.

[28] Subedi P, Subedi K and Thapa B 2013 Application of a hybrid cellular automaton-markov (CA-Markov) model in land-use change prediction: A case study of Saddle Creek Drainage Basin, Florida. *Applied Ecology and Environmental Sciences* **1** 6 126-132.

[29] Halmy M W A, Gessler P E, Hicke J A and Salem B B 2015 Land use/land cover change detection and prediction in the north-western coastal desert of Egypt using Markov-CA. *Applied Geography* **63** 101-112.

[30] Memarian H, Balasundram S K, Talib J B, Teh Boon Sung C, Mohd Sood A and Abbaspour K C. (2013). KINEROS2 application for land use/cover change impact analysis at the H ulu L angat B asin, Malaysia. *Water and Environment Journal* **27** 4 549-560.

[31] Calder I R 2007 Forests and water—ensuring forest benefits outweigh water costs. *Forest ecology and management* **251** 1-2 110-120.

[32] Zhang L, Nan Z, Yu W and Ge Y 2015 Modeling land-use and land-cover change and hydrological responses under consistent climate change scenarios in the Heihe River Basin, China. *Water Resources Management* **29** 13 4701-4717.

[33] Liu Y, Zhang J, Wang G, Liu J, He R, Wang H, Liu C and Jin J 2012 Quantifying uncertainty in catchment-scale runoff modeling under climate change (case of the Huaihe River, China). *Quaternary International* **282** 2012 130-136.

[34] Qin J, Ding Y, Wu J, Gao M, Yi S, Zhao C, Ye B, Li M and Wang S 2013 Understanding the impact of mountain landscapes on water balance in the upper Heihe River watershed in northwestern China. *Journal of arid land* **5** 3 366-383.

[35] Ragettli S, Pellicciotti F, Bordoy R and Immerzeel W W 2013 Sources of uncertainty in modeling the glaciohydrological response of a Karakoram watershed to climate change. *Water Resources Research* **49** 9 6048-6066.

[36] Seibert J and van Meerveld H J 2016 Hydrological change modeling: challenges and opportunities. *Hydrological Processes* **30** 26 4966-4971.

[37] Islam Z 2011 A review on physically based hydrologic modeling. *University of Alberta: Edmonton, AB, Canada*.

[38] Doocy S, Daniels A, Murray S and Kirsch T D 2013 The human impact of floods: a historical review of events 1980-2009 and systematic literature review. *PLoS currents*, **5**.

[39] Vashisht A K and Ranjan P 2020 Intermittent Multi-column Sand Filter: A Unique Solution to Multiple Applications. *Journal of The Institution of Engineers (India): Series A* **101** 1 69-75.

[40] Immerzeel W W, Beek L P and Bierkens M F 2010 Climate Change Will Affect the Asian Water Towers. *Science* **328** 1382-1385.

[41] Buytaert W 2012 Assessment of the impacts of climate change on mountain hydrology: development of a methodology through a case study in the Andes of Peru. *Mountain Research and Development* **32** 3 385-386.

[42] Eriksson M, Xu J, Shrestha A B, Vaidya R A, Santosh N and Sandstrom K 2009 *The changing Himalayas: impact of climate change on water resources and livelihoods in the greater Himalayas*. International centre for integrated mountain development (ICIMOD).

[43] IPCC 2013 / Change, I. P. O. C. (2014). *Climate change 2013: the physical science basis: Working Group I contribution to the Fifth assessment report of the Intergovernmental Panel on Climate Change*. Cambridge University Press.

[44] Lamsal P, Kumar L, Shabani F and Atreya K 2017 The greening of the Himalayas and Tibetan Plateau under climate change. *Global and Planetary Change* **159** 77-92.

[45] Anand J, Gosain A K, Khosa R and Srinivasan R 2018 Regional scale hydrologic modeling for prediction of water balance, analysis of trends in streamflow and variations in streamflow: The case study of the Ganga River basin. *Journal of Hydrology: Regional Studies* **16** 32-53.

[46] Shah T 2009 Climate change and groundwater: India's opportunities for mitigation and adaptation. *Environmental Research Letters* **4** 3 035005.

IOP Conf. Series: Materials Science and Engineering 1020 (2021) 012014 doi:10.1088/1757-899X/1020/1/012014

Reinforcing Livelihood by Interdependency of HR Capacity-ICT: Relevance to Multi-cropping practice system in Indian Eastern Himalaya

Jyoti Hatiboruah [a], Rajive Mohan Pant [b], Shibabrata Choudhury [c]

[a b c] North Eastern Regional Institute of Science and Technology,
Nirjuli, Arunachal Pradesh-791109, India.

jhbglt9@gmail.com

Abstract. The rural economy of Eastern Himalayan region of India is mostly agriculture based, dominated by traditional as well as community knowledge, has been using as source of livelihood since long years back. But with Green evolution and technological advancement, High Yielding variety of cash crop like Large Cardamom (Amomum subulatum Roxb.) and influence of skill-information-technology are emerging in a great extent respectively only to ensure sustainable livelihood. Since it is seen that only through large cardamom cultivation, it is difficult to ensure sustainability, hence concept of Multicropping is very much demanding. Here, this paper highlights the influence and interdependency of Human Resource Capacity and Information and Communication Technology (ICT) on Multicropping as well as on large cardamom cultivation to ensure sustainable livelihood in this particular region.

Key words: Large Cardamom, Multicropping, ICT, HR capacity building

1. Introduction

Multicropping practice as a part of Agricultural Development in the country like India where more than 121 cr. population is the utmost requirement to challenge sustainability, livelihood and food security. Here the agricultural development talks about all round development i.e. crop pattern, technology skills etc. and it can be understood as transition from cultural historical based rural production and marketing norms to structural farming of small and marginal farmers into a global market through integration [1-5]. In agricultural development information-intensive, knowledge-intensive production and marketing practices are the key elements [5-7]. On the other side, the technological advancement, the Information and Communication Technology (ICT) is introducing a new era in the field of agriculture specially in two broad areas, one is Market Efficiency and the other one is Knowledge Dissemination through mobile devices, remote sensors etc. [5-7], [24-25]. Market efficiency talks about the coordination on agricultural inputs, market information, financial transaction, e-commerce platform etc. whereas the Knowledge Dissemination focus on disseminating knowledge to small and marginal farmers on contemporary technologies which can uplift production practice, food security, sustainability and welfare [5], [7-10]. In the rural economy of Eastern Himalayan region of India, the cultivation of Large Cardamom (Amomum subulatum

Roxb) as a High Yielding Variety (HYV) of cash crop is much popular as one of the main sources of livelihood [11-15]. But these days due to varied challenging dimensions of land resource, climate, different crop diseases etc. compelling to go for multi-cropping practice along with the interdependency of ICT [16-17],[21], [23-25]. By focusing on Sustainability, it is very high in demand to coordinate among Large Cardamom cultivation along with Multicropping practice, ICT interdependency and Capacity building as only cultivation of Large Cardamom cannot ensure sustainability [1-2],[18],[24],[26]. Due to its high demand both in national and international market, its glamour among the cultivators is still very high [27]. If capacity building on manpower and ICT interdependency goes hands on together in agricultural sector, obviously it will bring a new era after the Green revolution [8], [19-20], [22]. This paper discusses necessity and importance of ICT interdependency along with interdependency of Human Resource (HR) Capacity Building with the objectives: i) To discuss the scenario of large Cardamom cultivation along with Multicropping practice in the region and ii) To discuss the interdependency of ICT as well as Capacity Building of manpower to ensure Sustainability. Here the paper is discussed in four sections, namely 1. Introduction 2. Research Methodology 3. The Case study 4. Conclusion. The first section "Introduction" part contains the current scenario of agricultural sector along with the need of some interdependency of ICT and Capacity building of HR to ensure sustainable livelihood. The "Research Approach" of this study includes Methodology, Study area, sampling and techniques. The Case study section is discussed in i) Current Scenario on livelihood practice in Himalayan region ii) In present context the Small and Marginal farmers' livelihood practice iii) Importance of ICT and Capacity Building of Manpower for cultivation of Large Cardamom iv) Planned outline reinforcing Livelihood of marginal farmers. The last section "Conclusion" summarizes the discussion of the paper in brief highlighting the future scope.

2. Research Methodology

2.1 Methodology

A qualitative approach is adopted for this research work which comprises both primary as well secondary data. For the collection of primary data, methods and tools that were adopted are structured questionnaire, personal interviews and for data validation purpose personal observation was used. The secondary data were collected from different sources like department of Horticulture, Government of India, spice board journal, annual reports of government of India: Ministry of Small and Medium Enterprises (MSME), Census India 2011, different business news, Ministry of Rural Development, United Nations Educational, Scientific and Cultural Organisation (UNESCO) etc. The combination of both judgement and convenient sampling techniques were used. The sample size for this study was 49 which represent almost 78% of total registered Large cardamom cultivators in Lower Subansiri.

2.2 Study area and sampling

Majority of the respondents which represent 80% of total sample size belong from Yachuli (Zero-II) as the majority habitants are aggressively associated with large cardamom cultivation and very few which represent only 20% of total sample size belong from Zero-I as the habitants are transforming themselves towards Kiwi cultivation gradually. The sample highlights a unique characteristic that involvement of young generation is not found active whereas the registered farmers belong from more than 35 years of age group. Though the farmers are associated with this cultivation from last many years but professional engagement only from last 3-9 years. A new ray is coming among the women who have registered as large cardamom cultivators though its size is small.

3. The Case Study

3.1 Current livelihood practice in Himalayan region

Most of the habitants of Lower Subansiri engage with vegetation of Maize, Pumpkin, Potato, Ginger, Chinese cabbage, Pisciculture along with rice cultivation etc. in large scale through Multicropping practice as major source of livelihood which are completely traditional practice based without any interdependency of ICT. Along with these cultivations almost every household maintain their own kitchen garden as a community practice. The traditional cum manual based cultivation is mostly done by the help of both hired labours as well as family members. Recently the emergence of formalized cooperative farming concept is grazing among the cultivators of Yachuli which can be used in an extensive manner to enhance the value chain in a global market.

3.2 In present context the Marginal farmers' livelihood practice

The marginal farmers do the Multicropping practice with vegetation and animal husbandry to meet the day to day expenditure but they go for large cardamom cultivation to meet the high demands of life i.e. of education of children and medical expenditure of family members.

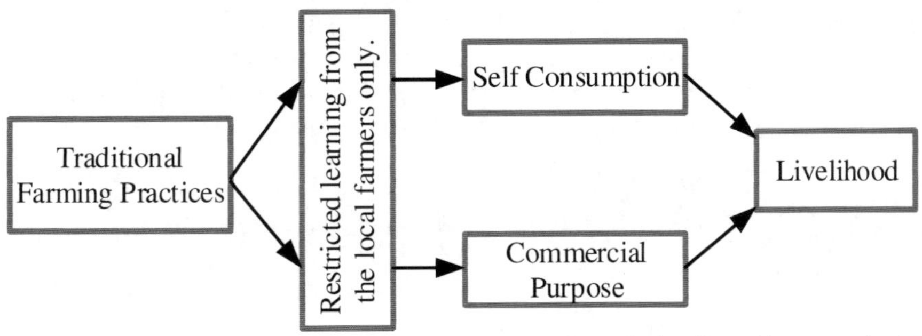

Figure 1. Diagrammatic representation of Existing practice of Livelihood by Marginal Farmers

Figure 1. highlights the traditional farming practice only that are being followed by the marginal farmers to ensure livelihood through Multicropping as well as large cardamom cultivation without any interdependency of ICT and change agent. With the advantage of geographical location and restricted learning from neighbour as well as from local farmers, the habitants go for Multicropping along with large cardamom cultivation. The ultimate production is used for both self - consumption and Commercial purpose as source of their livelihood. But gradually regions' climatic condition is changing due to slight increase in population, life style etc. Almost all the respondents of marginal farmers agreed and do believe that only to ensure sustainability they go for Large cardamom cultivation. Most importantly the marginal farmers are adopting its commercialization from last few couple of years which is great sign for them as well as for the economic condition of the place.

3.3 Importance of ICT and Capacity Building of Manpower for cultivation of Large Cardamom

The traditional as well as community-based knowledge and information from friends and relatives were the main sources of information based on which cultivation of Large cardamom is done commercially by the marginal farmers. But to sustain in the competitive global market with value chain it is very difficult without enhancing the use of ICT as well as building capacity of manpower on this regard. So, interdependency of ICT and Capacity building of manpower in the form of skill upgradation are the utmost requirements to challenge the global requirements either by the government agencies or by some non- government organisation by considering the gap. As it is coming to know that Government organise some kind of awareness programme on large cardamom cultivation in a very rare scale and some kind of training session only on cultivation process which is not even upgraded one only for the beneficiaries. So here it will be worth mentioning that interdependency of ICT will enhance the farmers on keeping updated on Market Efficiency like updated agricultural inputs, market information; scientifically proved cultivation practice to get more success and most importantly on financial transactions where farmers lack knowledge. In addition to this ICT helps in Knowledge Dissemination to marginal farmers on contemporary technologies which can uplift production practice. At the same time if capacity building of manpower on those areas, which are already being covered by ICT, are done practically it will enhance the farmers into a such level which will give them the strength to face global challenge.

3.4 Planned outline reinforcing Livelihood of marginal farmers

The existing practice model is not acquainted with ICT as well as with Capacity Building of manpower which make the marginal farmers weaker in accepting the global challenges with value chain. So, all the challenges can be categorised in four broad parts. They are- a) Challenges with the Cultivation i.e. in-depth physiological knowledge about the plant as well as its sapling, its cultivation, plant's various diseases based on climatic ailment etc. b) Challenges with Manpower Ability i.e. handling technique from plant to supply of raw material which includes acquiring entrepreneurial skill, branding the farmers and their product by themselves, pricing strategy in between etc. c) Challenges with Climatic Condition which talks about the changing pattern of required climate like temperature, rain water, soil pattern, shading pattern, land structure, forestry knowledge etc. d) Challenges with ICT and Capacity Building i.e. how much the marginal farmers are updated with latest Agri- input technologies, latest market information, financial resources and transactions, e-commerce platform etc. with the involvement of local community.

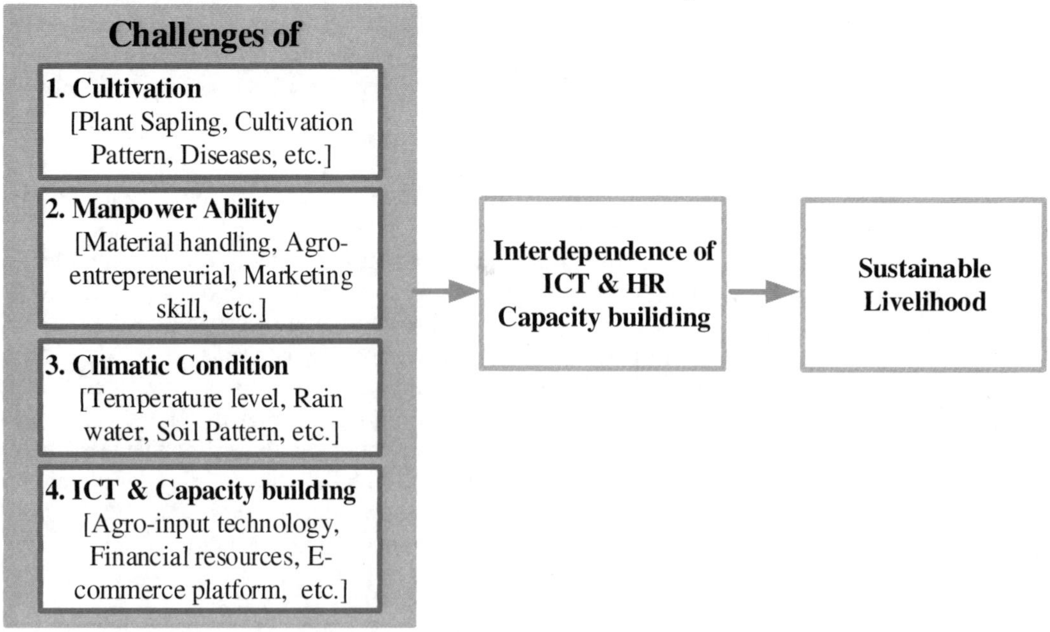

Figure 2. Proposed Planned outline reinforcing Livelihood of marginal farmers

Figure. 2 on Proposed outline on Sustainable Livelihood for marginal farmers is addressing on these issues so that sustainability of marginal farmers can be ensured through the interdependency of ICT and HR capacity building.

The proposed framework is talking about the high level of importance on the interdependency of ICT along with local community knowledge to enhance value chain on each and every stage of cultivation as well as to the stage of final product, and the interdependency of HR capacity building to enhance the ICT in reality through practical utilization in the field of agriculture more specially with the large cardamom cultivation. In short, if ICT and HR capacity building initiative are adopted parallelly either by Government agencies or by the Non- government agencies with the support of Government or the cooperative farmers group (since it is emerging in the proposed area) all the four mentioned challenges area can be addressed by the marginal farmers through a proper coordination and linkage.

4. Conclusion
This paper has highlighted the importance and influencing capacity of ICT and HR capacity building along with community-based knowledge for agricultural development for sustainability of rural economy. Effective coordination and implementation among the four dimensions of challenges of marginal farmers would help in strengthening the farmers towards the global competition, also can address the sustainability of livelihood through large cardamom cultivation and Multicropping practices in the proposed region with satisfactory result.

Acknowledgments
The authors are thankful to the funding agency: National Mission on Himalayan Studies (NMHS) for their support through project no. GBPNI/NMHS-2017-18/SG15.

References
[1] Pothula A. L., Singh, A.I. (2013). Postharvest Processing of Large Cardamom in the Eastern Himalaya: A review and Recommendations for increasing the sustainability of a Niche crop. *Mountain Research and Development (MRD),* vol. 33, no. 4, November, 2013, pp. 453-462, https://doi.org/10.1659/MRD-JOURNAL-D-12-00069.1

[2] Opolot N. H, Isubikalu et. all (2018). Influence of university entrepreneurship training on farmers' competences for improved productivity and market access in Uganda". *Cogent Food & Agriculture.*

[3] Eva D., Artemis K. (2007). Capacity Building in Agricultural Biotechnology in Turkey. *Food Reviews International*, vol. 19, no. 4, pp. 437-445, DOI: 10.1081/FRI-120025484.

[4] Durugkar S., & Poonia C. R. (2017). Optimum utilization of natural resources for home garden using wireless sensor networks. *Journal of Information and Optimization Sciences,* 38:6, 1077-1085.

[5] Karanasios S., & Slavova M. (2016). How to development actors do "ICT for development"? A strategy-as-practice perspective on emerging practices in Ghanaian agriculture. *Information Systems Journal,* 29:888-913.

[6] Zewge A., & Dittrich Y. (2017). Systematic Mapping Study of Information Technology for Development in Agriculture (The Case of Developing Countries). *The Electronic Journal of Information Systems in Developing Countries,* vol. 82, no. 2, pp. 1-25.

[7] Lindlom J., Lundstrom C., et. all. (2017). Promoting Sustainable intensification in precision agriculture: review of decision support systems development and strategies. *Precision Agriculture,* 18:309–331, DOI 10.1007/s11119-016-9491-4.

[8] Blasi et. all. (2015). Alternative food chains as a way to embed mountain agriculture in the urban market: the case of Trentino. *Agricultural and Food Economics,* DOI 10.1186/s40100-014-0023-0

[9] Ahmed et all. (2017). Cropping systems diversification, improved seed, manure and inorganic fertilizer adoption by maize producers of eastern Ethiopia. *Journal of Economic Structures,* DOI 10.1186/s40008-017-0093-8.

[10] Johnson et all. (2015). The Capacity Building Experience of Women Engaged in Determining the Cost and Affordability of Healthy Food in Nova Scotia, Canada. *Journal of Hunger & Environmental Nutrition,* DOI: 10.1080/19320248.2014.962769.

[11] Sen S. and Kansal A. (2019). Integrating value-chain approach with participatory multi-criteria analysis for sustainable planning of a niche crop in Indian Himalayas. *Journal of Mountain Science*, vol. 16, no. 10, pp. 2417-2434.

[12] Sharma E. (2000). A Boon for Mountain Populations: Large Cardamom Farming in the Sikkim Himalaya. *Mountain Research and Development,* Vol. 20, No. 2, pp: 108-111.

[13] Sharma G. (2016). Declining Large-Cardamom Production Systems in the Sikkim Himalayas Climate Change Impacts, Agro economic Potential, and Revival Strategies. *Mountain Research and Development,* Vol. 36, No. 3, pp: 286-298.

[14] Sharma G. (2002). Performance of an Age Series of Alnus-Cardamom Plantation in the Sikkim Himalaya: Productivity, Energetics and Efficiency. *Annals of Botany,* pp: 261-272

[15] Dauda A. K., (2017). Gender capacity building needs on soil fertility management practices among smallholder arable crop farmers in Kwara State, Nigeria.

[16] Ferreroa G., et. all (2019). Capacity building and training approaches for water safety plans: A comprehensive literature review. *International Journal of Hygiene and Environmental Health,* vol. 224, no. 4, February, pp. 615-627, doi: 10.1016/j.ijheh.2019.01.011

[17] Kamruzzaman M., Takeya H. (2008). Influence of Technology Responsiveness and Distance to Market on Capacity Building. *International Journal of Vegetable Science,* vol.14, no. 3, pp. 216-231, DOI: 10.1080/19315260802164848.

[18] Sendegeya A.M (2016). The role of academia in capacity building for sustainable energy development: the case of Namibia. *Energy,* pp: 218-222.

[19] Ahmed M. H., Mesfin H. M. (2017). The impact of agricultural cooperatives membership on the wellbeing of smallholder farmers: empirical evidence from eastern Ethiopia. *Agricultural and Food Economics.*

[20] Babu S. C. et. all. (2015). Strengthening Capacity for Agribusiness Development and Management in Sub-Saharan Africa. *Africa Journal of Management,* ISSN: 2332-2373

[21] Choudhury S., Pant R.M. et. all. (2016). Destination Branding of Ziro Through Potentiality of Bio-tourism. *Bioprospecting of Indigenous Bio-resources of North-East India*, pp. 329-337, DOI: 10.1007/978-981-10-0620-3_21

[22] Sendegeya A.M (2016). The role of academia in capacity building for sustainable energy development: the case of Namibia. *Energy,* pp: 218-222.

[23] Johnson C. P., Williams P. L. et. all (2016). The Capacity Building Experience of Women Engaged in Determining the Cost and Affordability of Healthy Food in Nova Scotia, Canada. *Journal of Hunger & Environmental Nutrition*, vol. 10, no. 3, 14th pp. 356-378. Journal Database, http://dx.doi.org/10.1080/19320248.2014.962769.

[24] Gangopadhyay P. K., et. all. (2019). Spatial targeting of ICT-based weather and agro-advisory services for climate risk management in agriculture. *Climatic Change,* 154:241-256, doi.org/10.1007/s10584-019-02426-5.

[25] Aker J. C., et. all. (2016). The promise (and pitfalls) of ICT for agriculture initiatives. *International Association of Agricultural Economists,* DOI: 10.1111/agec.12301.

[26] S. K. Ghosh, A. K. Das and S. Choudhury, (2020) ICT Interventions in Rural Development Schemes: A Case of Rupshi Development Block, Assam, *2020 8th International Conference on Reliability, Infocom Technologies and Optimization (Trends and Future Directions) (ICRITO), Noida, India, 2020, pp. 349-354, doi: 10.1109/ICRITO48877.2020.9197987.*

[27] J.H. Boruah, R.M. Pant, S. Choudhury, (2020), Sustainable livelihood by HR capacity building in multicropping practices of Indian Eastern Himalayan region: A case study, Journal of Statistics and Management Systems, 23, pp-277-284, https://doi.org/10.1080/09720510.2020.1724626

A Survey on Voice over Internet Protocol (VoIP) Reliability Research

Om Prakash Roy[1], Vinod Kumar[2]

[1, 2] Department of EE, NERIST, Arunachal Pradesh 791109, India

[1]oproy61@gmail.com, [2]vinodnerist@gmail.com

Abstract. VoIP technology deals with the real-time data communication for voice transfer in the form of digital packets through internet communication and facilitates public to make use of internet for video and phone calls. The voice data packets are transferred from source to the destination and vice versa. VoIP require high speed internet connection for data transfer on data network. It is necessary to arrive the data packet from its source to the destination with high level of reliability. It is very important to analyse link failure, packet loss, delay and jitter during the data communication. Components involved in data communication should be reliable. This paper provides a survey for step wise development and use of reliability techniques for gaining high quality of voice in VoIP network.

Keywords: VoIP, Reliability, availability, QoS

1. Introduction

VoIP technology deals with the real-time data communication for voice transfer in the form of digital packets through internet communication. The voice data packets are transferred from source to the destination and vice versa. For the purpose of communication of data between two nodes, first the voice signal which is in the analog form is converted to digital form as data packets then sent over the internet using codec. The process is repeated at the receiving end but in reverse order, first digital data is converted to analog signal then communicated to the destination node [1]. VoIP is a component based system which handle critical applications and evaluation of reliability and security of this becomes very important [2].

1.1. PSTN Vs VoIP

Public Switched Telephone Network (PSTN) is the oldest communication service for providing end to end switched circuit communication for voice signal transfer for more than a hundred year. With the advancement of data networks, the service is being avoided because of its cost and data transfer inefficiency. Voice and data transfer over the same network becoming popular because of involvement of less communication infrastructure like cables, exchange branch etc. which ultimately reduces cost, this encouraged the people to use VoIP [3].

1.2. Architecture of VoIP

Making a phone call using broadband internet connection with packet switching network is very fast and cost saving. VoIP is the best known for transfer of audio and video using packet network. The architecture of VoIP differs from PSTN. There are some differences like centralization of audio services which is possible only by using VoIP systems [4]. The voice transmission in VoIP network is shown in Figure 1.

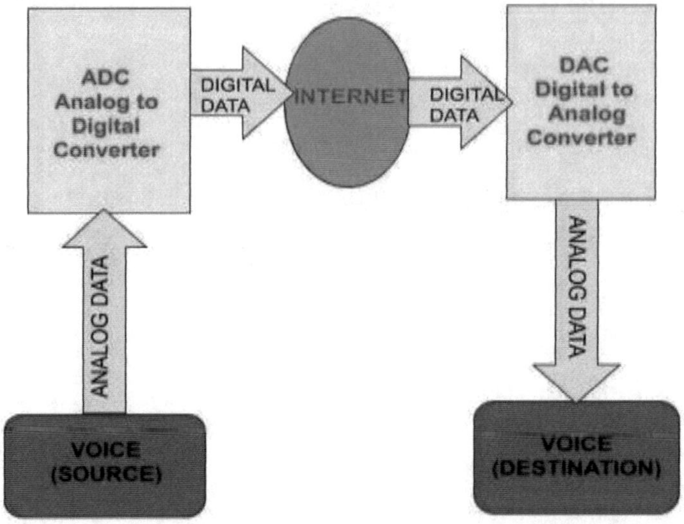

Figure 1. Voice Transmission in VoIP Network.

This paper is organized in four sections, introduction part covers introduction to VoIP along with architecture of VoIP. Literature survey is presented in section 2 along with year wise reliability analysis techniques in table format. Further, section 3 describes about reliability of VoIP, reliability matrices and reliability issues in VoIP system. Finally, section 4 concludes the paper.

2. Literature Survey

Authors surveyed the VoIP challenges, discussed about the implementation with some factors considerations, protocols requirements and necessary components for connection set up [5]. The researchers surveyed VoIP technology with brief introduction. They discussed about protocols, jitter, delay, loss of packet and feasibility of this technology over the satellite links [6]. Modelling and computation for availability was performed on the basis of corresponding signalling and bearer path. They have shown accuracy of Simplified Availability Modelling Worksheet (SAMOW) after comparison with Markov model [7]. Authors investigated performance and dependability modelling for services engaged in transfer of audio data and used Stochastic Petri Net (SPN) for modelling approach and evaluated the dependability of audio data communications [8]. Worked on end to end service reliability for VoIP and developed models for network elements like links, routers, switches gateways, call servers [9]. Highlighted the development of soft switches for Voice over Internet Protocols and Asterisk (PBX) free software tested as fault tolerant and had shown the possibility of improvement of the service availability [10]. Presented widely-used concept of fault-tolerance and found that mapping a design at system level is possible for Quality of Service performance like fault-tolerance needs at the level of application compared to Open System Interconnection (OSI) model [11]. Research covers the issues of Voice over IP impacting the reliability and integrity and described, delay, packet loss, and jitter. Moreover, analysed the hardware component level reliability by using Markov Model [12]. Presented the Voice over IP architecture on the basis of node to node rules. The peer-to-peer technology is applied by integrating load balancing and requirements for failover [13]. Surveyed recent areas for routing protocols developments, Quality of Service (QoS) techniques, packet related mechanisms, standards for wireless communication and further surveyed QoS parameters affecting the quality of audio or video in voice over IP communication in wireless mesh networks e.g. loss of packet, jitter and delay [14]. Presented modelling for packet loss which is one of the QoS parameters and performed analysis for jitter and packet loss for VoIP traffic [15]. Proposed Gilbert–gamma topology for packet-loss in speech quality transmitted over Internet Protocol (IP) networks. In the topology, they shown introduction of packet-loss process for the speech recognition systems in VoIP [16]. Evaluated algorithms for the purpose of analysis of reliability and availability of VoIP. The methodology was based on IP Markov at component level and constituted tools for reliability and availability of complex networks [17]. Evaluated availability for VoIP networks services in internet using measurement data and metrics for availability [18]. Proposed Rapid Availability Prototyping for Testing Operational Readiness (RAPTOR), a software tool for gaining availability during life cycle [19]. Utilized a network simulator known as OPNET Modeler 14.5 for reliable VoIP network and assured the gain of network reliability [20]. Proposed SATCOM (SATellite COMMunication) links as reliable and efficient way to transmit IP data for voice communication and as a backup transmission link for overall resilience improvement [21]. Proposed a hierarchical model by combining performance and reliability, failure of more than one component was considered for modelling reliable VoIP network in wireless environment. Markov chain was utilized for constructing a reliability model and reliability metrics like Mean Time To Failure (MTTF) is recorded in VoIP network [22]. Proposed an approach for Voice over IP and discussed challenges. Lagrange's interpolation was applied for reliability and fault tolerance. But for effective stego-message communication reliability issues need addressing [23]. Reviewed PSTN for five nines reliability and observed the achievement of same level for voice over IP. Using Linux open source, they experimented for providing higher level of availability. But the work is in continuation for theoretical analysis of availability [24]. Explored VoIP for transport protocols and presented a Reliable Signalling Transport Protocol (RSTP). They presented RSTP advantages for reliability. But in this direction an efficient algorithm is required [25]. Studied the main VoIP components like servers, firewalls, routers etc., which can affect the VoIP service. The modelling used Continuous Time Markov Chain (CTMC) for software rejuvenation and was used for reliability and availability analysis in two cases one with rejuvenation and other without rejuvenation [26]. Followed Markov process for degraded states in modelling software rejuvenation in a VoIP system. Moreover, checking the control states triggering for rejuvenation was tested [27]. Presented an easy-to-use software tool, called NetExpert. The reliability and availability model were used in the software. They presented a comparative analysis of network topologies for fiber network and found NetExpert as tool for the design of different network configurations [28]. Proposed an E-model for Thai environment and presented solution for delay in packet transfer. They defended E-model for reliability and accuracy. The authors suggested the findings as latest evidence for language identification in VoIP system. But only Thai voice samples were considered, moreover this bias factor need further investigation [29]. Presented reliability evaluation of business VoIP proposed an approach which identified two VoIP design one about reliability standards and secondly matrices for measurement of enterprise VoIP services [30]. Mentioned about VoIP reliability including IP networking infrastructure. Reliability matrices at element and service level was mentioned. Authors found matrices as an essential requirement tool for measurement of VoIP reliability [31]. Measured and evaluated QoS for Voice over IP (VoIP). Mean Opinion Score (MOS), latency, packet loss and jitter remained the main criteria for evaluation [32]. Detected link failure over the Wide Area Network (WAN) by linking two tunnel one primary and another secondary tunnel. Data path change during the call for packets transmission over non-failure tunnel [33]. A brief description about year wise development of reliability analysis techniques for VoIP are shown in Table 1 below.

Table 1. An Overview of VoIP Reliability Analysis Techniques.

Author	Year	Techniques used for reliability analysis
Hou *et al*	2002	Markov analysis and Reliability Block Diagram [7].
Yoma *et al*	2005	ordinary Gilbert model [16].
Sharma *et al*	2007	SHARPE software for reliability analysis [2].
Koutras *et al*	2007	Semi-Markov modelling framework [27].
Koutras *et al*	2009	Continuous Time Markov Chain (CTMC) [26]
Castanon et al	2009	Stochastic Petri Net [28].
Palade I	2010	Reliability block diagram - IP RBD and Markov chains [17].
Guimares *et al*	2010	Stochastic Petri Net [8]
Hamdaqa *et al*	2011	Lagrange Interpolation [23].
Pal *et al*	2011	Linux in conjunction with a load balancer and ENUM [24].
Mahmood *et al*	2012	OPNET Modeller 14.5 is used as network simulator [20].
Palade I	2012	RAPTOR, Monte Carlo simulation [19].
Toral-Cruz *et al*	2013	2-state and 4-state Markov chains [15],
Chhabra *et al*	2015	Mean Opinion Score and R-Score in Wireless VoIP [37].
Bensalah *et al*	2019	Riverbed Modeller simulator [32].
Murphy *et al*	2019	RAPTOR [38].
Dasari *et al*	2019	NS-3 version 3.27,adhoc network topology [39].
Hu *et al*	2020	MOS and QoE for packet [40].
Malekzadeh	2020	The NS3 simulation tool implement the model [41].
Ganesan *et al*	2020	Virtual Box, Mininet VM, POX Controller, Wireshark [42].

3. Reliability of VoIP System

Reliability is probability of success. Reliability of a system or a device or a component is the probability of performance of its intended function for a given period of time with certain working condition. For example, a system having reliability of 0.9999 for one year has a continuously 99.99% functioning probability without any failure for the same year [10].

3.1. Reliability and Availability

Reliability effects availability, 99.99% reliability is not equal to 99.99% availability. Reliability measures ability of a system to function without failure, whereas availability is measurement of ability for providing service level continuously to customers. The reliability may be improved in VoIP by considering redundant hardware equipment within the networks including connections. The access gateways, signalling gateway and media server etc. can be made fault tolerant. Software can be upgraded to avoid the service loss [4]. Availability is the key metric for applications. Customers requires a high level of services. A small loss of service is acceptable but failure at system level is not tolerable [34].

3.2. Reliability and Availability of Data Networks

It is very difficult to measure the reliability and availability of data networks because of involvement of large number of equipment in a network including hardware and software. Therefore, it is impossible for data networks to catch the reliability bar defined by VoIP providers. As a measure for reliability, it requires calculation of Mean Time Between Failures (MTBF) [17]. In case of hardware and software repairable components of a network such as VoIP, if we assume the constant failure rate. then the reliability function become

$$R(t) = e^{-\lambda t} = e^{-t/MTBF} \quad \text{where } \lambda = 1/MTBF \tag{1}$$

Availability measure failures that comes out appears within the time interval (0, t). In case of irreparable system, the definition to measure A(t) is equivalent with R(t). Availability ‘A(t)’ that can be measured as below.

$$A(t) = MTBF/(MTBF+MTTR) \tag{2}$$

3.3. Reliability Matrices
Communication networks like VoIP system consist of many electronic components and when a failure occurs unaffected components are not powered down. A very small increase in availability of system is very important [10, 26, 27]. Considering the important matrices to measure the reliability are mentioned as below.

3.3.1. Uptime
It is the time measurement of a system in hours for its operational state. For example, uptime of a VoIP system is 5000 hours. In case of reliability calculations, it is same as Mean Time Between Failures (MTBF).

3.3.2. Downtime
Average time of the equipment in hours when it is not in working mode. For example, downtime of VoIP equipment is 10 hours. In case of reliability calculations, it is same as MTTR.

3.3.3. Mean Time To Repair (MTTR)
It is the used by a person to repair the equipment for proper functioning. For calculating availability for any system MTBF and MTTR are required.

3.3.4. Mean Time Between Failures (MTBF)
MTBF is a parameter for measuring reliability and it is average time interval counted in hours between the failures.

3.3.5. Hardware MTBF
It is the mean time between hardware component failures. Hardware component redundancy avoids hardware equipment failures.

3.3.6. System MTBF
System MTBF is the mean time between system failures. The failure predicts the outage and may be restored by repairing. The MTBF of hardware equipment for a system can be increased by adding redundant components.

3.3.7. Failure Rate
Failure rate of hardware equipment is inverse of its MTBF and similarly, failure rate of a system is inverse of its MTBF. When Failure Rate becomes higher, MTBF will be smaller. Failure rate = 1/MTBF.

3.4. Reliability Issues in VoIP System
The VoIP networks are prone to cyber-attacks than the circuit switched networks. Voice quality and network issues need reliability assessment [30]. VoIP network faces many problems such as packet loss, jitter, latency, voice quality, software and hardware equipment. Different types of reliability issues exist in the present developments of VOIP technologies and are mentioned below.

3.4.1. Software Reliability

Software reliability has an important role for proper functioning of the system without failure because failure of any system component may increase other related problems. Security of VoIP system operation is possible with reliable software. The operation of the communication network and the availability of the files needed has an important role in reliability analysis [35].

3.4.2. Link Failure

There are several reason for link failure for example, equipment failure, cable cut, power failure, switch or router failure, virus attack, packet lost from source to destination and Denial of Service (DoS) attack. Link failure can occur suddenly and may remain for long time. Link can be re-established if the problem is caused by non-availability of suitable path because reliable routers can find another possible path. Link failure impacts the quality of speech in VoIP network [14].

3.4.3. Packet Loss

Packet loss in VoIP system is caused by non-availability of suitable bandwidth on the network system. Data packets travel through router in the network follow different paths from source to destination and some packets may have lost in between. Packet loss causes the degraded audio and video quality which is not acceptable in standard and business communications. Moreover, loss of few packets in VoIP communication is not a big problem but in case of a loss of data packet carrying confidential information passing through the network may raise the question on integrity of secret information [16].

3.4.4. Network Availability and Outages

Network systems availability is not possible all the time because of involvement of lots of components in between such as hardware, software and link. VoIP system failure in business system may lead to a big loss of money because of communication failure. The service providers are required to find a suitable solution. Outage occur when one main system is down and reliability standards for VoIP are necessary such as PSTN standards [36].

3.4.5. Quality of Service (QoS)

Due to the involvement of different types of network applications in VoIP system quality of service is a big demand. The voice packets must reach from source to destination along with quality of transmission. Packet loss, jitter, latency and delay are the main issues which remain in VoIP network. Real time communication such as VoIP requires a low level of latency in network for the purpose of performance of high level of voice and video quality. QoS can influence the Quality of Experience (QoE) in VoIP communication. To evaluate the reliability of VoIP traffic QoE from the VoIP users is very critical to analyse [32].

3.4.6. Network Design

Network design is responsible for reliability of the VoIP system. If VoIP network is not designed properly than reliability of the system will be affected. Latency, packed loss and jitter are the key issues in network design. When VoIP is implemented over wireless networks, it becomes critical to design the network and issues such as reliability and QoS cannot be ignored. For the purpose of development of 5G network, network design is emerging as difficult area. The performance of network systems like VoIP communication systems require consideration in designing stable environment. Indices like reliability and availability may be important to measure the performance of a system [11].

3.4.7. Power Requirements

Power supply interruption creates the outages in VoIP network. The power required depends on specifications of the equipment. Sometime equipment consumes a very less power than the actual supply. Power supply rating cannot be considered to be same as power consumption of network equipment. Most of the network equipment requires either normal AC or normal DC supply. VoIP phone gets power through Ethernet via unused wires. In case of analog phone and digital phone, power is supplied by PBX switch [36].

4. Conclusions

In the past year communication development of VoIP has taken place at a large extent with the reliability issues. It is tried to survey reliability techniques used by researchers with discussion on existing reliability issues. Further, an overview of reliability analysis techniques is presented in table format along with important reliability matrices to be considered in VoIP development. It is observed that future of VoIP depends on improving the quality of voice by solving the reliability issues. Whereas number of solutions has been deployed by the researchers but the attention should be given on solving Quality of Service (QoS) problems such as link failure, packet loss, delay and jitter during the VoIP data communication. Moreover, Reliability improvement for components involved in data communication should be given importance for betterment of voice quality in VoIP systems.

References

[1] Jalendry S and Verma S 2015 A detail review on voice over internet protocol (voip) *Int. J Eng. Trends Technology* **23(4)** 161-166

[2] Sharma V S and Trivedi K S 2007 Quantifying software performance, reliability and security: An architecture-based approach *Journal of Systems and Software* **80(4)** 493-509

[3] Sinaeepourfard A and Hussain H M 2011 Comparison of VoIP and PSTN services by statistical analysis *IEEE Student Conference on Research and Development* 459-461

[4] Raj R, Gagneja A and Singal R 2012 Voice over Internet Protocol–The Technology and Its Application *International Journal of Scientific and Engineering Research* **3(5)** 1-6

[5] Pourghasem J, Karimi S and Edalatpanah S 2012 A Survey of Voice Over Internet Protocol (VOIP) Technology *IJCMSA* **6(3-4)** *53– 62*

[6] Raheja T and Munjal D 2015 A Comprehensive Survey on Voice over Internet Protocol (VoIP). *IJARCET* **4(4)** 1552-1557

[7] Hou W and Okogbaa O G 2002 A simplified availability modeling tool for networks with 1: 1 redundant software-hardware systems *IEEE Annual Reliability and Maintainability Symposium Proceedings (Cat. No. 02CH37318)* 556-562

[8] Guimares A P, Maciel P R M and Matias R 2010 Dependability and performability modeling of voice and data networks *IEEE Latin-American Conference on Communications* 1-6

[9] Conway A E and Khasnabish B 2006 End-to-end network reliability modeling of enterprise VoIP services *IEEE/IFIP Network Operations and Management Symposium* 404-413

[10] Gorti A 2006 A fault tolerant VoIP implementation based on open standards *IEEE Sixth European Dependable Computing Conference* 35-38

[11] Al-Kuwaiti M, Kyriakopoulos N and Hussein S 2009 A comparative analysis of network dependability, fault-tolerance, reliability, security, and survivability *IEEE Communications Surveys and Tutorials* **11(2)** 106-124

[12] Ali S R 2019 Reliability Analysis of VoIP System *Next Generation and Advanced Network Reliability Analysis* (Springer Cham) 211-244

[13] Fiedler J, Kupka T, Magedanz T and Kleis M 2006 Reliable VoIP services using a peer-to-peer intranet *Eighth IEEE International Symposium on Multimedia (ISM'06)* 121-130

[14] Meeran M T, Annus P and Le Moullec Y 2016 The current state of voice over internet protocol in wireless mesh networks *IEEE International Conference on Advances in Computing, Communications and Informatics* 2567-2575

[15] Toral-Cruz H, Pathan A S K and Pacheco J C R 2013 Accurate modeling of VoIP traffic QoS parameters in current and future networks with multifractal and Markov models *Mathematical and Computer Modelling* **57(11-12)** 2832-2845

[16] Yoma N B, Busso C and Soto I 2005 Packet-loss modelling in IP networks with state-duration constraints *IEE Proceedings-Communications* **152(1)** 1-5

[17] Palade I 2010 Reliability and availability engineering for VoIP communications systems *IEEE 8th International Conference on Communications* 309-312

[18] Jiang W and Schulzrinne H 2003 Assessment of voip service availability in the currentinternet *Proceedings of the 4th International Workshop on Passive and Active Network Measurement (PAM 2003)* 1-7

[19] Palade I 2012 A new approach in using RAPTOR to evaluate reliability and availability for VoIP systems *IEEE 9th International Conference on Communications* 217-220

[20] Mahmood M K, Al-Naima F M and Uzunoglu N K 2012 Design and simulation of a data transmission network for industrial control system subject to reliability improvement *IEEE International Conference on Future Communication Networks* 23-28

[21] Gierszal H, Sdongos E, Kiedrowski Ł, Korzeniowki, D Pluciński, K Brajer, Ł and Jackson J 2016 Reliability of SATCOM transmission for PPDR usage *IEEE 12th International Conference on Wireless and Mobile Computing, Networking and Communications* 1-8

[22] Gupta V and Dharmaraja S 2013 Reliability and performance modelling of VoIP system with multiple component failures *International Journal of Reliability and Safety* **7(1)** 58-74

[23] Hamdaqa M and Tahvildari L 2011 ReLACK: a reliable VoIP steganography approach *IEEE Fifth International Conference on Secure Software Integration and Reliability Improvement* 189-197

[24] Pal S, Gadde R and Latchman H A 2011 On the reliability of Voice over IP (VoIP) telephony *SPRING 9th International Conference on Computing Communications and Control Technologies* (Orlando, Florida, USA) 1-6

[25] Bai G, Long K, Wang W and Cheng S 1999 A new reliable signalling transport protocol RSTP for voice over IP *IEEE Fifth Asia-Pacific Conference on... and Fourth Optoelectronics and Communications Conference on Communications* **(1)** 263-266

[26] Koutras V P, Salagaras C P S and Platis A N 2009 Software rejuvenation for higher levels of VoIP availability and mean time to failure *IEEE Fourth International Conference on Dependability of Computer Systems* 99-106

[27] Koutras V P and Platis A N 2007 VoIP availability and service reliability through software rejuvenation policies *IEEE 2nd International Conference on Dependability of Computer Systems* 262-269

[28] Castanon G, Sarmiento A M, Ramirez-Velarde R and Aragón-Zavala A 2009 Software tool for network reliability and availability analysis *Wire Journal International* **74** 1-13

[29] Daengsi T and Wuttidittachotti P 2013 VoIP quality measurement: enhanced E-model using bias factor *IEEE Global Communications Conference* 1329-1334

[30] Chu C H K, Pant H, Richman S H and Wu P 2006 Enterprise VoIP reliability *IEEE 12th International Telecommunications Network Strategy and Planning Symposium* 1-6

[31] Johnson C R, Kogan Y, Levy Y, Saheban F and Tarapore P 2004 VoIP reliability: a service provider's perspective *IEEE Communications Magazine* **42(7)** 48-54.

[32] Bensalah F, Bahnasse A and El Hamzaoui M 2019 Quality of Service Performance Evaluation of Next-Generation Network *IEEE 2nd International Conference on Computer Applications and Information Security* 1-5

[33] Datta S, Ragula B and Balasubramaniam P 2015 U.S. Patent No. 8,995,252 *U.S. Patent and Trademark Office* (Washington, DC) 1-14

[34] Vargas E and Blue Prints S 2000 High availability fundamentals. Sun Blueprints series Part no . 806-6857-10 1-19

[35] Chen M J, Wen C C, Lin H C and Chu Y S 2016 ASIC design and implementation for VoIP intrusion prevention system *IEEE* International Conference on Applied System Innovation 1-4

[36] Rippon W J 2006 Threat assessment of IP based voice systems 1st IEEE Workshop on VoIP

Management and Security 19-28
[37] Chhabra A and Singh D 2015 Assessment of VoIP E-model over 802.11 wireless Mesh network *IEEE International Conference on Advances in Computer Engineering and Applications* 856-860
[38] Murphy K, Carter C, Grimes E and Malerich A 2019 RAPTOR 7.0 Tutorial Workbook 10-6
[39] Dasari V, Alexander D, Brown S, Brooks J, Panneton B and Su S 2019 Analysis of High fidelity ns-3 simulations to study real-time application performance in tactical wireless networks *IEEE International Conference on Big Data* 3955-3959
[40] Hu Z, Yan H, Yan T, Geng H and Liu G 2020 Evaluating QoE in VoIP networks with QoS mapping and machine learning algorithms *Neurocomputing* **386** 63-83
[41] Malekzadeh M 2020 Combination of Access Categories and Frame Aggregation Schemes For Bandwidth Efficiency of VHT 802.11 AC *Journal of Engineering Science and Technology* **15(1)** 617-635
[42] Ganesan N and Thangaraju B 2020 Pox Controller based Qos Evaluation for 5G Systems-Network Slicing *IEEE 7th International Conference on Signal Processing and Integrated Networks* 394-398

IOP Conf. Series: Materials Science and Engineering 1020 (2021) 012016 doi:10.1088/1757-899X/1020/1/012016

Evaluation of energy efficiency and moisture diffusivity for convective drying of large cardamom

Soumya Dash[*a], Kshirod Kumar Dash[*b], Shibabrata Choudhury[*a]
[a]Centre for Management Studies, North Eastern Regional Institute of Science and Technology, Nirjuli, Arunachal Pradesh, India, 791109
[b]Department of Food Processing Technology, Ghani Khan Choudhury Institute of Engineering and Technology Malda, West Bengal, India, 732141

*Corresponding author: soumyadash2003@gmail.com , kshirod@gkciet.ac.in, shibabrata.choudhury@gmail.com

Abstract: The drying behavior of freshly harvested large cardamom capsules were studied in a hot air dryer within the temperature range of 50-80°C. The study on drying rate showed that the constant rate drying period was very short, and capsules mostly followed the falling rate drying period. The drying during falling rate drying was followed unsteady state mass transfer and was governed by Fick's diffusion law. The drying rate was faster initially when the moisture content in the product was high but with a decrease of moisture content, the drying rate was found to be decreased. Nine different drying kinetics models were fitted to find the best suitable model to predict the drying kinetics of large cardamom capsules. The two-term model was the best-suited model out of the nine different models, representing the drying properties of large cardamom with the highest R^2 and the lowest RMSE and χ^2 values. The maximum effective moisture diffusivity D_{eff} value of 3.898×10^{-10} was obtained at a drying temperature of 80°C, while the lowest value was 1.949×10^{-10} at 50°C. The dependency of effective moisture diffusivity with drying temperature was correlated by Arrhenius equation. The activation energy showing the minimum energy needed to extract moisture from a solid matrix of large cardamom capsule was calculated as 32.54 kJ/mol and the pre-exponential factor of Arrhenius equation was calculated at 7.6×10^{-4} m^2/s.
Keywords: large cardamom; thin layer drying; effective moisture diffusivity; activation energy.

1. Introduction

Drying is basically a simultaneous heat transfer and mass transfer process. The heat needed for the evaporation of moisture was supplied from the hot air circulating around the product. During the process, moisture from the product was transferred to the surrounding hot air during the drying process. Diffusion is a characteristic activity of drying materials and the transfer of water vapor from the inside of the food material to the outer surface during drying is controlled or regulated by the diffusion process. Drying requires the simultaneous transfer of heat and mass. Drying is primarily aimed at reducing moisture present in the product to enhance storage stability. Therefore, drying of fruits and vegetables has been done as a way of enhancing storability.

Large cardamom has a leading role among different spices and has immense commercial significance. Large cardamom (*Amomum subulatum Roxb*. Family: Zingiberaceae) is the value-added cash crop and the primary source of cash income for eastern Himalayan farmers. It is also known as black cardamom, and reinforces and intensifies the recipes' taste. The main production centers in India are the sub-Himalayan ranges of Arunachal Pradesh, a part of Nagaland, Sikkim and Darjeeling district of West Bengal. Drying is the most critical operation after harvesting the large cardamom. The cardamom capsule moisture content has to be reduced to a safe level for a longer storage period. Proper drying will maintain the quality of the product and minimize losses. Delays in drying, inadequate drying, or ineffective drying may decrease the consistency of the capsule and cause losses.

Drying results in a substantial decrease of space demand for storage of the commodity for longer length. The different drying methods implemented for drying of fruits and vegetables includes tray drying, vacuum drying (Khawas et al., 2016), microwave drying (Dash and Das, 2020; Dash and Das, 2019a), microwave vacuum drying (Raj and Dash, 2020; Dash, Chakraborty, and Singh, 2020), osmotic drying (Prithani, and Dash, 2020; Dash and Balasubramaniam, 2018), and fluidized bed drying (Dash and Das, 2019b). The prediction of drying behavior can be obtained from suitable drying models. The drying of different food products was predicted through different drying model in different food products such as onions (Singh and Sodhi, 2000), green bean (Doymaz, 2005), hawthorn fruit (Sousa et al., 2018), ear corn (Friant et al., 2004), apricots (Doymaz, 2004a), and mulberry (Doymaz, 2004b) are available. Each food material has unique behavior due to the difference in structure and composition. The drying kinetics study can be implemented to design suitable dryers and optimize the drying parameters (Kingsly et al., 2006). The different thin-layer drying kinetics model studied for drying of food materials are toria seeds (Rangroo and Rao, 1992), dates (Hassan and Hobani, 2000), chilli pepper (Tunde-Akintunde and Ajala, 2010), millet samples (Ojendiran and Raji, 2010), srilankan paddy (Syamali et al., 2009), sesame seeds (Khazaei and Daneshmandi, 2007), amaranth grain (Ronoh et al., 2010), and parboiled wheat (Debabandya et al., 2005). The different kinetic models were implemented to evaluate the effective moisture diffusivity of the product. The effective moisture diffusivity obtained at different temperature were implemented to evaluate the activation energy in different products such as star fruit (Dash, Gope, Sethi, and Doloi, 2013), kachkal banana peel (Khawas et al., 2014), ripe Jackfruit (Saxena and Dash, 2015), culinary banana slices (Khawas et al., 2015), and elephant apple (Nag and Dash, 2016). However, a very limited study had been conducted on the drying of freshly harvested spices like large cardamom. Based on this, the objective of the study was to observe the drying behavior of freshly harvested large cardamom. Different thin layer model was fitted to the drying data to obtain the most suitable model to represent the kinetics behavior of the drying process. The effective moisture diffusivity and activation energy for the process was evaluated.

2. Materials and method

2.1. Drying procedure

The freshly harvested large cardamom capsules of Arunachal Pradesh, Itanagar, were collected from local farmers. Freshly harvested cardamom capsule has a moisture content of 78.59±3.65 percent (wet basis). The raw material was dried immediately by hot air drying until it reaches a moisture content (8.2±0.5 percent wet basis). The drying experiments were conducted using a laboratory tray dryer at a specific temperature. The moisture loss due to drying was recorded during the experiment with the aid of an electronic balance.

2.2. Mathematical modeling

The drying of large cardamom was conducted at four different temperatures, such as 50, 60, 70, and 80°C. The moisture content during the drying was obtained at an interval of 30 min. The obtained moisture content was expressed in dimensionless moisture ratio as presented in Eqn. (1).

$$\mathrm{MR} = \frac{Mt-Me}{Mo-Me} \qquad (1)$$

In Eqn. (1) the parameterMR, M_t, M_o, M_e are expressed in terms of kg water/kg dry matter) and the parameters are defined as follows.

$\mathrm{MR} =$ Moisture Ratio
$\mathrm{M_t} =$ Moisture content at time $\mathrm{t} = \mathrm{t}$
$\mathrm{M_o} =$ Initial moisture content at time $\mathrm{t} = 0$
$\mathrm{M_e} =$ Equilibrium moisture content at time $\mathrm{t} = \infty$

The dimensionless moisture ratio was fitted with different models present in Table (1).

2.3. Estimation of Effective Moisture Diffusivity

The drying of biological material such as fruits and vegetables follows the falling rate period. The drying process follows the falling rate period was generally described by Fick's second Law of mass transfer, as shown in Eqn. (2)

$$\partial M/\partial t = D_{eff}\, \partial^2 M/\partial x^2 \qquad (2)$$

The large cardamom capsule was considered as geometrically spherical in shape and hence, the solution of diffusion models for spheres was applied. The different assumptions were:

i. Initial moisture content across the fresh capsule was uniform.
ii. The capsule was believed to be homogeneous and isotropic solids and the transition of mass to the core was symmetric.
iii. The moisture was assumed to be dispersed uniformly across the cardamom capsule in the initial state (i.e., at time t=0)
iv. Thermal equilibrium occurs between the capsule surface and the drying air.
v. The coefficient of diffusion is constant, and the shrinkage is negligible.
vi. Migration of moisture is by diffusion only.

The solution of Eqn. (2) for spherical geometry is presented in given Eq. (3) (Crank, 1975).

$$MR = \frac{6}{\pi^2}\sum_{n=1}^{\infty}\frac{1}{(n)^2}\, exp^{\frac{[-(n)^2\pi^2 D_{eff}\, t]}{r^2}} \qquad (3)$$

Where r = radius of the cardamom capsule. On neglecting the higher order terms, the equation (3) reduces to Eqn. (4).

$$MR = \frac{6}{\pi^2}\, exp^{\frac{[-\pi^2 D_{eff}\, t]}{r^2}} \qquad (4)$$

Taking natural logarithm on both sides of Eqn. (4) the obtained equation is presented in Eqn. (5).

$$\ln MR = \ln\frac{6}{\pi^2} - \frac{[\pi^2 D_{eff}\, t]}{r^2} \qquad (5)$$

The slope of the graph plotted between ln(MR) and drying time would be used to calculate the effective moisture diffusivity of the product during drying.

$$slope = \frac{[-\pi^2 D_{eff}]}{r^2} \qquad (6)$$

$$D_{eff} = \frac{-slope\, r^2}{\pi^2} \qquad (7)$$

The effective moisture diffusivity was correlated with different drying temperatures by using the Arrhenius equation to obtain the activation energy (Wang et al., 2018).

$$De = Do\, exp\left(\frac{E_a}{RT}\right) \qquad (8)$$

$$\ln De = \ln Do - \frac{E_a}{RT} \qquad (9)$$

In Eqn. (8) and (9), the parameters E_a, R, T, and D_o are defined as following.

$E_a = Energy\ of\ activation\ (kJ/mol)$

$R = universal\ gas\ constant\ (8.3143\ kJ/mol)$

$T = absolute\ air\ temperature\ (K)$

$D_o = preexponential\ factor\ of\ the\ Arrhenius\ equation\ (m^2/s)$

The activation energy represented the minimal energy to be supplied in order to disrupt the interactions between water-solid or water-water and transfer water molecules in the solid from one point to another point within the solid. The smaller activation energy indicated that the water molecule could pass in the sample more readily. The activation energy needed for the initiation of drying was determined using the Arrhenius equation.

2.4. Statistical analysis

The different statistical parameters such as coefficient of determination (R^2), chi-square ($\chi 2$), and root mean square error (RMSE) were evaluated to select the best suitable model to represent the drying kinetics. The best suited model was chosen based on the highest R^2 and lowest ($\chi 2$) values (Khawas et al.,2016; Togrul and Pehlivan, 2002). R^2 value indicates the correctness in fitting and χ^2 and RMSE value indicates bias error, i.e., deviation in the fitting.

Table 1: The Different mathematical models applied to the experimental data

Model No.	Model Name	Model	References
1	Newton	$MR = exp(-kt)$	Westerman et al., 1973
2	Page	$MR = exp(-kt^n)$	Page, 1949
3	Modified page	$MR = exp[-(kt)^n]$	Yaldiz et al., 2001
4	Henderson and Pabis	$MR = a\,exp(-kt)$	Yagcioglu et al., 1999
5	Logarithmic	$MR = a\,exp(-kt) + c$	Yaldiz and Ertekin, 2001
6	Verma et al.	$MR = a\,exp(-kt) + (1 - a)exp(-gt)$	Verma et al., 1985
7	Wang and Singh	$MR = Mo + at + bt^2$	Ozdemir and Devres,1999
8	Two term	$MR = a\,exp(-k_0 t) + b\,exp(-k_1 t)$	Rahman et al., 1998
9	Two term exponential	$MR = a\,exp(-kt) + (1 - a)exp(-kat)$	Yaldiz et al., 2001

3. Results and Discussions

3.1. Drying behavior of large cardamom

The variation of moisture ratio of large cardamom at four different drying temperatures is 50, 60, 70, and 80°C, as shown in Fig. (1). It was observed that the moisture content of the product was reduced at a faster rate during the initial drying period. The moisture content curves were complex in nature. The rate of moisture loss was higher with an increase in drying air temperature, causing a reduction in drying time to attain a particular level of moisture content. The faster moisture vaporization at higher temperatures was due to an increase of vapor pressure with the product at an elevated temperature level. In addition to this, the relative humidity of drying air also decreases with an increase in temperature, causing an increased moisture removal rate. The equilibrium moisture content of the cardamom capsules within the temperature was 50-80°C were 6.29-8.75 kg/kg dry matter. With an increase in drying air temperature, the relative humidity of drying air was decreased. Lowe relative humidity at higher temperatures was able to reduce the equilibrium moisture content to a lower level. The time taken to attend a particular moisture content was decreased by 16.4 % and 11.2 % when the drying temperature was increased from 50 to 60°C and 60 to 70°C, respectively.

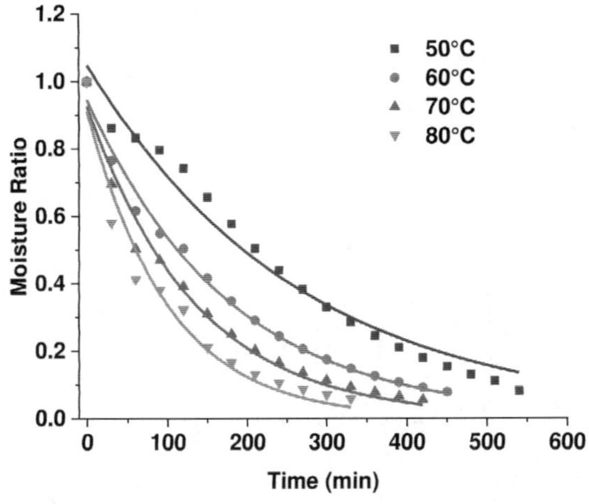

Fig.1: Plot of moisture ratio and time of large cardamom at different temperatures

3.2. *Fitting of drying model for drying of large cardamom*

The drying constants, root mean square error (RMSE), reduced chi-square value (χ^2) and coefficient of determination (R^2) were calculated for all the nine models at 50, 60, 70, and 80°C. The obtained values of R^2 and χ^2 and RMSE for different models are presented in Table (2). The most suitable model capable of predicting the drying of large cardamom was selected based on the highest R^2 and lowest RMSE and χ^2 values.

Fitting moisture ratio data in Newton model within the temperature range of 50-80°C had the R^2, χ^2 and RMSE values of 0.952-0.987, 0.003-0.021, and 0.03-0.06 respectively. The k value for newton model was found to increase from 0.0036 to 0.01108 with an increase of temperature from 50-80°C. Moisture ratio data fitting in the 50-80°C temperature range for Page model had the R^2, χ^2 and RMSE values of 0.971-0.994, 0.0106-0.0314, and 0.0190-0.0249 respectively. The k value for the Page model was found to rise from 0.0007 to 0.0483 with a temperature increase from 50-80°C. Moisture ratio results fitting in the Modified Page model within the 50-80°C temperature range resulted in the R^2, χ^2 and RMSE values of 0.951-0.994, 0.0104-0.0260, and 0.0190-0.0249 respectively. The k value for Modified Page model was found to rise from 0.00359 to 0.01243, with a temperature increase from 50-80°C. Fitting of moisture ratio data in Henderson and Pabis model within the temperature range of 50-80°C had the R^2, χ^2, and RMSE values of 0.960-0.991, 0.0106-0.0213, and 0.0238-0.0520 respectively. The k value for Henderson and Pabis model was found to increase from 0.00379 to 0.01001 with an increase of temperature from 50-80°C. Moisture ratio fitting of the Logarithmic model fitting within the 50-80°C temperature range had R^2, χ^2, and RMSE values of 0.9621-0.9916, 0.0019-0.0203, and 0.0231-0.0432, respectively. The k value for the Logarithmic model was observed to increase from 0.00214 to 0.01326 with temperature increase from 50-80°C. Fitting of moisture ratio data in Verma et al. model within the temperature range of 50-80°C had the R^2, χ^2 and RMSE values of 0.9541-0.9972, 0.0106-0.0211, and 0.0133-0.0409 respectively. The k value for Verma et al. model was found to increase from 0.00392 to 0.00779 with an increase of temperature from 50-80°C. Fitting of moisture ratio data in Wang and Singh model within the temperature range of 50-80°C had the R^2, χ^2 and RMSE values of 0.873-0.984, 0.0097-0.0227, and 0.0215-0.1937 respectively. The k value for Wang and Singh model was found to increase from 0.00277 to 0.00768 with an increase of temperature from 50-80°C. Fitting of moisture ratio data in the two-term exponential model within the temperature range of 50-80°C had the R^2, χ^2 and RMSE values of 0.7593-0.9967, 0.0015-0.0176, and 0.0150-0.1371 respectively. The k value for the two-term exponential model was found to increase from 0.00523 to 0.0478 with an increase of temperature from 50-80°C.

Table 2: Drying kinetics parameters of convective drying of large cardamom

Drying models	Temperature	Parameters	R^2	χ^2	RMSE
Newton	50	$k = 0.0036$	0.9738	0.0022	0.0474
	60	$k = 0.00602$	0.9872	0.0109	0.0305
	70	$k = 0.00809$	0.9752	0.0218	0.0426
	80	$k = 0.01108$	0.9526	0.0036	0.0600
Page	50	$k = 0.000724703$ $n = 1.2841$	0.9831	0.0106	0.0236
	60	$k = 0.01255$ $n = 0.86081$	0.9847	0.0204	0.0190
	70	$k = 0.02433$ $n = 0.78172$	0.9946	0.0314	0.0191
	80	$k = 0.0483$ $n = 0.69076$	0.9710	0.0107	0.0249
Modified page	50	$k = 0.00359$ $n = 1.28846$	0.9731	0.0260	0.0236
	60	$k = 0.00618$ $n = 0.86231$	0.9947	0.0116	0.0190

	70	$k = 0.00862$ $n = 0.78309$	0.9646	0.0104	0.0191
	80	$k = 0.01243$ $n = 0.69176$	0.9510	0.0107	0.0249
Henderson and Pabis	50	$k = 0.00379$ $a = 1.049$	0.9767	0.0120	0.0435
	60	$k = 0.00567$ $a = 0.94643$	0.9916	0.0106	0.0238
	70	$k = 0.00746$ $a = 0.92707$	0.9821	0.0213	0.0349
	80	$k = 0.01001$ $a = 0.91298$	0.9608	0.0130	0.0520
Logarithmic	50	$k = 0.00214$ $a = 1.36948$ $c = -0.36734$	0.9621	0.0107	0.0246
	60	$k = 0.00608$ $a = 0.93017$ $c = 0.02483$	0.9916	0.0106	0.0231
	70	$k = 0.00887$ $a = 0.9006$ $c = 0.05042$	0.9854	0.0211	0.0302
	80	$k = 0.01326$ $a = 0.87654$ $c = 0.07347$	0.9701	0.0123	0.0432
Verma et al.	50	$k = 0.00392$ $a = 1.08004$ $g = 2.2$	0.9781	0.0019	0.0409
	60	$k = 0.00532$ $a = 0.89145$ $g = 2.2$	0.9972	0.0112	0.0133
	70	$k = 0.00656$ $a = 0.82127$ $g = 2.2$	0.9958	0.0203	0.0161
	80	$k = 0.00779$ $a = 0.72266$ $g = 2.2$	0.9541	0.0104	0.0191
Wang & Singh	50	$a = -0.00277$ $b = 0.00000196201$	0.9843	0.0105	0.0215
	60	$a = -0.00477$ $b = 0.00000630586$	0.9628	0.0227	0.0502
	70	$a = -0.00584$ $b = 0.00000901659$	0.9194	0.0159	0.1739
	80	$a = -0.00768$ $b = 0.0000154222$	0.8730	0.0097	0.1937
Two term	50	$k_0 = 0.00392$ $k_1 = 1.9681$ $a = 1.08006$ $b = -0.08006$	0.9966	0.0020	0.0039
	60	$k_0 = 0.0053$ $k_1 = 0.10317$ $a = 0.8878$ $b = 0.11226$	0.9970	0.0002	0.0132

	70	$k_0 = 0.00635$ $k_1 = 0.06257$ $a = 0.7914$ $b = 0.20936$	0.9962	0.0003	0.0147
	80	$k_0 = 0.09586$ $k_1 = 0.00764$ $a = 0.29451$ $b = 0.70558$	0.9936	0.0005	0.0188
Two term exponential	50	$k = 0.00523$ $a = 1.82125$	0.9932	0.0106	0.0235
	60	$k = 0.04785$ $a = 0.11136$	0.9967	0.0102	0.0150
	70	$k = 0.00525$ $a = 0.99784$	0.7593	0.0176	0.1278
	80	$k = 0.04408$ $a = 0.2039$	0.9801	0.0015	0.1371

Among the different models discussed, two-term model had a most significant fitting of data in terms of average R^2 (0.998) and χ^2 (1.17×10^{-3}) values and hence proposed as the best model describing the drying behavior of large cardamom. Fitting of moisture ratio data in two-term model within the temperature range of 50-80°C had the R^2, χ^2 and RMSE values of 0.9936-0.9970, 0.0002-0.020, and 0.0039-0.0188 respectively. It was observed that among the four drying constants of two-term model, 'k_0' and 'b' increased with temperature while 'k_1' and 'a' decreased with temperature. The 'k_0' and 'b' value for the two-term model was found to increase in the range of 0.00392 - 0.0958 and 0.08006-0.70558 respectively whereas the 'k_1' and 'a' value was found to decrease in the range of 1.9681-0.00764 and 1.08006-0.70558 with the increase of temperature from 50-80°C.

3.3. *Effective moisture diffusivity for drying of large cardamom*

Predicted moisture ratios were plotted against time at different temperatures as 50, 60, 70, and 80°C to obtain the effective moisture diffusivity. From the respective slope, moisture diffusivity values were determined using Eqns. (6 and 7). The highest effective diffusivity (D_{eff}) value of 3.898×10^{-10} was obtained at drying temperature 80°C while the lowest value of 1.949×10^{-10} at 50°C (Table 3). The diffusivity for drying of different food was found to be in the same range as obtained for the convective drying of large cardamom. Effective diffusivity in the temperature range of 50 -120°C for rough rice and brown rice with the moisture content of 15 % (db) was observed to be 1.3×10^{-10} - 3.2×10^{-9} and 4.5×10^{-10} - 3.9×10^{-9} $m^2.s^{-1}$ respectively (Bakshi and Singh, 1980). Drying analyses of two types of millet i.e. SOSAT C88 and EXBORNO with an initial moisture content of 31.6±1% (wb) and 22.75±2.25% (wb) was conducted at drying temperatures of 40, 50, 60, and 70°C. The effective moisture diffusivity within the temperature range of 40-70°C was increased and the obtained values were in the range of $2.86\times10^{-9} – 5.27\times10^{-9}$ and $1.17\times10^{-9} – 2.98\times10^{-9}$ $m^2.s^{-1}$ respectively (Ojediran and Raji, 2010). A study of effective diffusivity of barley grain assuming the shape of the grain as spherical geometry showed that the diffusivity was varied in the range of 1.99×10^{-11} -5.31×10^{-11} $m^2.s^{-1}$ (Markowski et al., 2010). The hot air drying and vacuum drying of bee pollen within the temperature range of 40-50°C revealed that the two-term model was the most suitable model to predict the moisture ratio values. The effective moisture diffusivity values for drying bee pollen were varied between 8.40×10^{-11} - 6.29×10^{-10} m^2/s (Kayacan et al., 2018). The moisture diffusivity of red pepper was studied under three drying conditions such as hot air drying (HAD), infrared-assisted hot air drying (IAHD), and pulsed vacuum drying (PVD). The diffusivity values for HAD, IAHD, and PVD were varied between 1.33 –5.83×10^{-10}, 1.38–6.87×10^{-10}, and 1.75–8.97×10^{-10} m^2/s, respectively (Deng et al., 2018).

Table 3: Effective moisture diffusivities of black cardamom capsule at different temperatures

Temperature	D_{eff}	Ln D_{eff}	R^2	Chi Square
50°C	1.949×10^{-10}	-22.3583	0.866	0.0281
60°C	2.729×10^{-10}	-22.0219	0.913	0.0763
70°C	3.508×10^{-10}	-21.7706	0.953	0.0459
80°C	3.898×10^{-10}	-21.6652	0.959	0.0174

3.4. Activation energy for drying of large cardamom

In order to determine the pre-exponential component (D_o) and activation energy (E_a) of the Arrhenius equation (Eqns. 8-9), the logarithm of effective moisture diffusivities (Deff) was plotted against the reciprocal of absolute temperature (Fig. 2). From the intercept of linear fit equation (Eqn. 2), the value of D_o for large cardamom was evaluated as 7.61×10^{-4} m^2 s^{-1} and the value of E_a was found to be 32.55 $kJ.mol^{-1}$.

$$D_{eff} = (7.61 \times 10^{-4}) \exp\left(-\frac{32.55 \times 10^3}{RT}\right) \quad (10)$$

The amount of energy barrier that has to overcome i.e., the threshold barrier to activate the diffusion of moisture is the activation energy and the estimated activation energy value was 32.55 kJ / mol. The activation energy can be decreased by increasing the temperature and the drying rate, but the product quality can be degraded. The drying of *Phyllanthus amarus* and *Phyllanthus niruri* showed that an activation energy values of 22.828 and 43.129 kJ/mol, respectively (Sousa et al., 2018). The obtained results are similar to the results obtained in the drying of various food products such as apple with activation energy value 19.96 – 22.62 kJ mol^{-1} (Teixeira and Tobinaga, 1998), carrot with activation energy value 22.43 kJ mol^{-1} (Togrul, 2006) and red pepper with activation energy value 24.76 kJ mol^{-1} (Kaleemullah and Kailappan, 2005).

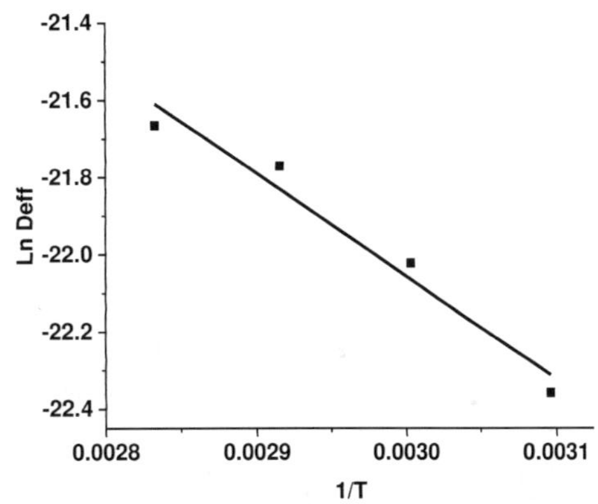

Fig.2: Relationship between the effective moisture diffusivities and temperatures

4. Conclusion

The drying behavior of larger cardamom was applied to nine different models in the literature to establish the drying kinetic model for the drying of large fresh cardamom. The two-term model delivered the best results among these models and showed good agreement with the experimental data obtained from the experiments, including the thin layer drying process. When the effects of drying air temperature on the constants and coefficients of the two-term model were examined, the R^2, $\chi 2$ and RMSE values were found to be in the range of 0.9936-0.9970, 0.0002-0.020, and 0.0039-0.0188 respectively. Therefore, the two-term model adequately described the drying behavior of large cardamom in the drying process within the temperature range of 50-80°C. The effective diffusivities increased with the drying temperature and varied from 1.949×10^{-10} to 3.898×10^{-10} m^2/s. The temperature dependence of diffusivity follows the Arrhenius relationship, and the activation energy for the diffusion of moisture was found to be 32.55 kJ/mol.

Acknowledgments

The authors are thankful to the funding agency: National Mission on Himalayan Studies (NMHS) for their support through project no.GBPNI/NMHS-2017-18/SG15.

References

[1] Bakshi, A. S., & Singh, R. P. (1980). Kinetics of water diffusion and starch gelatinization during rice parboiling. *Journal of Food Science*, *45*(5), 1387-1392.

[2] Crank, J. (1975). The mathematics of difiusion. *2nd end.*

[3] Dash, K. K., & Balasubramaniam, V. M. (2018). Effect of high pressure on mass transfer kinetics of granny smith apple. *Drying Technology*, *36*(13), 1631-1641.

[4] Dash, K. K., & Das, S. K. (2019a). Puffing quality and sensory evaluation of reduced sodium puffed rice. *Quality Assurance and Safety of Crops & Foods*, *11*(8), 757-768.

[5] Dash, K. K., & Das, S. K. (2019b). Optimization of fluidized bed preconditioning for microwave puffed rice using integrated artificial neural network and genetic algorithm approach. *Journal of Food Process Engineering*, *42*(6), e13158.

[6] Dash, K. K., & Das, S. K. (2020). Modeling and optimization of microwave puffing of rice using artificial neural network and genetic algorithm. *Journal of Food Process Engineering*, e13577.

[7] Dash, K. K., Chakraborty, S., & Singh, Y. R. (2020). Modeling of Microwave Vacuum Drying Kinetics of Bael (Aegle marmelos L.) Pulp by Using Artificial Neural Network. *Journal of The Institution of Engineers (India): Series A*, 1-9.

[8] Dash, K. K., Gope, S., Sethi, A., & Doloi, M. (2013). Study on thin layer drying characteristics star fruit slices. International Journal Agriculture and Food Science Technology, 4(7), 679-686.

[9] Debabandya, M., &Srinivas, R. P. (2005). A thin layer drying model of parboiled wheat. Journal of Food Engineering, 66, 513-518.

[10] Deng, L. Z., Yang, X. H., Mujumdar, A. S., Zhao, J. H., Wang, D., Zhang, Q., Wang, J., Gao, Z. J., & Xiao, H. W. (2018). Red pepper (Capsicum annuum L.) drying: effects of different drying methods on drying kinetics, physicochemical properties, antioxidant capacity, and microstructure. *Drying technology*, *36*(8), 893-907.

[11] Deng, L. Z., Yang, X. H., Mujumdar, A. S., Zhao, J. H., Wang, D., Zhang, Q., & Xiao, H. W. (2018). Red pepper (Capsicum annuum L.) drying: Effects of different drying methods on drying kinetics, physicochemical properties, antioxidant capacity, and microstructure. *Drying Technology*, *36*(8), 893-907.

[12] Doymaz, I. (2004a). Effect of pre-treatments using potassium metabisulphide and alkaline ethyl oleate on the drying kinetics of apricots. Biosystems Engineering, 89(3), 281–287.

[13] Doymaz, I. (2004b). Pretreatment effect on sun drying of mulberry fruits (Morus alba L.). Journal of Food Engineering, 65, 205–209.

[14] Doymaz, I. (2005). Drying behaviour of green beans. Journal of Food Engineering, 69(2), 161-165.

[15] Friant, N. R., Marks, B. P., & Bakker-Arkema, F. W. (2004). Drying rate of ear corn. Transactions of the ASAE, 47(5), 1605-1610.

[16] HASSAN, B. H., & HOBANI, A. I. (2000). Thin-layer drying of dates. Journal of Food Process Engineering, 23(3), 177-189.

[17] Kaleemullah, S., & Kailappan, R. (2005). Drying kinetics of red chillies in a rotary dryer. *Biosystems Engineering*, *92*(1), 15-23.

[18] Kayacan, S., Sagdic, O., & Doymaz, I. (2018). Effects of hot-air and vacuum drying on drying kinetics, bioactive compounds and color of bee pollen. *Journal of Food Measurement and Characterization*, *12*(2), 1274-1283.

[19] Khawas, P., Das, A. J., Dash, K. K., & Deka, S. C. (2014). Thin-layer drying characteristics of Kachkal banana peel (Musa ABB) of Assam, India. *International Food Research Journal*, *21*(3).

[20] Khawas, P., Dash, K. K., Das, A. J., & Deka, S. C. (2015). Drying characteristics and assessment of physicochemical and microstructural properties of dried culinary banana slices. *International Journal of Food Engineering*, *11*(5), 667-678.

[21] Khawas, P., Dash, K. K., Das, A. J., & Deka, S. C. (2016). Modeling and optimization of the process parameters in vacuum drying of culinary banana (Musa ABB) slices by application of artificial neural network and genetic algorithm. Drying Technology, 34(4), 491-503.

[22] Khawas, P., Dash, K. K., Das, A. J., & Deka, S. C. (2016). Modeling and optimization of the process parameters in vacuum drying of culinary banana (Musa ABB) slices by application of artificial neural network and genetic algorithm. *Drying Technology*, *34*(4), 491-503.

[23] Khazaei, J., & S. Daneshmandi. (2007). Modeling of thin layer drying kinetics of sesame seeds: Mathematical and neural networks modeling. *International Agrophysics*, 21, 335-348.

[24] Kingsly, A. R. P., Singh, D. B., Manikantan, M. R., & Jain, R. K. (2006).Moisture dependent physical properties of dried pomegranate seeds (Anardana). Journal of Food Engineering, 75, 492–496.

[25] Markowski, M., Białobrzewski, I., & Modrzewska, A. (2010). Kinetics of spouted-bed drying of barley: Diffusivities for sphere and ellipsoid. *Journal of Food Engineering*, *96*(3), 380-387.

[26] Nag, S., & Dash, K. K. (2016). Mathematical modeling of thin layer drying kinetics and moisture diffusivity study of elephant apple. *International Food Research Journal*, *23*(6).

[27] Ojendiran, J. O., &Raji, A. O. (2010). Thin layer drying of millet and effect of temperature on drying characteristics. International Food Research Journal, 17, 1095-1106.

[28] Prithani, R., & Dash, K. K. (2020). Mass transfer modelling in ultrasound assisted osmotic dehydration of kiwi fruit. *Innovative Food Science & Emerging Technologies*, 102407.

[29] Raj, G. B., & Dash, K. K. (2020). Microwave vacuum drying of dragon fruit slice: Artificial neural network modelling, genetic algorithm optimization, and kinetics study. *Computers and Electronics in Agriculture*, *178*, 105814.

[30] Raji, A. O., & Ojediran, J. O. (2011). Moisture sorption isotherms of two varieties of millet. *Food and bioproducts processing*, *89*(3), 178-184.

[31] Rangroo, S., &Rao, D. G. (1992). Drying of toria (*Brassica campistris*): Part 2—Drying conditions. Journal of food engineering, 17(1), 59-68.

[32] Ronoh, E. K., Kanali, C. L., Mailutha, J. T., &Shitanda, D. (2010). Thin layer drying kinetics of Amaranth grains in a natural convection solar tent dryer. *African Journal* of *Food Agriculture*, *Nutrition* and Development, 10(3), 2218-2233.

[33] Saxena, J., & Dash, K. K. (2015). Drying kinetics and moisture diffusivity study of ripe Jackfruit. *International Food Research Journal*, *22*(1), 414.

[34] Singh, H., &Sodhi, N. S. (2000). Dehydration kinetics of onions. Journal of Food Science and Technology, 37(5), 520–522.

[35] Sousa, A. D., Ribeiro, P. R. V., Canuto, K. M., Zocolo, G. J., Pereira, R. D. C. A., Fernandes, F. A. N., & Sousa de Brito, E. (2018). Drying kinetics and effect of air-drying temperature on chemical composition of Phyllanthus amarus and Phyllanthus niruri. Drying technology, 36(5), 609-616.

[36] Sousa, A. D., Ribeiro, P. R. V., Canuto, K. M., Zocolo, G. J., Pereira, R. D. C. A., Fernandes, F. A. N., & Sousa de Brito, E. (2018). Drying kinetics and effect of air-drying temperature on chemical composition of Phyllanthus amarus and Phyllanthus niruri. *Drying technology*, *36*(5), 609-616.

[37] Syamali, A. K., Jayaweera, H. H. E., &Ariyaratne, T. R. (2009). Thin layer drying of some Srilankan paddy varieties under low humid conditions. Proceedings of Technical Sessions, Institute of Physics, Srilanka, 25, 36-44.

[38] Toğrul, H. (2006). Suitable drying model for infrared drying of carrot. *Journal of food engineering*, *77*(3), 610-619.

[39] Togrul, I.T., &Pehlivan, D. (2002). Mathematical modeling of solar drying of apricots in thin

layers. Journal of Food Engineering, 55(3), 209–216.
[40] Tunde-Akintunde, T. Y., &Ajala, A. (2010). Air drying characteristics of chili pepper. International Journal of Food Engineering, 6(1).
[41] Wang, J., Law, C. L., Nema, P. K., Zhao, J. H., Liu, Z. L., Deng, L. Z., Gao, Z. J. & Xiao, H. W. (2018). Pulsed vacuum drying enhances drying kinetics and quality of lemon slices. *Journal of Food Engineering*, *224*, 129-138.

Piece-wise linear regression: A new approach to predict COVID-19 spreading

Apurbalal Senapati[1], Soumen Maji[2] and Arunendu Mondal[3]

[1] Department of Computer Science & Engineering, Central Institute of Technology Kokrajhar, Assam-783370, India.
[2] Department of Civil Engineering, Central Institute of Technology Kokrajhar, Assam-783370, India.
[3] Department of Chemistry, Central Institute of Technology Kokrajhar, Assam-783370, India.

Email: a.senapati@cit.ac.in

Abstract. The coronavirus disease 2019 (COVID-19) pandemic is the most rapidly evolving global emergency since March 2020 and one of the most exercised topics in all aspects of the world. So far there are numerous articles that have been published related to COVID-19 in various disciplines of science and social context. Since from the very beginning, researchers have been trying to address some fundamental questions like how long it will sustain when it will reach the peak point of spreading, what will be the population of infections, cure, or death in the future. To address such issues researchers have been used several mathematical models from the very beginning around the world. The goal of such predictions is to take strategic control of the disease. In most of the cases, the predictions have deviated from the real data. In this paper, a mathematical model has been used which is not explored earlier in the COVID-19 predictions. The contribution of the work is to present a variant of the linear regression model is the piece-wise linear regression, which performs relatively better compared to the other existing models. In our study, the COVID-19 data set of several states of India has been used.

Keywords: Piece-wise linear regression, COVID-19, Mathematical model, Prediction, Machine learning

1. Introduction

The COVID-19 pandemic gradually deteriorated day by day since the beginning of the year 2020 around the world. Several countries like Vietnam, New Zealand, South Korea, Cuba, Germany, France, etc. survived successfully and the spread of the disease is declining now. However, on the other hand, in countries like the USA, India, Brazil, Russia, etc. the disease is spreading very fast. At present in India, the daily positive cases are identified as more than 80,000 on average. Almost around the world imposed the lockdown as a common strategy to control the spreading of the disease up to some extent. The different countries change their lockdown strategies from time to time for their socio-economical situations and that also affects the spreading of a pandemic. Fig. 1 shows the trend of spreading positive COVID-19 cases in some of the selected states in India whereas Fig. 2. shows a comparison on the daily averaged Covid-19 positive cases during lockdown and unlock-1.

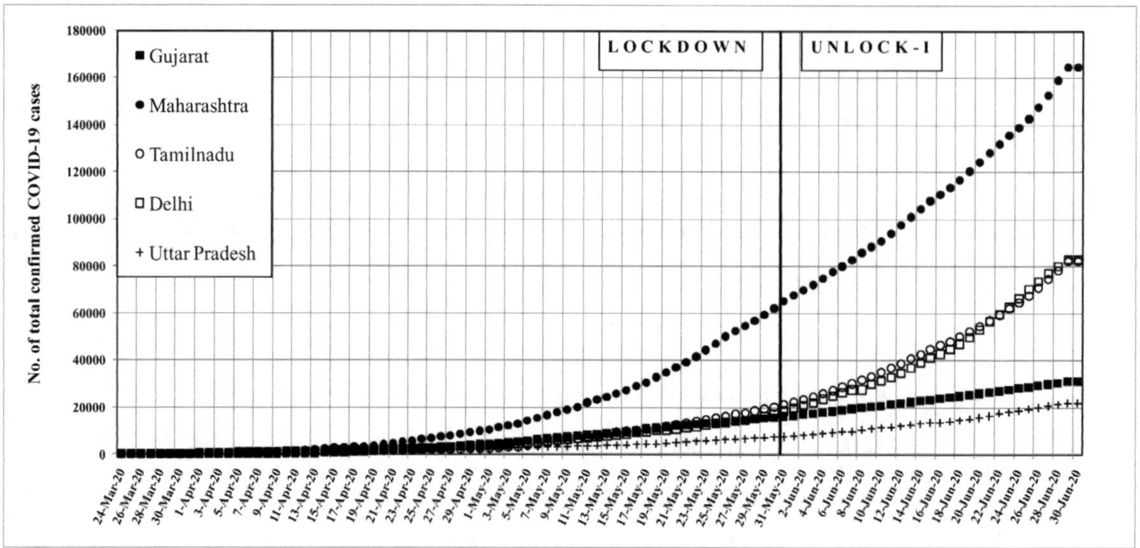

Fig. 1: Trend of spreading positive COVID-19 cases in Gujarat, Maharashtra, Tamil Nadu, Delhi, and Uttar Pradesh from 24th March to 30th June 2020.

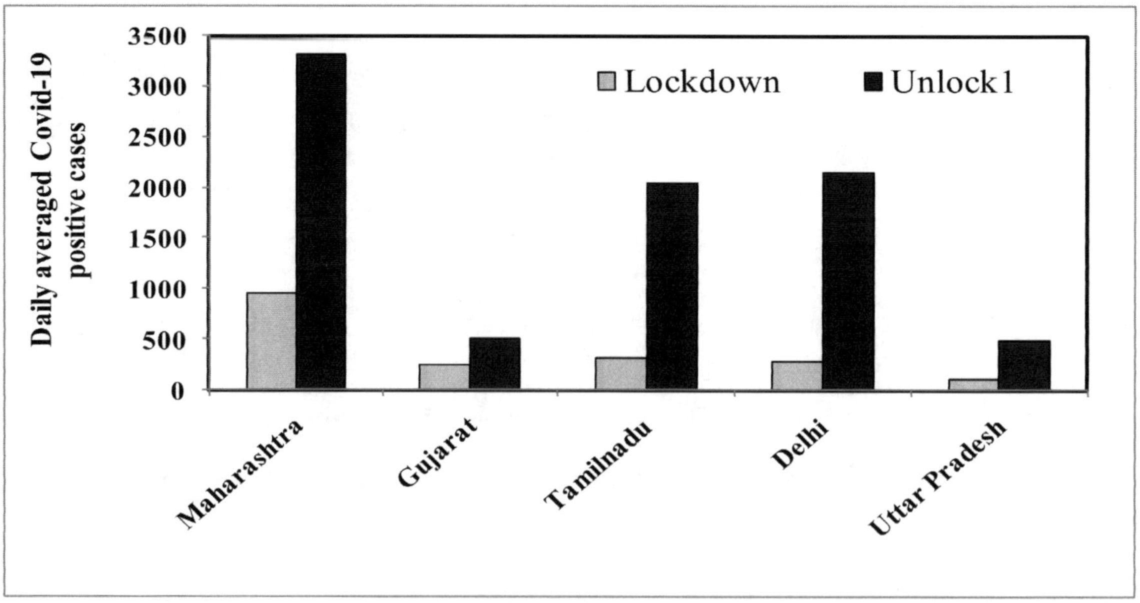

Fig. 2. Daily averaged Covid-19 positive cases shown during lockdown and unlock-1 in most affected states like Maharashtra, Gujarat, Tamil Nadu, Delhi, and Uttar Pradesh.

2. Existing predicted models:

History shows the outbreaks of infectious diseases in several civilizations is quite past [1], [2]. Some standard mathematical models were established for infectious disease [3], [4]. A recent one was Middle East Respiratory Syndrome (MERS) (2012), avian influenza (H7N9), and Ebola [5]. Bernoulli et al. [6] used a mathematical model to analyze the mortality of smallpox in England and later on different types [7] of models come. Since the last few months from a computational point of view, several data-driven statistical models have been adopted to predict the infection, recovery, death, time interval of the pandemic, etc. of the COVID-19. So many models like the non-linear regression framework are used to address the cumulative and daily death rate [8]. A data-driven technique is also used where the mobile GPS tracker is used to find out the social distancing behavior [9]. Yang et al. [10] used an AI-based SEIR model to predict the trend of pandemic COVID-19. Apart from these, many researchers across the globe have used different statistical models to forecast COVID-19 related outcomes and their interpretations [11], [12] and [13]. Some of these predictions are good to some extent but most of them are far from reality [7], [8] and [14]. Although attempts have been made [15] to improve the existing model with the recent availability of the updated data and external policies on the predictions as well [15], [16] and [17]. Interestingly, Luo [18] used the heuristic approach to address the problem and discussed the challenges of such kind of models, and explained their complexity. These reports do not conclude that all such models or analyses are reliable, rather these should be considered as predictions based on the available data, and these predictions are highly dependent on the reliability of these data. So the accuracy of the models depends on the accuracy of the collected data although there will be an error and which will not be the weakness of the models, but a part of the model. Since there are several reports already predicted based on the data available in the past, hence now it is the time to verify their predictions. Few such results are discussed here.

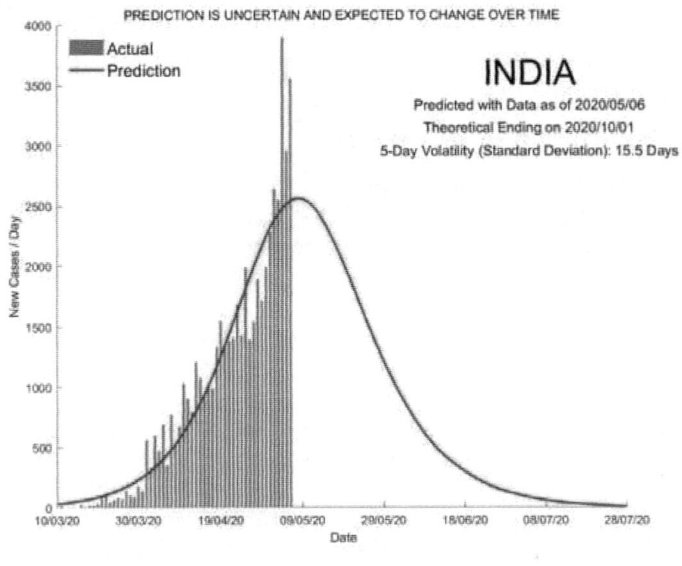

Fig. 3. Prediction of Covid-19 pandemic in India (Luo 2020)

Luo [18] has used the "SIR epidemic model" using the data from different countries to predict the key dates of transition during the coronavirus lifecycle across the globe. For India, they have predicted that 97% of the pandemic will be ended by the 3rd week of May 2020 and at the end of May 2020, it will over 99% with their declaration that, it may contain errors. Fig.3 shows their predictions with data as of dated 06-May-2020.

Xu [19] tried to model the COVID-19 pandemic by the Farr Law. It says that the epidemics like AIDS, SARS, Ebola, etc. tend to rise and fall in a roughly symmetrical pattern or bell-shaped curve. But, the practical scenario does not follow exactly the symmetric pattern rather it is skewed on one side in the case of COVID-19.

On the other hand, Ben-Israel [20] concluded, based on simple statistical analysis that, the spread of COVID-19 follows a particular pattern of repetition in a curve. He also says that the growth and decline in the new cases follow the same trend across countries around the world. Particularly, he stressed that it reaches a peak after about 40 days and flattened to almost zero after 70 days and it is independent of the other external factors. But, the actual scenario is far from this conclusion.

Some researchers [21], [22] attempted to predict the impact of lockdown in the spreading of COVID-19 in India and did similar studies for the other countries. Schueller et. al. [22] considered the first 21 days of the lockdown in March 2020 and also considered several parameters like incubation period, the infected population has no symptoms or mild symptoms, the age distribution of the populations, etc. They predicted, with moderate and hard lockdown may reduce the transmission of the disease. But this prediction is difficult to verify. Because based on the clinical expert it is obvious that the lockdown reduces the spreading of infection but the actual measure is quite difficult. Singh [21] tried to understand the spreading characteristic of COVID-19 in India taking into account the physical distancing and lockdown. They have used differential based mathematical modeling. Singh [21] used the modified logistic growth model in his prediction using the truncated information on novel corona confirmed cases in India. There are several conditions applied in his model and predicted that at the end of April 2020 increasing growth will start to decline and there will be no newer cases at the end of July 2020. Whereas a similar study of Mandal et al. [23] has shown that, the impact of social distancing and that can reduce the maximum cases up to 62 percent.

Chatterjee et al. [24] predicted the COVID-19 epidemic in India based on a stochastic model. Their study shows that uninterrupted epidemic in India could result a minimum of 364 million positive cases with the death of 1.56 million and the highest acceleration of the pandemic by mid-July. To verify the prediction yet has to observe.

Ghosh et al. [25] performed the state-wise analysis in India. They built three growth models (exponential model, the logistic model, and the Susceptible Infectious Susceptible (SIS) model) to predict infected people in the states in the next coming 30 days. They used the parameter daily infection-rate (DIR) values for each state. This model furnished the three predictions for the states, they have interpreted the results combining from the models rather than individually. They claimed that over 126 thousand people will be affected in COVID-19 by May 31, 2020. They also observed people are not following the lockdown because of the essential work in several places and results spreading the infection.

Menon et al. [26] described several models in their ongoing projects. They have incorporated features like asymptomatic, symptomatic infections, the effect of migration, quarantine, etc. and have also addressed several subjective type questions and the predictions of the number of deaths, the number of infections, etc.

Senapati et al. [27] tried to overcome the limitations of the linear regression model using the piecewise regression model. They have done their experiment from the data of some selected states of India. The results are compared with the other regression model and show their results bid others.

3. Linear regression model and its limitation

Linear Regression is a supervised machine learning algorithm used for a regression task. The regression model is used for prediction value based on independent variables. It is mostly used for finding out the relationship between variables and forecasting. There are variations of regression models, they differ based on – the kind of relationship between the dependent and independent variables, they also considered the number of independent variables.

The linear regression performs the task to predict a dependent variable (y, considered as an output) based on a given independent variable (x, considered as an input). So, the regression technique finds out a linear relationship between input and output. Geometrically this is interpreted as a best fitted straight line. Mathematically it can be represented as in equation (1)

$$y = mx + c \tag{1}$$

Where the unknown values m and c are to be estimated by the gradient decedent technique from the given data i.e. from the training data is given as in equation (2)

$$\{(x_i, y_i): i = 1, 2, \ldots, n\} \quad (2)$$

3.1 Limitation of the model

Since the linear regression model is based on the principle of the relationship of the input and the output data which is linear but in real-life data, the linear relationship does not hold. Sometimes only partitions of the data hold the linearity (Fig. 4) in that case, a single line is not enough to fit the whole data. On the other hand, the outlier data makes huge effects on the regression.

3.2 Proposed model

Since we are dealing with the COVID-19 data and based on our observation is that the COVID-19 data does not follow the linear property. Rather it follows the linear property for a short period and then it changes its direction. In this case, it is not well suited for the linear regression and if still it is used the linear regression the predictions are far from the actual. To overcome this problem several researchers, use polynomial regressions (Gupta et al. 2020). In our study, the piecewise regression model [28] has been used.

From the COVID-19 data, it has been observed that it follows the linearity within a small interval (window). Then it shifts its directions, again follows the linearity, again shifts its trends, and so on. Under this condition, one line simply is not enough to fit the entire data set, then the concept of piecewise linear regression comes to overcome such limitation. When the data set follows different linear trends over the different partitions of data, then we should model the regression function in pieces. Each linear regression is corresponding to a partition i.e., the pieces, and whether the pieces are connected or not connected depends, on the data and the problem. The connecting points are known as the breakpoints, i.e., the locations where the slope changes. Fig 4 shows the actual data, and Fig 5 shows the piecewise correlation regression lines.

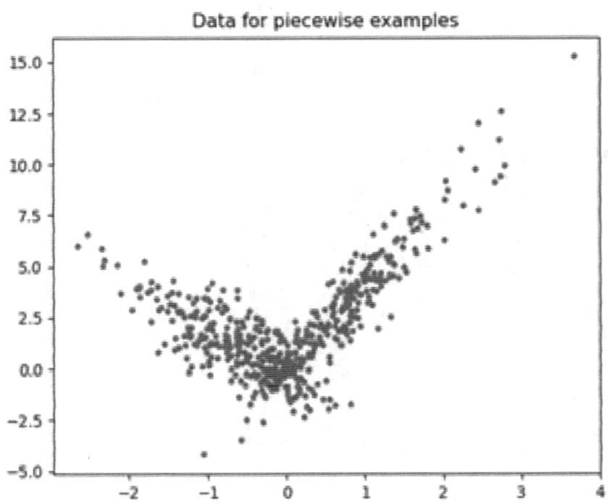

Fig. 4. Actual data points

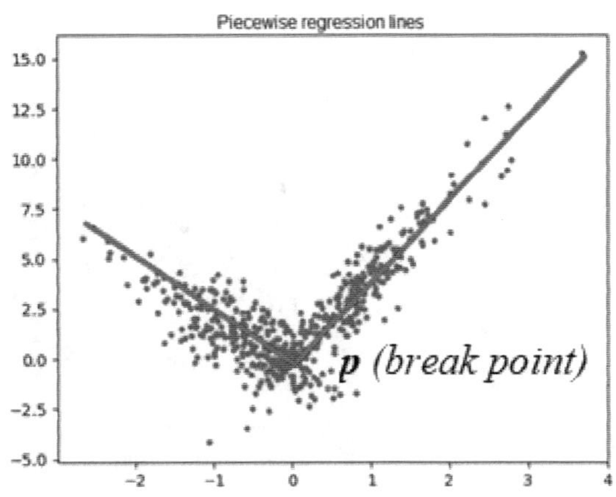

Fig. 5. Piecewise correlation regression lines

The mathematical representation of the piecewise linear regression, consider the lines in Fig. 5 represented by equation (3)

$$y = \begin{cases} m_1 x + c_1 & when,\ x \leq p \\ m_2 x + c_2 & when,\ x > p \end{cases} \quad (3)$$

The point at x = p is the joining point of two lines, i.e., a breakpoint. Find out the breakpoints is also a challenging problem. In our case, we have chosen heuristically, and it is considered a seventh length window, i.e., increasing rate follows the linearity for seven days, and then it changes.

4. System evaluation

The system is evaluated in two different aspects. First, its output is compared with the actual output and it is also compared with the other existing systems.

To check the performance of the *piece-wise regression model* past COVID-19 positive confirm cases have been used. The system generated data have been compared with the actual data. Particularly in this evaluation the COVID-19 positive data (29) of the state Maharashtra, the most spreading state in India have been used from 1st, 6th, 12th, 18th, 24th, and 30th August 2020. Table 1 shows the comparison. The column **Actual** is the actual positive confirm cases and the column **Predicted** represents prediction by the *piece-wise regression model*.

Table 1: Comparison of the system generated value with the actual value

Date	Actual	Predicted
01-08-2020	42,2118	41,9238
06-08-2020	46,8265	46,7689
12-08-2020	53,5601	53,6944
18-08-2020	59,5865	60,8557
24-08-2020	67,1942	68,4708
30-08-2020	74,7995	75,9658

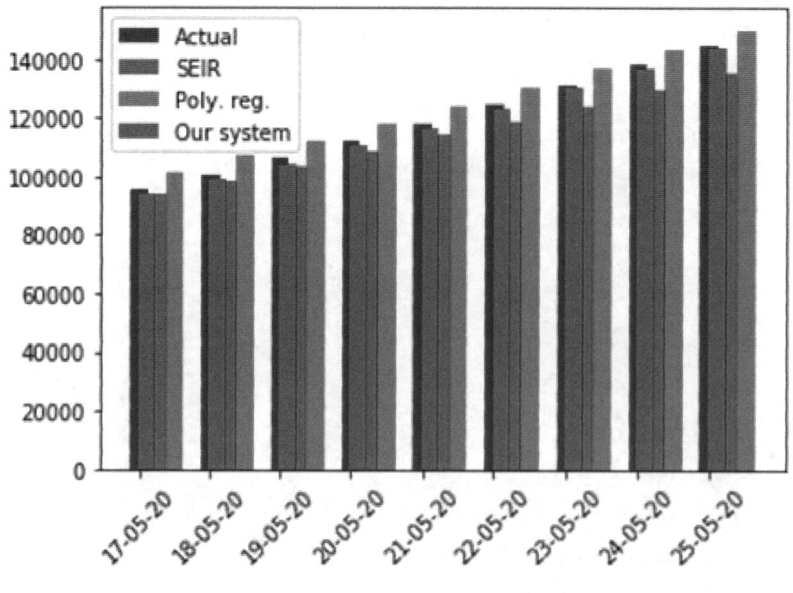

Fig. 6. Comparison of our system with the existing systems (SEIR and Polynomial regression) and actual value

Now the system is compared with the other systems. The comparison is done by the SEIR model and a Polynomial regression model [30]. The comparison is only meaningful when it is considering the same data for the same dates. Hence, we have considered the same data set for the same date. Data have been used in the positive confirm cases from 17th May 2020 to 25th May 2020. Fig 6. shows the graphical representation of the comparison. The result (Fig 6) shows that our system (piece-wise linear regression) is very close to the actual value compared to the other existing system.

5. Conclusions

In this study, a *piece-wise linear regression* model has been used for the COVID-19 predictions which are not done earlier. A comparative study shows that it performs better than the other existing model and hence can be used for future prediction in better accuracy. From the mathematical point of view, the model performs better in local data i.e. in a small partition. Hence there needs further investigation about the optimal partition to improve further.

6. References

[1] Siro Igino, T 2007 The biblical plague of the Philistines now has a name, tularemia. *Med Hypotheses.* **69**:1144e1146. https://doi.org/10.1016/j.mehy.2007.02.036.

[2] Acuna-Soto R, Stahle DW, Therrell MD, Gomez Chavez S, Cleaveland MK 2005 Drought, epidemic disease, and the fall of classic period cultures in Mesoamerica (AD 750-950). Hemorrhagic fevers as a cause of massive population loss. *Med Hypotheses.* **65**(2):405-9. doi: 10.1016/j.mehy.2005.02.025. PMID: 15922121.

[3] William Ogilvy, K., McKendrick A. G. (1927). A contribution to the mathematical theory of epidemics. *Proc. R. Soc. Lond.* A115700–721 http://doi.org/10.1098/rspa.1927.0118.

[4] Anderson RM, May RM 1992 Infectious Diseases of Humans: Dynamics and Control. Oxford, New York: *Oxford University Press*; 1992.

[5] Emergencies preparedness, response (WHO), Disease outbreaks by year 2020 Available at : https://www.who.int/csr/don/archive/year/en/ (Accessed April 23, 2020)

[6] Bernoulli D, Blower S 2004 An attempt at a new analysis of the mortality caused by smallpox and of the advantages of inoculation to prevent it. *Rev Med Virol.* **14**:275e288. https://doi.org/10.1002/rmv.443.

[7] Siettos CI, Russo L 2013 Mathematical modeling of infectious disease dynamics. *Virulence.* **4**:295e306. https://doi.org/10.4161/viru.24041.

[8] Christopher JL M 2020 IHME COVID-19 health service utilization forecasting team. Forecasting the impact of the first wave of the COVID-19 pandemic on hospital demand and deaths for the USA and European Economic Area countries *medRxiv* 2020.04.21.20074732; doi: https://doi.org/10.1101/2020.04.21.20074732

[9] Woody S, Tec MG, Dahan M., Gaither K, Lachmann M., Fox S, Meyers LA, G Scott J 2020 Projections for first-wave COVID-19 deaths across the US using social-distancing measures derived from mobile phones *medRxiv* 2020.04.16.20068163; doi: https://doi.org/10.1101/2020.04.16.20068163

[10] Yang Z, Zeng Z, Wang K, Wong S S, Liang W, Zanin M, Liu P, Cao X, Gao Z, Mai Z, Liang J, Liu X, Li S, Li Y, Ye F, Guan W, Yang Y, Li F, Luo S, Xie Y, … He J 2020 Modified SEIR and AI prediction of the epidemics trend of COVID-19 in China under public health interventions. *Journal of thoracic disease* **12**(3), 165–174. https://doi.org/10.21037/jtd.2020.02.64

[11] COVID-19 Daily Epidemic Forecasting by the *Institute of Global Health* 2020 https://renkulab.shinyapps.io/COVID-19-Epidemic-Forecasting (Accessed on 25 May 2020).

[12] COVID-19 Confirmed and Forecasted Case Data 2020 https://covid-19.bsvgateway.org/. (Accessed on May 25, 2020).

[13] COVID-19 Projections 2020 https://covid19.healthdata.org/projections. (Accessed on May 25, 2020).

[14] Jewell NP, Lewnard JA, Jewell BL 2020 Caution Warranted: Using the Institute for Health Metrics and Evaluation Model for Predicting the Course of the COVID-19 Pandemic. *Ann Intern Med.* DOI: https://doi.org/10.7326/M20-1565.

[15] Marchant R, Samia NI, Rosen O, Tanner MA. Cripps S 2020 Learning as We Go: An Examination of the Statistical Accuracy of COVID19 Daily Death Count Predictions. *medRxiv* 2020.04.11; doi: https://doi.org/10.1101/2020.04.11.20062257

[16] Grey S, Mac Askill A 2020 Special Report: Johnson listened to his scientists about coronavirus - but they were slow to sound the alarm. *Reuters. A.* https://reut.rs/2Rk3sa7. (Accessed on May 25, 2020).

[17] Resnick B 2020 The White House projects 100,000 to 200,000 Covid-19 deaths. *Vox*. https://www.vox.com/science-and-health/2020/3/31/21202188/us-deaths-coronavirus-trump-white-house-presser-modeling-100000. (Accessed on May 25, 2020).
[18] Luo J 2020 Predictive Monitoring of COVID-19. *SUTD Data-Driven Innovation Lab*. https://ddi.sutd.edu.sg/when-will-covid-19- end/.
[19] Xu J, Cheng Y, Yuan X, Li W V, Zhang L 2020 Trends and prediction in daily incidence of novel coronavirus infection in China, Hubei Province and Wuhan City: an application of Farr's law. American *Journal of Translational Research*, 12(4), 1355.
[20] Ben-Israel I 2020 The end of exponential growth: The decline in the spread of coronavirus. *The Times of Israel*. https://www.timesofisrael.com/the-end-of-exponential-growth-the-decline-in-the-spread-of-coronavirus/. (Accessed on May 25, 2020).
[21] Singh B P 2020 Forecasting Novel Corona Positive Cases in India using Truncated Information: A Mathematical Approach. *medRxiv*. Doi: https://doi.org/10.1101/2020.04.29.20085175.
[22] Schueller E, Klein E, Tseng K, Kapoor G, Joshi J, Sriram A, Laxminarayan R 2020 COVID-19 in India: Potential Impact of the Lockdown and Other Longer Term Policies. Retrieved from Washington D.C. https://cddep.org/publications/covid-19-india-potentialimpact-of-the-lockdown-and-other-longer-term-policies/.
[23] Mandal S, Bhatnagar , Arinaminpathy N, Agarwal A, Chowdhury A, Murhekar M, Sarkar S 2020 Prudent public health intervention strategies to control the coronavirus disease 2019 transmission in India: A mathematical model-based approach. *Indian Journal of Medical Research*, **151**(2), 190.
[24] Chatterjee K, Chatterjee Ko, Kumar A, Shankar S 2020 Healthcare impact of COVID-19 epidemic in India: A stochastic mathematical model, *Medical Journal Armed Forces India*. Doi: https://doi.org/10.1016/j.mjafi.2020.03.022.
[25] Ghosh P, Ghosh R, Chakraborty B 2020 COVID-19 in India: State-wise Analysis and Prediction. *medRxiv preprint*. Doi: https://doi.org/10.1101/2020.04.24.20077792.
[26] Menon A, Rajendran N K, Chandrachud A, Setlur G 2020 Modelling and simulation of COVID-19 propagation in a large population with specific reference to India. *medRxiv preprint*. Doi: https://doi.org/10.1101/2020.04.30.20086306.
[27] Senapati A, Nag A, Mondal A, Maji S 2020 A novel framework for COVID-19 case prediction through piecewise regression in India. *International Journal of Information Technology*. Doi: https://doi.org/10.1007/s41870-020-00552-3
[28] McZgee V E, Carleton W T 1970 Piecewise regression. *Journal of the American Statistical Association*, 65(331), 1109-1124.
[29] SRK Novel Corona Virus Disease 2019 in India 2020 covid_19_india.csv, Version 205, https://www.kaggle.com/sudalairajkumar/covid19-in-india?select=covid_19_india.csv. (Accessed on Sept 28, 2020).
[30] Gupta R, Pandey G, Chaudhary P 2020 Machine Learning Models for Government to Predict COVID-19 Outbreak. Doi: https://doi.org/10.1145/3411761

IOP Conf. Series: Materials Science and Engineering 1020 (2021) 012018 doi:10.1088/1757-899X/1020/1/012018

Adopting Sustainable Farming: Implications of Renewable Energy and ICT

S. Dash[a], S. Choudhury[b]

[a,b] North Eastern Regional Institution of Science and Technology, Nirjuli, Arunachal Pradesh, India.

Email: soumyadash2003@gmail.com

Abstract: Contemporary sustainable farming balances higher productivity with a focus on green technological intervention. The green intervention such as renewable energy and information communication technology (ICT) though have been widely accepted by other sectors, farming community yet to yield the dividend from these two tools. In this paper, a modified technology acceptance model has been presented with a focus towards the adoptability of actual sustainable farming. Further, a success story of sustainable farming with the intervention of ICT is discussed. Finally, interdependency of renewable energy and ICT is elaborated with a focus to formulate a generalized theme towards feasibility of actual sustainability farming is presented.

Key Words: Renewable Energy, ICT, Sustainable Farming.

1. Introduction

Information and Communication Technology (ICT) and Renewable energy has played a pivotal role to accomplish holistic growth objectives of both developing and developed economy. The World Summit on the Information Society in 2003 addressed e-agriculture and featured the application of ICTs as a priority in agricultural development. ICT help to provide adequate knowledge to many farmers. It offers predictive information regarding weather, fertilizers and pest intake, as well as to bring transperancy in processes like land litigation, and market pricing. E-governace has empowered farmers of interior location to access adequate information with all convenience (Ghosh and et al., 2020). Agriculture is the foundation and backbone of the Indian economy and 68% of Indian population is chiefly dependent on agriculture for their livelihood and stake in gross domestic product (GDP) is found to be nearly 20 percent. The agricultural sector in India is lagging behind in many aspects and characterized by poor infrastructure in connectivity along with disintegrated and diversed market. The fragmented land ownership of farmers with less adoptability of technology in farming often often faces challenges in accessing reliable and timely information. ICT has enabled the reform in the traditional practices of agriculture for augmented productivity and sustainability (Gangopadhyay et al., 2019; Karanasios and Slavova, 2019; Lindblom et al., 2017). Appropiate applications of ICT in farming sectot is expected resolve the issues of food security. Customised ICT plays crucial role in facilitating marginal farmers for optimal utilization of available limited resources. ICT enables in defining, storing, and retrieving data for dissemination in agriculture sector with the support of both public and private agencies. Thus farming community are able to access information with the help of user friendly need based information of their respective geographic location. Hence, ICT in agriculture has become an important field of research and application related to e-agriculture.ICT has great impact in agriculture sector because it provides information about business activities related to agriculture supply chain, profitability of farmers as well as value to the customers. Further it helps in retaining of existing stakeholders, as well as acquiring new dimension for financing, and investment avenues. Tools and techniques of ICT helps in accessing formulating market strategy for sourcing and delivery systems. Internet is a significant source for gathering the present information on every part of agriculture and transforming agricultural sector into modern digital agriculture.

In various rural development projects, such as rural cluster electrification and enhanced agricultural technology, information and communication technologies have been introduced to make agriculture more sustainable (Choudhury et al., 2016, Ghosh et al., 2020). Renewable energy systems are those that are not prone to degradation and use primary energy supplies. Examples of green energies include solar, wind, geothermal, and biomass. Solar energy is the light that comes right from the Sun. It is the most concentrated source of energy on Earth, including nuclear weapons. Photovoltaic cell, which transforms sunlight directly into electricity, is the fastest growing form of renewable energy, rising at 50 per cent a year. The Sun provides more than 10 000 times the electricity actually consumed by humans per year. The renewable energy was implemented in different agriculture system and successful results were achieved (Khan et al., 2018; Liu et al., 2017; Chel et al., 2011). The renewable energy can be implemented to compensate the energy shortfall in a cost-effective and environmental sustainable manner (Choudhury et al., 2018). Different cost-effective approaches have been developed by various researchers using solar power, wind power, hydroelectric power for use in various industrial sectors, technical institutions and underdeveloped rural areas (Parida and Chatterjee, 2016; Parida et al., 2016a; Parida et al., 2016b). The different energy sources, e.g. solar, wind, hydraulic, biomass, organic waste, biofuels, and combined heat and electricity, provide the electric generator with a simple, reliable and efficient solution. Conserving important non-renewable natural fuels without causing degradation to the atmosphere. Solar energy can be used in its different ways, e.g. solar photovoltaics, direct solar thermal, and green energies and wind can provide the solution to the world's energy challenges and potentially making the atmosphere safe for future generations by reducing air emissions from the use of fossil fuels.

The Technology Acceptance Model (TAM) is an information system theory for using a modern technology which simulates how users embrace and use a technology different studies on implementation of TAM user acceptance of word processors spreadsheets (Mathieson, 1991), e-mail (Szajna, 1996), voice mail (Straub, 1995) , and telemedicine technology (Hu, 1999). The Technology Acceptance Model has been tested empirically and applied in multiple applications (Mathieson, 1991; Taylor and Todd, 1995; Venkatesh, 2000; Moon and Kim, 2001; Liu et al., 2010; Bagozzi, 2007; Davis 1989; Lee et.al., 2003; Venkatesh and Davis, 1996; Venkatesh et.al., 2003). The overall objective of the research was to analyze the linkage of the Renewable Energy, information and communication technology to agriculture sector for sustainable farming. The specific objectives of the research were to implement Renewable Energy and Information and Communication Technology (ICT) for Adopting Sustainable Farming and evaluate the efficiency of the process. The proposed paper has also outlined the implication of Technical Acceptance Model (TAM) to achieve the same.

2. Research model and hypothesis

2.1. Application of information and communication technologies for sustainable farming

Poor farmers lack credit for the purchase of farm inputs such as seeds, fertilizers, crop protectants. Smaller or more distant farmers always have fewer trading opportunities and they have to rely on a few traders with less attractive credit conditions, or may not at all have access to credit from the commodity market. ICT is one of these alternatives, which has recently unlocked tremendous capacity to directly strengthen agriculture in developed countries. ICT has found a foothold in the booming mobile, telecom and Internet markets, including in vulnerable smallholder farms and their operations. Computer-based technologies and networking mechanisms such as social media, data archives of information (online or offline), digital imagery and film, and cell phones are all included in ICTs.

It enables accessing requisite information embedded with Just-in-Time (JIT) for the beneficiaries located at distant geographic location at an affordable cost. This also guarantee gender equality and involve women in the course of decision making. ICTs have the option of receiving immediate and accurate reports of the software, operation, applied technology or information. It helps to increase efficiency and quick accomplishment of work. The mechanism will reinforce the seed industry, such as timely and updated reports on seed-related issues such as the introduction of new crops, false seeds and seed price information, etc. Sustainable growth of rural & agricultural areas is also supported by ICT. This also helps to empower rural netitizens with the conservation of environment by avoiding unwanted exploration through implementation of appropriate technology, and productive strategies for development, markets, banking and financial services. ICT is concerned with recording, extracting and distributing digital data. ICT makes the system more competitive, efficient and attentive to meet the demands.

2.2. Application of Technology Acceptance Model for sustainable farming

The technology acceptance model (TAM) was implemented to predict the acceptability of a mechanism and to define the improvements that may be added to the system. As per TAM, perceived usefulness (PU) and perceived ease of use (PEOU) are decisive factors regarding the acceptability of any information system. In addition to this the technology acceptance model suggests that the use of an information system is determined by the behavioral intention.

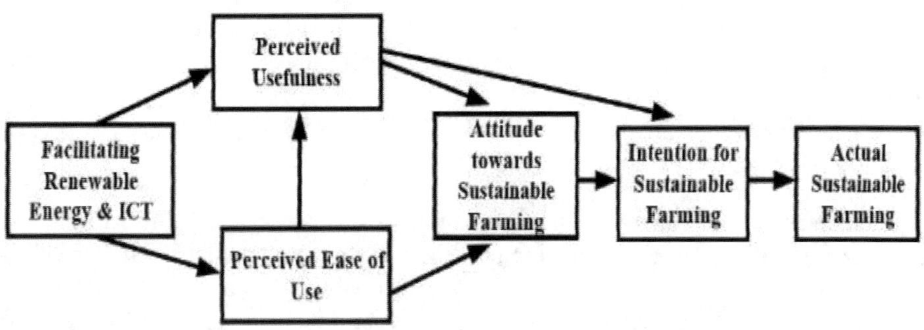

Figure-1: Proposed technology acceptance model for sustainable farming

For sustainable agriculture the behavioral intention is dependent on 3 factors such as attitude towards sustainable farming (AS), intention for sustainable farming (IS), and confidence for implementation of

sustainable farming (CIS). The technology acceptance model for the sustainable agriculture is shown in Fig1. PU denotes the degree of the conviction of a person that a certain scheme is being used can boost its efficiency. Whereas, PEOU indicates affinity of an individual belief system for easy usage of a specific creative information system. AS is defined as the degree to which a farmer implements ICT and renewable energy for sustainable farming. The intention of sustainable farming (IS) implies the intent of individual users and is characterized as a type of external psycho-motor response quantified by the actual course of action of individual users.
In this study the relationship between ICT and renewable energy and sustainable farming was established. ICT and renewable energy is meant for the implementation for farming system by the farmers. Although there are various real and observable benefit steps a simple approach has been created to quantify and calculate the implements of ICT and renewable energy for sustainable farming. In the selection of the necessary details for the design, a method of survey was implemented. A standardized questionnaire consisting of a series of questions was taken as a testing tool. The methodology of questionnaire surveying was used in order to collect primary results. A major portion of the questionnaire consisted of semi-closed questions such that each respondent could select from the alternatives presented and also share their opinion. The TAM questionnaires were implementing renewable energy and ICT, perceived usefulness, perceived ease of use, attitude towards sustainable farming, intention for sustainable farming, and confidence for implementation. The participants were asked to responds the questions asked above using the 5 points (1-5) Scale. The five point scale was designated as1 as lowest point with denoting 'strongly disagree' and 5 as 'strongly agree' and subsequently other interval number denotes all other interval values. Twenty elderly people (over 51 years of age), twenty middle-aged people (between 31 and 50 years of age) and twenty younger adults (between 18 and 30 years of age) from the local population were chosen for the study. Every member of the group should be able to use smartphones and computers. An introductory overview on ICT and green energy was presented to all participants. In order to test the model parameters, they were expected to finish a survey and attempt all the given questionnaire and the score had to be provided on the five-point scale.

3. Results and Discussions

In agriculture system ICT and Renewable energy are used in weather forecasting, taking expert opinion, and for farming techniques by illustrative video. Renewable energy was implemented for generating and storage of electricity which further can be implemented for electricity based water pump and storage of farm output. The Indian farmers are lesser interested for implementation of ICT and renewable energy in agriculture due to their low level of education, and higher initial cost of investment for implementing the technology. The farmers often risk introducing the technique because they are not sure about the yield. ICT is mainly dependent on electricity. Privatized mega-power plant can't afford less efficient rural electrification, so green energy can help marginal farmers make it sustainable by partnership at village level.

3.1. Descriptive statistic for parameters of sustainable farming

The descriptive statistics pertaining to the different factors of sustainable agriculture are shown in Table 1. The mean score of the variables implementing renewable energy and ICT (IRE-ICT), perceived usefulness, perceived ease of use, attitude towards sustainable farming, intention for sustainable farming, and confidence for implementation were 3.43, 3.70, 3.31, 3.60, 3.76, and 4.06 respectively. The score of each variables was higher than average which indicated that the strongly agree with each parameter of sustainable agriculture.

Table 1 Descriptive statistics of the variables for implementation of sustainable agriculture

Variable	*Minimum*	*Maximum*	*Mean*	*SD*
Implementing renewable energy & ICT	*1.00*	*5.00*	*3.43*	*1.43*
Perceived usefulness	*1.00*	*5.00*	*3.70*	*1.29*
Perceived ease of use	*1.00*	*5.00*	*3.30*	*1.44*
Attitude towards sustainable farming	*1.00*	*5.00*	*3.60*	*1.40*
Intention for sustainable farming	*1.00*	*5.00*	*3.77*	*1.22*
Confidence for implementation	*1.00*	*5.00*	*4.07*	*1.11*

3.2. Correlation between variables of sustainable farming

The correlation coefficients among the various variables are summarized in Table (2). According to the findings obtained there is a positive and significant relationship between the variables implementing renewable energy and ICT, perceived usefulness, perceived ease of use, attitude towards sustainable farming, intention for sustainable farming, and confidence for implementation. It was observed that implementing renewable energy and ICT was strongly correlated with attitude towards sustainable farming ($p < 0.01$, $r = 0.604$) and the correlation was found to be significant.

Table 2 Correlation coefficients between the Technology Acceptance Model parameters

Variables	*IRE-ICT*	*PU*	*PEOU*	*AS*	*IS*	*CIS*
IRE-ICT	*1*					
PU	*0.558***	*1*				
PEOU	*0.537***	*0.55***	*1*			
AS	*0.604***	*0.426**	*0.572***	*1*		
IS	*0.415**	*0.522***	*0.354*	*0.566***	*1*	
CIS	*0.393**	*0.135*	*0.202*	*0.172*	*0.113*	*1*

*Correlation is significant at the 0.05 level (2-tailed).
**Correlation is significant at the 0.01 level (2-tailed).

The correlation coefficients of implementing renewable energy and ICT with perceived usefulness, perceived ease of use, attitude towards sustainable farming, intention for sustainable farming and confidence for implementation with implementing renewable energy and ICT were 0.558, 0.536, 0.604, 0.414, and 0.393 respectively. Therefore, the use renewable energy such as solar, hydro-power or wind-farms in sustainable agriculture is highly beneficial for the farmers. Farmers can use renewable based rural electrification technology for electrical fencing of agriculture land. Irrigation for optimal watering by electric pumps. Post harvest activity can also be achieved by the usage of electric dryer. For this purpose, marginal farmers can use solar panel based irrigation for distantly located farmland. At the same time established farmers can rely upon geothermal heat pumps. The correlation between perceived usefulness and perceived ease of use, attitude towards sustainable farming, intention for sustainable farming, and confidence for implementation were evaluated. Based on the results obtained there was positive and significant relationships between perceived usefulness and perceived ease of use ($p < 0.01$, $r = 0.550$) and also relationships between perceived usefulness and intention for sustainable farming was found to be highly significant ($p < 0.01$, $r = 0.522$). The correlation between confidence for implementation was found to be significant for implementing renewable energy & ICT whereas the correlation between confidence for implementation and perceived usefulness, perceived ease of use, attitude towards sustainable farming, intention for sustainable farming was found to be insignificant ($p > 0.05$). The findings of correlation experiments have shown that the model variables are related to each other.

4. Conclusion

This paper analyzed the use of the TAM as deterministic for the acceptance of ICT and renewable energy in sustainable farming. The study has a novel addition of implementation of renewable technologies leading to sustainable farming. This study was conducted to classify and correlate the parameters implementing renewable energy and ICT, perceived usefulness, perceived ease of use, attitude towards sustainable farming, intention for sustainable farming, and confidence for implementation towards sustainable agriculture technologies. There is a significant positive relationship observed between the variables implementing green energy and ICT, perceived utility, perceived ease of use, attitude towards sustainable farming, intention for sustainable farming, and confidence for implementation, according to the findings obtained. ICT would enhance interchange of their experience with an interactive ICT based systems for both tangible and intangible causes. Farmers with minimal access to resources, energy and technology can access information via computer-based apps and networking platforms such as social media, digital database archives, and digital imagery and recording, as well as mobile phones. Farmers with minimal access to resources, energy and technology can access information. The analysis offers valuable perspectives for farmers and decision makers of implementing ICT and renewable energy for sustainable agriculture.

Acknowledgments

The authors are thankful to the funding agency: National Mission on Himalayan Studies (NMHS) for their support through project no.GBPNI/NMHS-2017-18/SG15. The authors would also like to acknowledge Technology Intervention for Tribal Empowerment (TITE) scheme of Department of Science and Technology, Government of India (Project no: SEED/TITE/001/2017(C).

References

[1] A. Parida, D. Chatterjee, (2016) A solar PV augmented hybrid scheme for enhanced wind power generation through improved control strategy for grid connected doubly fed induction generator, Cogent Engineering, 3: 1147337. doi.org/10.1080/23311916.2016.1147337

[2] A. Parida, S. Choudhury, S. Chatterjee, D. Chatterjee, (2016a) A cost effective trade-off based renewable power augmented energy efficient load model for manufacturing industries for demand side management", Cogent Engineering, 3: 1239299. doi.org/10.1080/23311916.2016.1239299

[3] A. Parida, S. Choudhury, S. Chatterjee, D. Chatterjee, (2016b) A rule based electrical energy distribution model for energy efficiency in technical institutions: A case study, Energy and Buildings, 126, 508-516. doi.org/10.1016/j.enbuild.2016.05.069

[4] Bagozzi, R. P. (2007). The legacy of the technology acceptance model and a proposal for a paradigm shift. Journal of the association for information systems, 8(4), 3.

[5] Chel, A., & Kaushik, G. (2011). Renewable energy for sustainable agriculture. Agronomy for Sustainable Development, 31(1), 91-118.

[6] Davis, F. D. (1989). Perceived usefulness, perceived ease of use, and user acceptance of information technology. MIS quarterly, 319-340.

[7] Gangopadhyay, P. K., Khatri-Chhetri, A., Shirsath, P. B., & Aggarwal, P. K. (2019). Spatial targeting of ICT-based weather and agro-advisory services for climate risk management in agriculture. Climatic change, 154(1-2), 241-256.

[8] Hu, P. J., Chau, P. Y., Sheng, O. R. L., & Tam, K. Y. (1999). Examining the technology acceptance model using physician acceptance of telemedicine technology. Journal of management information systems, 16(2), 91-112.

[9] Karanasios, S., & Slavova, M. (2019). How do development actors do "ICT for development"?

A strategy-as-practice perspective on emerging practices in Ghanaian agriculture. Information Systems Journal, 29(4), 888-913.
[10] Khan, M. T. I., Ali, Q., & Ashfaq, M. (2018). The nexus between greenhouse gas emission, electricity production, renewable energy and agriculture in Pakistan. Renewable Energy, 118, 437-451.
[11] Lee, Y., Kozar, K. A., & Larsen, K. R. (2003). The technology acceptance model: Past, present, and future. Communications of the Association for information systems, 12(1), 50.
[12] Lindblom, J., Lundström, C., Ljung, M., & Jonsson, A. (2017). Promoting sustainable intensification in precision agriculture: review of decision support systems development and strategies. Precision Agriculture, 18(3), 309-331.
[13] Liu, I. F., Chen, M. C., Sun, Y. S., Wible, D., & Kuo, C. H. (2010). Extending the TAM model to explore the factors that affect Intention to Use an Online Learning Community. Computers & education, 54(2), 600-610.
[14] Liu, X., Zhang, S., & Bae, J. (2017). The nexus of renewable energy-agriculture-environment in BRICS. Applied energy, 204, 489-496.
[15] Mathieson, K. (1991). Predicting user intentions: comparing the technology acceptance model with the theory of planned behavior. Information systems research, 2(3), 173-191.
[16] Moon, J. W., & Kim, Y. G. (2001). Extending the TAM for a World-Wide-Web context. Information & management, 38(4), 217-230.
[17] S. Choudhury, A. K. Patra, A. Parida, S. Chatterjee, (2018) Renewable energy capacity estimation for Indian energy sector using energy demand forecasting through Fuzzy Time Series, Advances in Smart Grid and Renewable Energy, Springer, 551-558. doi.org/10.1007/978-981-10-4286-7_54
[18] S. Choudhury, A. Parida, R. M. Pant, S. Chatterjee, (2016) GIS augmented computational intelligence technique for rural cluster electrification through prioritized site selection of micro-hydro power generation system, Renewable Energy, 142, 487-496. https://doi.org/10.1016/j.renene.2019.04.125
[19] S. K. Ghosh, A. K. Das and S. Choudhury, (2020) ICT Interventions in Rural Development Schemes: A Case of Rupshi Development Block, Assam, 2020 8th International Conference on Reliability, Infocom Technologies and Optimization (Trends and Future Directions) (ICRITO), Noida, India, 2020, pp. 349-354, doi: 10.1109/ICRITO48877.2020.9197987.
[20] Straub, D., Limayem, M., & Karahanna-Evaristo, E. (1995). Measuring system usage: Implications for IS theory testing. Management science, 41(8), 1328-1342.
[21] Szajna, B. (1996). Empirical evaluation of the revised technology acceptance model. Management science, 42(1), 85-92.
[22] Taylor, S., & Todd, P. (1995). Decomposition and crossover effects in the theory of planned behavior: A study of consumer adoption intentions. International journal of research in marketing, 12(2), 137-155.
[23] Venkatesh, V. (2000). Determinants of perceived ease of use: Integrating control, intrinsic motivation, and emotion into the technology acceptance model. Information systems research, 11(4), 342-365.
[24] Venkatesh, V., & Davis, F. D. (1996). A model of the antecedents of perceived ease of use: Development and test. Decision sciences, 27(3), 451-481.
[25] Venkatesh, V., Morris, M. G., Davis, G. B., & Davis, F. D. (2003). User acceptance of information technology: Toward a unified view. MIS quarterly, 425-478.

Saliency Detection in Text Documents using Policy-Driven Reinforcement Learning Methodologies

Gaurav Meena, Sarika Choudhary, Ravi Raj Choudhary

Department of Computer Science, Central University of Rajasthan, Ajmer, Rajasthan, India-305817

E-mail: gaurav.meena@curaj.ac.in, www@curaj.ac.in

Abstract. As the amount of information grows, it is challenging to find concise information. Thus it is necessary to build a system that could present human quality summaries. Saliency detection is a tool that provides abstracts or keywords of a given document. In this paper, three different approaches have been implemented for saliency detection. In all these three approaches, sentences are represented as a feature vector. In the first approach, features like root words, vocabulary intersections, words, and inclusion of numerical data use. This model is trained by using general Algorithms, Like Porter's Stemmer, Spell check. In the second approach, apart from the features used in the first approach, TF-IDF scores, Mean, Standard Deviation, and a Threshold value of a word is also used as features. In the third approach, Maximal Marginal Relevance (MMR) algorithm is used to generate a summary.

Keywords : Saliency, Summaries, Porter's Stemmer, Spell check, TF-IDF, Mean, Standard Deviation, Threshold value, Maximal Marginal Relevance (MMR).

1. Introduction

With the happening to the data transformation, electronic reports are turning into a standard media of business and scholarly data. A large number of electronic archives are created and made accessible on the web every day. Be that as it may, it is difficult to peruse every single archive. So there is a need for introducing the data as notable words so that there is no compelling reason to peruse the entire record. An enormous sum of information may be accessible around net-like. Anyhow retrieving the right what more critical information is is not time permits. Search engines, for example, such that Google, Yahoo. Try to serve similarity as a data get agenize. They recover what is more rank the data similarity as for every those user's appeal. In any case, due to those absences of profound comprehensions of standard dialect also humankind's intelligence, they best perform shallow string. Saliency identification in content reports is the way toward distinguishing the most critical data from the archive or set of related records and speaking to the principle thought in significantly lesser space, usually by a variable of three to ten than the original content. On a basic level, Saliency location in content archives is conceivable on account of the usually happening repetition in content and because imperative or striking data is spread unevenly in literary reports information. Those data recovery will be at distant more excellent over that client camwood handles what is more oversee. Those clients need to dissect the searched outcomes one by one until the palatable majority of the data may be required.

It is a chance devouring and wasteful, Assuming that report camwood be speaking to for fewer expressions with key vital words, after that, this aggravates our existence more straightforward.

Automatic Text Summarization [1] may be a quick mining undertaking that extracts first sentences with a blanket just about every last one of the main ideas of those archives.

The main benefit of this is that the time consumed by the user to read the whole document to find the useful information reduces. With record outline judgment available, a client might undoubtedly choose it is pertinence on their investment also procure wanted documents for a significant part very less mental load. The main challenge is to identify redundancy.

2. Literature Review

2.1. Pioneering Works:

Luhn et al., 1958 [2] Excerpts of specialized papers that are more magazine articles that fill those needs of regular digests had been made entirely via programmed implies. In the exploratory investigation depicted, the total content of an article to machine-discernable state is examined by an IBM 704 information transforming machine and broke down as per a standard program.

Edmundson et al., 1969 [3] portray new strategies for naturally extricating reports to screening reasons for existing, i.e., the PC choice of sentences hosting the best possibility to passing on to the individual using the substance of the report card. The exploration strategy incorporates strategies to those courses of action of the required lexicons, those sets of the control parameters, and the near assessment of the programmed separates with physically created removes.

Baxendale et al., 1958 in [4] tell about Machine strategies for lessening specialized foul documents on their segregating indices, which investigated. The PC program mimicked human checking designs in choosing "point sentences" and expressions made out of things and modifiers. The measure of buildup coming about because of every technique and the relative consistency in files are analyzed. It demonstrated that the organized file gave by the expression is more significant and segregating.

2.1.1. Statistical Method Salton et al., 1988 in [5] demonstrated that that pertinence from claiming record terms is conversely proportional to the amount from claiming documents in the corpus holding those expressions. The mathematical statement for term pertinence assessment given by tfi x idfi, where tfi is the recurrence of term i in the report, and idfi is the reversed recurrence of archives containing this term. Sentences might get a chance to be in this way scored, for example, by summing the significance of terms in the sentence.

Kupiec et al., 1995 in [6] said that to condense is to lessen in many-sided quality, and consequently long while holding a portion of the fundamental characteristics of the first. This work principally centered on record extricates, a specific sort of registered archive rundown. Report extricates comprising about 20 percent of the first can be as enlightening as the full content of a record, which recommends that even shorter concentrates might be valuable characteristic synopses.

Eduard Hovy et al., 1999 [7] gave a condition on SUMMARIST, and it is an endeavor to make a robust computerized content synopsis framework, because of the condition: rundown = point distinguishing proof + translation + era. Each of these stages contains a few free modules, vast numbers of them prepared on extensive corpora of content.

3. Methodology and Implementation

Machine learning is [8] may be a method for data examination that mechanizes legitimate model creation. Using calculations that iteratively pick up starting with data, machine learning in permits machines to find shrouded encounters without being unequivocally programmed the place with taking a look. Arthur Samuel (1959). Machine Learning: field of study that provides for Pcs those limits to take without being explicitly programmed.

Tom Mitchell, Well-postured learning in Problem: a workstation system gets similarly as a matter for knowledge (E), concerning a portion undertaking (T), What is more, a percentage execution measure (P). In its execution for T, Likewise measured Toward P, enhances with knowledge E. Two of the majority comprehensively grasped machine learning in methodologies used as supervised learning and unsupervised learning. In any case, there are also separate strategies for machine learning in the different methods may be reinforcement learning.

3.1. Supervised learning
Algorithms would be readied utilizing denoted cases, for instance, a piece of information where the place the desired yield may be referred. For example, a touch for mechanical assembly might bring information F (Failed) or R (runs).

3.2. Unsupervised learning
Unsupervised learning utilizes information that needs no chronicled denote. Those structures will not be completely frank the "right reply. "Those computations must fathom what is shown up. The objective is to investigate the information also discover a few structures inside. Unsupervised learning in capacities amazingly on regard built information.

3.3. Semi-supervised learning
Semi-supervised learning utilized for vague requisitions from supervised learning. Over whatever case, it employments both checked and unlabelled information to getting to prepared a minimal measure from claiming named information with a tonal of unlabelled information (in light of the reality that unlabelled information will be more sensible furthermore less push will get).

3.4. Reinforcement learning
Reinforcement learning now uses for robotics, gaming, and route. With help taking in, those figuring figures through experimentation which activities yield the best remunerates. This sort of learning has three essential parts: those master (the learner alternately boss), that world (everything those director interfaces with), and exercises (what the driver can do).

3.5. Input Dataset
We input data in the program for processing in the form of XML (Extensible Markup Language) files. From the XML files, we need the only text portion of the file for the further processing of the program. For extracting the text from the XML files, we use the JAXB API of the JAVA. The input dataset considered from the DUC2001 dataset.

3.6. Exclude List
Prepare an Exclude List, which contains all those words which are not available in the dictionary like the name of persons, countries, cities. It is a manually created list.

3.7. Dictionary
We are using a Dictionary from WordNet for the checking of the spellings of words extracted from the input documents.

3.8. Stopwords
Stopwords are expressions that are separated when alternately following preparing information (text). If stop expressions generally allude to the relatable point expressions in a language, there is no single widespread rundown about stop expressions utilized by all tools. In fact, not know devices utilize such a rundown. Any arrange for talk maybe pegged concerning illustration those stop expressions to a habituated plan.

3.9. Tokenizing

Tokenizing is a treat to alter an edict into root words. The sentence segmentation revolves around the composing delimiter, which is space (" "). The scheme in this regard deal with will then be continued with stemming.

3.10. Stemming

Stemming is a process of transforming a word to return it to its root word by applying specific rules so that each word has the same representation. Stemming is in this research uses Porter's Stemmer Algorithm (1980). This Algorithm deals with confix, prefix, and suffix [9].

3.11. TF-IDF

Tf-idf remains for term frequency-inverse document frequency, and the tf-idf weight is a weight regularly utilized done data recovery and mining. This weight is an accurate measure used to assess how critical a statement will be within a document to an accumulation or corpus. Those significance builds proportionally to structure about times a saying gives the idea in the report; however, is wipe out separately normality of clearing in the corpus. Varieties of the tf-idf weighting plan need aid regularly utilized by search engines as a vital apparatus for scoring and positioning a document's pertinence provided for a client inquiry.

3.11.1. TF: Term Frequency, which measures how habitually a term happens to a document. Since without a unique case point of interest may be uncommon over a piece, it will be accessible that expression might appear significantly all the more introduce long documents over abbreviated ones. Thus, those expression frequency is much-branched detail limit (those aggregate amount from claiming terms in the document) separately similarly as an approach of normalization:

$$TF(t) = \text{(Number of times term t shows up in an document)/(Total number of terms in the document)}$$

3.11.2. IDF: Inverse Document Frequency, which measures how vital expression is. Same time registering TF, the sum terms need aid viewed as just as imperative. Nonetheless, it may be referred to that certain terms, for example is, of, and *that*, might show up a considerable measure of times, however, need little significance. Accordingly, we have to weigh down the incessant terms same time scale up those extraordinary ones, by registering those following:

$$IDF(t) = \log \frac{Totalnumberofdocuments}{Numberofdocumentswithtermtinit}$$

TF-IDF Weighting: We now consolidate the definitions of term frequency and inverse document frequency, to prepare a composite weight for each term in each document. That tf-idf weighting plan will term t a weight to document d provided for by

$$tf - idf_{t,d} = tf_{t,d} * idf_t \quad (1)$$

Clinched alongside other words, $tf - idf_{t,d}$ assigns to term t a weight over document d that is

- Most elevated the point when t happens a lot of people times inside a little number of documents (thus loaning helter-skelter segregating control with the individual's documents);
- Bring down the point when the expression happens fewer times for a document, or happens in large portions documents (thus putting forth a less maintained pertinence signal);
- Least the point when those term happens clinched alongside virtually every last bit of documents.

3.12. Cosine Similarity

Cosine similarity [10] is a basic calculation in order to obtain a similarity score between two vectors. Cosine similarity calculation consists of two stages which are:

- Calculating relevance between document and query Calculating Cosines of two vectors which are W (weight) of every document with W (weight) of the query.
- Calculating similarity between documents Calculating Cosine angle of two vectors which are W (weight) of a word and W (weight) of another word (another document.

Cosine similarity equation is as follows:

$$SIM(S_1, S_2) = \frac{\sum_i t_{1i} t_{2i}}{\sqrt{\sum_i^{t_{1i}^2}} \sqrt{\sum_i^{t_{2i}^2}}} \tag{2}$$

Information:
S_1 = the vector of the candidate query or document
S_2 = the vector of the document beside the candidate
t_i = weight of word

3.13. Maximal Marginal Relevance (MMR)

The vast majority of modem IR search engines available a positioned rundown for retrieved documents requested by the declining importance of the user's inquiry. In vary, we inspired those requirements to "relevant novelty" as a prone possibly predominant paradigm. An initial close estimation should correct mapping variety think as of farthest point importance and variety independently and provide a straight mix as those metrics. We bring those straight consolidation "marginal relevance" - i.e., a document needs minor secondary importance if it may be both fittings of the inquiry and incorporates nominal analogy to formerly chose documents [11].

$$\begin{aligned} MMR &= ARG_{max} C_{i \epsilon R/S} [\lambda Sim_1(Q, Di) - (1-\lambda)_{max} \\ &\quad \times C_{i \epsilon S} Sim_2(Di, Dj)] \end{aligned} \tag{3}$$

Where, D_i: Documents collection, Q: Query, R: Relevant documents, S: Current result set. Provided for those over definition, MMR [11] computes incrementally those standard relevance-

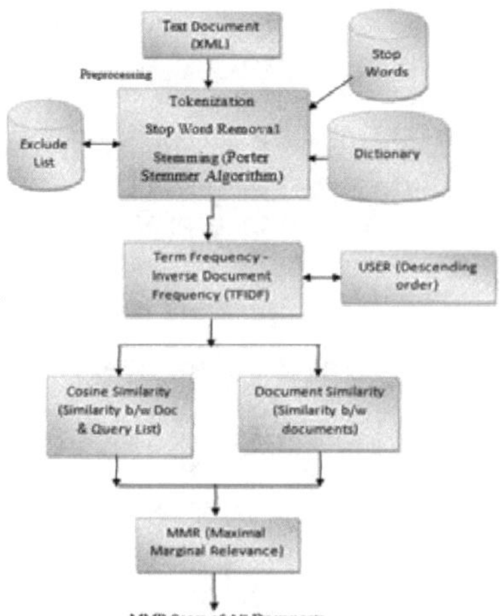

Figure 1. MMR Score for all documents

IOP Conf. Series: Materials Science and Engineering 1020 (2021) 012019 doi:10.1088/1757-899X/1020/1/012019

ranked rundown when those parameterλ =1, and computes a maximal differences positioning around those documents in R when λ =0. To intermediate qualities from claiming λ in the interim [0, 1], a linear combination from claiming both criteria may be optimized. Clients longing to example those majority of the data space around the query, ought set λ at an more modest value, and the individuals longing will centring done around different possibly covering or reinforcing important documents, ought further bolstering situated λ to an worth closer to 1. We found that an especially successful search method (reinforced by those client ponder examined below) may be on begin for a little λ (e.g. λ = 0. 3) in place will comprehend the majority of the data space in the locale of the query, and on concentrate on those most vital parts utilizing an reformulated inquiry (possibly through relevance feedback) and a bigger quality for λ(e.g. λ = 0. 7).

4. Result and Discussion

In the pre-processing part, we get the list of words with their TF-IDF scores in document wise as well as in descending order also. So, we use the document wise score in calculating the similarity between the documents and the score, which is in descending order for creating the Query List. Furthermore, find the similarity between the Query List and each input document.

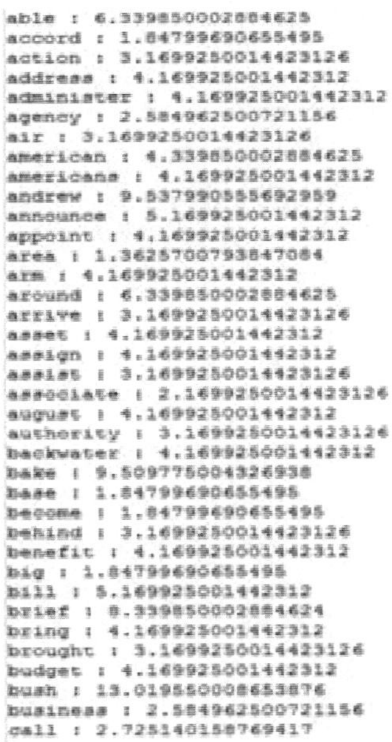
able : 6.339850002884625
accord : 1.847996906655495
action : 3.1699250014423126
address : 4.169925001442312
administer : 4.169925001442312
agency : 2.584962500721156
air : 3.1699250014423126
american : 4.339850002884625
americans : 4.169925001442312
andrew : 9.537990555692959
announce : 5.169925001442312
appoint : 4.169925001442312
area : 1.3625700793847084
arm : 4.169925001442312
around : 6.339850002884625
arrive : 3.1699250014423126
asset : 4.169925001442312
assign : 4.169925001442312
assist : 3.1699250014423126
associate : 2.1699250014423126
august : 4.169925001442312
authority : 3.1699250014423126
backwater : 4.169925001442312
bake : 9.509775004326938
base : 1.847996906655495
become : 1.847996906655495
behind : 3.1699250014423126
benefit : 4.169925001442312
big : 1.847996906655495
bill : 5.169925001442312
brief : 8.339850002884624
bring : 4.169925001442312
brought : 3.1699250014423126
budget : 4.169925001442312
bush : 13.019550008653876
business : 2.584962500721156
call : 2.7251401588769417

Figure 2. TF-IDF score list document wise

maff 37.529325012980806
gajdusek 33.359400011538385
nih 33.359400011538385
brown 29.189475010096185
kuru 28.529325012980813
victims 22.189475010096619
defect 20.849625007211156
president 18.4799690655495
feed 17.359400011538385
mitrova 16.67970000576925
saudi 16.67970000576925
won 16.67970000576925
prion 15.849625007211563
commissar 15.849625007211563
cattle 14.039100017307775
bush 13.019550008653876
disease 13.0
ban 12.935978345588465
ministry 12.935978345588465
loss 12.935978345588465
germane 12.924812503605778
politics 12.67970000576925
outbreak 12.67970000576925
catastrophe 12.67970000576925
miles 12.67970000576925
clinton 12.509775004326936
fema 12.509775004326936
relief 12.509775004326936
concluded 12.509775004326936
magazine 12.509775004326936
conduit 12.509775004326936
encephalopathies 12.509775004326936
patient 12.509775004326936
biology 12.509775004326936
cluster 12.509775004326936
doses 12.509775004326936
fascine 12.509775004326936

Figure 3. TF-IDF score list score wise

We performed a pilot test with seven documents for news articles. The reason for examining to figure out if they might advise those contrast the middle of standard ranking technique and MMR. Starting with these scores should be obvious there will be no vast measurable contrast in the middle of those λ =0. 7 and λ =0. 3 scores. This is often demonstrated by instances the place the λ =1 summary judgement neglected to lift up a bit from claiming important majority of the data and the re-ranking for λ =0. 7 or 0. 3 might.

Those MMR-flow choice techniques to summarization meet expectations surpass for longer documents. MMR may be likewise beneficial to plunge of excursions from multiplex points harshly those same topics. News stories cool an excellent deal repeat of history in-formation.

IOP Conf. Series: Materials Science and Engineering 1020 (2021) 012019 doi:10.1088/1757-899X/1020/1/012019

Table 1. MMR(15%) Scores at different values of λ

Documents	$\lambda = 0.0$	$\lambda = 0.1$	$\lambda = 0.2$	$\lambda = 0.3$	$\lambda = 0.4$
D1	-0.984561744	-0.796186568	-0.607811392	-0.419436215	-0.231061039
D2	-0.995201172	-0.813461284	-0.631721397	-0.449981509	-0.268241621
D3	-0.968851794	-0.785248336	-0.601644877	-0.418041419	-0.23443796
D4	-0.976726974	-0.798918083	-0.621109191	-0.443300299	-0.265491408
D5	-0.986422112	-0.804535412	-0.622648712	-0.440762011	-0.258875311
D6	-0.991420629	-0.806195342	-0.620970556	-0.435744769	-0.250519483
D7	0	0.085755928	0.171511855	0.257267783	0.34302371

Table 2. MMR(15%) Scores at different values of λ

Documents	$\lambda = 0.5$	$\lambda = 0.6$	$\lambda = 0.7$	$\lambda = 0.8$	$\lambda = 0.9$	$\lambda = 1$
D1	-0.042685863	0.145689313	0.334064489	0.522439665	0.710814841	0.899190017
D2	-0.086501734	0.095238154	0.276978041	0.458717929	0.640457817	0.822197704
D3	-0.050834502	0.132768956	0.316372415	0.499975873	0.683579332	0.86718279
D4	-0.087682516	0.090126376	0.267935267	0.445744159	0.623553051	0.801361942
D5	-0.076988611	0.104898089	0.28678479	0.46867149	0.65055819	0.83244489
D6	-0.065294196	0.119931091	0.305156377	0.490381664	0.67560695	0.860832237
D7	0.428779638	0.514535565	0.600291493	0.68604742	0.771803348	0.857559275

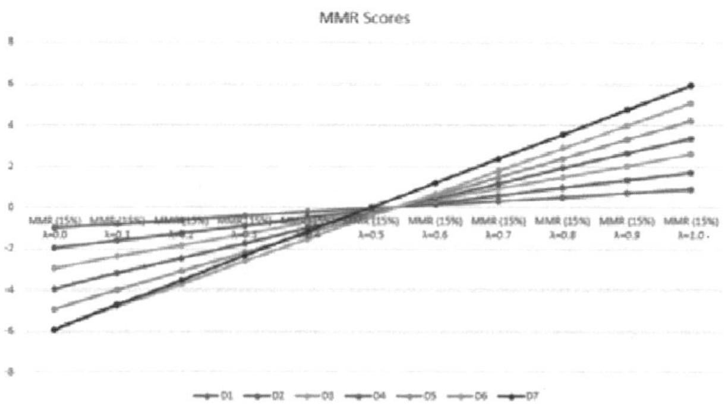

Figure 4. Graph of MMR scores

5. Conclusion

Our System is for saliency detection in text documents based on the extraction method. This salient expression alternately expressions extraction will increase a chance to be a tremendous sufficient for a spectator to comprehend the primary clue of a document. Machine learning methods help in saliency detection in content archives frame firmly related fields, for example, information retrieval or text mining.

As future work we can survive this work for all types of saliency detection in text documents. Integrating the user's prior knowledge into models will be a decisive factor in saliency detection in text documents. We plan to incorporate Reinforcement Learning Methodologies or dynamically improving the algorithm performance on quasistationary data. In particular, frequently occurring concepts will finalize, and novel ideas will include over a timeline additionally first-order Markov Chain. Assumptions can attempt, and Q- Learning algorithms will be incorporated.

References

[1] Neto, Joel Larocca et. al. "Automatic text summarization using a machine learning approach." Brazilian Symposium on Artificial Intelligence. Springer Berlin Heidelberg, 2002.
[2] Luhn, Hans Peter. "The automatic creation of literature abstracts." IBM Journal of research and development 2.2 (1958): 159-165.
[3] Edmundson, Harold P. "New methods in automatic extracting." Journal of the ACM (JACM) 16.2 (1969): 264-285.
[4] Baxendale, Phyllis B. "Machine-made index for technical literature—an experiment." IBM Journal of Research and Development 2.4 (1958): 354-361
[5] Salton, Gerard, and Christopher Buckley. "Term-weighting approaches in automatic text retrieval." Information processing and management 24.5 (1988): 513-523.
[6] Kupiec, Julian, Jan Pedersen, and Francine Chen. "A trainable document summarizer." Proceedings of the 18th annual international ACM SIGIR conference on Research and de-velopment in information retrieval. ACM, 1995.
[7] Hovy, Eduard, and Chin-Yew Lin. "Automated text summarization and the SUMMARIST system." Proceedings of a workshop on held at Baltimore, Maryland: October 13-15, 1998. Association for Computational Linguistics, 1998.
[8] Yulita, Winda, and Feddy Setio Pribadi. "The Implementation of Maximum Marginal Rel-evance Method on Online National and Local News Portal." Proceeding of Engineering In-ternational Committee 4 (2015).
[9] Carbonell, Jade, and Jade Goldstein. "The use of MMR, diversity-based reranking for re-ordering documents and producing summaries." Proceedings of the 21st annual international ACM SIGIR conference on Research and development in information retrieval. ACM, 1998.
[10] Huang, Anna. "Similarity measures for text document clustering." Proceedings of the sixth new zealand computer science research student conference (NZCSRSC2008), Christchurch, New Zealand. Vol. 4. 2008.
[11] Carbonell, Jaime G., and Jade Goldstein. "The use of MMR, diversity-based reranking for reordering documents and producing summaries." SIGIR. Vol. 98. 1998.

IOP Conf. Series: Materials Science and Engineering 1020 (2021) 012020 doi:10.1088/1757-899X/1020/1/012020

Security and Challenges in Voice over Internet Protocols: A Survey

Vinod Kumar[1], Om Prakash Roy[2]

[1,2]Department of EE, NERIST, Arunachal Pradesh 791109, India

[1]vinodnerist@gmail.com, [2]oproy61@gmail.com

Abstract. Voice over Internet Protocol (VoIP) an emerging new technologies is based on real-time communication of data by sending and receiving digital signals using Internet Protocols (IP). VoIP has become popular in business organization and scientific communities. Voice coder (Codec) named as coder and decoder, play an important role in packet switch network for transfer of audio and video data. A signalling protocol known as Session Initiation Protocol (SIP) is used for data communication between nodes in VoIP network. In this paper we survey on security related issues in VoIP communication system and covered several security attacks. We proceed with survey on all the problems that have big effects on VoIP security and challenges such as data manipulation, substitution attack, power backups, jitter, latency and delay. Finally, we conclude the paper with further research on security of VoIP communication system.

Keywords: VoIP, Security, SIP, RTP, Codec

1. Introduction

Voice over Internet Protocol (VoIP) an emerging new technologies is based on real-time communication of data by sending and receiving digital signals using Internet Protocols (IP). VoIP communication system is different than Public Switched Telephone Network (PSTN). PSTN is plain old telephone system based on circuit switching technology whereas in VoIP technology use protocols such as Session Initiation Protocol (SIP) in which text-based encoding format is used for real time data communication. Data communication with security is the demand of the present day world. Four types of communications are possible in VoIP. Firstly, computer to computer, secondly computer to phone, thirdly phone to computer and lastly, phone to phone. Simple application software like Softphone, Skype, Instant messenger need to download and installation in the computer system. Computer user need to use hardware device such as microphone. Skype is the best example for such type of service [1].

1.1. VoIP Compression Algorithms(Codecs)

Voice coders (Codecs) are named as coders and decoders. These Codecs are used in packet switch network for transfer of audio and video data. Modem is an example for codec which modulates and demodulates the audio signals, this means conversion of analog signal into digital signal and vice versa. Codecs are defined by ITU-T for real time voice and video communication. Table 1 below give an example of codecs with various parameters. G.711 is best in quality and widely used codec. But one disadvantage of G.711 is that it consumes more power than other codecs [3].

Table 1. Example of Codecs with various parameters.

Codec Type	Packitization Delay (ms)	Payload (bytes)	Tx Rate (pkt/s)	Bit Rate (kbps)
G.7.11	20	160	50	64
G.729	20	20	50	8
G.723	20	20	34	5.3

1.2. VoIP Gateways

VoIP gateways, known as MEDIA gateways. These devices play the role of mediator between phone networks and VoIP networks. Gateway has contained minimum one phone port and one Ethernet port for each device. Gateways like Analog FXO, ISDN, Cellular Phone etc. are the best examples [4].

1.3. Protocols for VoIP Systems

Protocols are responsible for real time data transfer over the internet, for example use of protocols for multimedia, online television and web camera. Many signalling protocols are applicable to a VoIP solution such as SIP, TRP, RTCP, SIP-T, BICC and CMSS. Protocol selection depends on services required and equipment availability. For SIP phone access, a network with SIP protocol is mandatory [4].

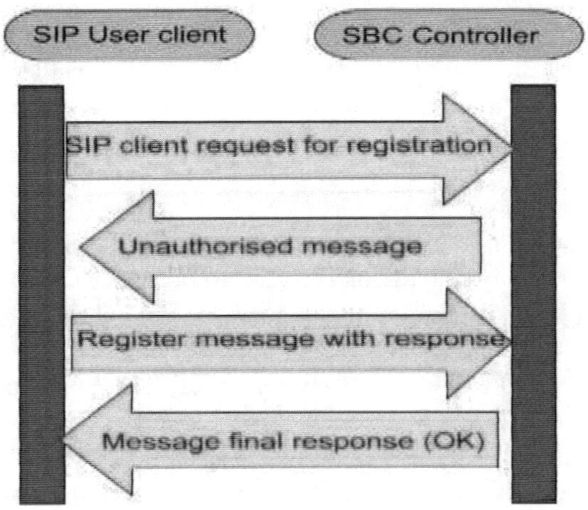

Figure 1. SIP messages registration and exchange.

1.3.1. Session Initiation Protocol (SIP)

SIP is used for making phone calls using Internet. This protocol can establish multimedia sessions, user location, availability, capabilities, session setup and management. For example, in case of communication between SIP user agent and Session Border Controller (SBC). There are two number of SIP phones running on any computer system connected to internet and both can identify each other on the network. The very first message from the source is known as Register message. It is mandatory to receive message by SIP client. Figure 1 as above shows standard registration exchanged SIP message. SIP is used by the service providers such as SIP trunking for PSTN replacement, and IP Multimedia Subsystem. Additionally, SIP can also be used in enterprise communications for Instant

Messaging and Unified Communication. Furthermore, SIP can be used for as transport such as TCP, TLS, UDP and SCTP. Moreover, SIP can be defined as a Web Socket transport [5].

1.3.2. RTP and RTCP

Real Time Protocol (RTP) protocol along with signalling protocol is used to build up connections over IP network. Generally, RTP is made for voice and video transfer. There are two sessions for video transfer in real time with separate SSRC identifiers, one for audio and another for video transfer. One drawback of RTP is that it does not assures the data packets transfer accurately. Real Time Control Protocol (RTCP) is applicable along with real-time protocols for transfer of data over the large network. The Quality of Service (QoS) parameters like jitter, delay and packet loss can be observed by RTCP. RTCP is not responsible for late arrival of packets. The packet structure for RTCP is defined in the RFC 3550. Real Time Control Protocol Extended Reports (RTCP-XR) is an example for improved version of RTCP [6].

1.3.3. H.323 and VoIP gateways Protocols

H 323 is an internet protocols for transfer of audio, video and digital data over the internet. This protocol operates at one step upward from the transport layer of network. VoIP gateways can be controlled and handled by various protocols. Examples of VoIP protocol are such as Asterisk, TDMoE, LTP, Megaco, MGCP, SIP and TGCP [4, 6].

1.3.4. Transmission Control Protocol (TCP)

TCP is a reliable transport protocol which has congestion control. Also, TCP is used to transport web (HTTP) traffic, but not suitable to carry real-time communication traffic such as RTP and retransmission used to implement reliability [7].

1.3.5. Transport Layer Security(TLS)

TLS is an older version security layer between the TCP and application that provides authentication and confidentiality services. In addition, TLS can run over UDP and DTLS which can be used to generate keys for SRTP also known as DTLS-SRTP [7].

1.4. *Advantages*

a) Low Cost - Cost of operation of VOIP system is low to as compare to traditional PBX.
b) One terminal - A VOIP system may be integrated into one PC system for web access to more than one customers. For example, web access and multimedia session can take place at a time in single system.
c) Portability - VoIP can be used anywhere in any corner of the world and important for the users with high mobility [4].
d) Upgradation - Upgradation of VoIP technology is very easy because it is based on software.
e) Bandwidth utilization - Bandwidth is utilized effectively of compression of voice calls because it does not transport 60% silence speech.

1.5. *Disadvantages*

a) High Initial Cost - Because of complexity in installation, moreover, VoIP do not have single standard.
b) Complexity - Voice and data security is a big challenge because VOIP systems and data networks works together, intruders may attack a VOIP system [8].
c) Low QoS - Quality of Service is not enough due to congestion on the network.
d)Power problem: If internet connection is lost, VoIP service is not available.

1.6. *VOIP Softwares*

a) Vox Bridge - For improving the quality of VoIP communication with data encryption and compression features.
b) VoIP client and server software - NAT traversal software for IP communications.
c) Improlabs - VOIP solution provider with cloud based bandwidth optimization service, mobile softphone and instant messaging solutions.
d) NetExpert - NetExpert is a modern network resource planning system with new network technologies such as IP and VoIP. NetExpert, a Network Planner is designed to simplify the network planning, preparing network configurations [9].

1.7. *Paper Organization*

This paper is organized in six sections, introduction part covers introduction to VoIP along with necessary codecs, gateways, protocols, advantages, disadvantages and necessary software required for VoIP system. In section 2 literature survey and section 3 covers security issues in VoIP system. Section 4 and 5 describes the challenges and problems respectively in VoIP communication. Finally, section 6 concludes the paper.

2. Literature Survey

Speech quality estimation models were analysed by considering set of input features and considered distortion types for different network to produce models predicting the quality degradation. The method of detecting impairment factors for equipment was produced and models were tested on VoIP data transfer [10]. A method for shifting a model to 7-kHz network is mentioned and authors noticed the change in parameters. But further study is needed with numerical validation full band voice sample [11]. The most relevant SIP-related threats and security mechanisms is presented to overcome the SIP security issues using asterisk software for multimedia applications [12]. Authors examined some VoIP security related problems in data networks. They proposed that digital password and signature more secure in VoIP communication and protect the system from threats [13]. Speaker identification was tested on VoIP traffic and found 70–75% in accuracy. But calculated data for their findings cannot be used as full identity [14]. Portable devices like mobile phone and laptop can run multimedia applications such as VoIP. These applications require sufficient power for communication between devices on networks running the applications. The battery lifetime is important factor. The authors proposed adaptive algorithm and presented experimental studies on different codecs [2]. Problem solving of strong peer's authentication was presented through VoIP system. Voice based applications and models were used based on SIP/SRTP VoIP protocols. Some more work is pending on this type of investigations and analysis [15]. A E-Model is proposed for voice quality evaluation as per the recommendation by the International Telecommunication Union- Telecommunication(ITU-T). Authors gives a review for various codecs used in voice transmission along with Mean Opinion Score (MOS) [3]. Authors explained 2-types of Denial of Service attacks, one flooding attack and another coordinated attacks. They suggested a model for studying SIP operation and detection of DoS attacks and described algorithms to identify DoS attacks by using the model [16]. Authors proposed several methods for VoIP security and Intrusion Prevention System(IPS) for VoIP with two modules. They found VoIP protected from spoofing attack and SIP flooding attack [17]. QoE in VoIP systems is evaluated as a MOS using an end-of-call "problem token questionnaire" (PTQ) and proven to be useful in practice [18]. An approach based on ELK and PDCA for IoT' security risks was presented and found preserving confidentiality, integrity and availability for business. Authors produced work for utilizing a tool within the IoT followed by repot but need further explore [19]. Performance evaluation method for Abstract Wireless Intrusion Detection and Prevention System (WIDPS) is proposed along with three principles of IT security as confidentiality, integrity, and availability were considered important. But perfect Bayesian Nash equilibrium for signalling games remained the future research [20]. PenQuest model for viewing information system attacks was introduced. The prototype design utilized taxonomies and security standards for cyber-attacks simulation works [21]. The earlier findings on security is analyzed for IoT devices along with tools to test these devices. They introduced a platform to identify weak points in IoT communications and found that IoT devices need analysis for security [22]. The passwords up to twelve characters long is analysed for maximum combinations. They tried to prove that password may be computed and revealed about the password not as a legitimate defence for privacy. [23]. The cybersecurity for health related data is analysed and proposed the actions in preserving the Confidentiality, Integrity, and Availability (CIA) [8].

3. Security issues in VoIP system

Security of VoIP system depends on security of its hardware and software components, means physical security like physical control of system by the people, network infrastructure and servers etc. Software components like any threats which may disturb the smooth functioning of installed software. Security issues as availability, confidentiality and integrity are defined by Central Intelligence Agency (CIA). Due to connectivity of VoIP systems with data network and sharing of software or hardware equipment, intruders gets more chances of attacks to a VOIP system as compared to other voice system such as traditional switching voice system [12]. Some of the security issues associated with VoIP system are discussed below and the types of security issues along with problems and possible solutions are shown in Table 2 below.

Table 2. Security issues in VoIP system.

Type of Security Issue	Problems	Possible solutions
o Vulnerabilities for attackers	Physical connection and gain access to network	User authorization implementation properly.
o Eavesdropping	Privacy loss	Encrypting the data
o Rogue Sets	Threat to confidentiality	Network lock down mechanism
o Spoofing attack	Unauthorised access	Firewall configuration
o Denial of Services (DoS) attacks	False SIP invite messages and memory exhaustion	Packet's source IP address on sending replies
o Flooding-based DoS attacks	Consumption of memory and processing resources.	Designing of mitigation scheme
o VoIP Media DoS attack	Non availability of transmission	Firewall at hardware and software level

3.1. Vulnerabilities for attackers

VoIP network sub systems has possibilities of vulnerabilities for attackers. Some of possibilities are same as in PSTN system. For example, in case of PSTN any physical connection can be made inside the network by the attackers to gain access to calls. Similarity in case of VoIP system attacker can gain access to physical transmission lines by using transmission control protocols (TCP/IP). VoIP system is based on computer hardware and software with operating systems like windows, Linux, Unix, Mac etc. These operating systems remain the target of hackers, ultimately VoIP systems are the part of it. Vulnerability is a big weak point in the system and a threat to a voice network. A proper user authorization implementation is necessary for protection from such type of attacks [12, 24].

3.2. Eavesdropping

Eavesdropping is related to observation and monitoring of communication between two or more devices by the attacker. In VoIP system audio and video data, fax document and credit card passwords may be captured by the hackers for further misuse. PSTN user remains comfortable about the security of conversation. VoIP user also expect the same level of security. Encrypting the data in VoIP communication is important [13].

3.3. Rogue Sets

Rogue sets is related to gaining access to VoIP network resources, intruders can install more applications to existing VoIP networks such as malicious VoIP applications. Rogue sets attacks is a big security threats to confidentiality and integrity due to unauthorized access to VoIP network. The network administrator may initiate the lock down mechanism for allowing the installation of applications only by the authorized person [13].

3.4. Spoofing attack

In such types of attacks attacker try to access to a network by showing that the message is delivered by a trusted system by "spoofing" Internet Protocols address of concern system. This type of fraud creates confusion for nodes inside VoIP system and the attacker tries to capture credentials to claim as a legitimate system user. To prevent such spoofing problem routers must be configured by not allowing the incoming packets with the source address of local domain [5].

3.5. Denial of Services (DoS) attacks

SIP is susceptible to DoS attacks. Intruders by sending false SIP INVITE messages in bulk creates problem in VoIP system management. Computation resources like processor time and bandwidth can be exploited by attackers without acknowledgement of messages. Data flow in VoIP wireless network without any encryption results in sniffing and make data trapping easy. This is known as voice trapping and an attack against confidentiality. As a bogus registration attacker can send harmful INVITE messages to the SIP server and redirect all messages to attacker's node. Such type of attacks causes an availability and confidentiality attack. To avoid such attack administrator may use the packet's source IP address to discard Domain Name System (DNS) resolution overhead on messages replies [24].

3.6. Flooding-based DoS attacks

SIP use INVITE and REGISTER messages and need much processing time. The main reason behind such attack is consumption of memory and processing resources. Such type of attack can also exhaust the link bandwidth of proxy server resulting DoS attack. Designing of mitigation scheme is necessary for protection of VoIP system [16].

3.7. VoIP Media DoS attack

The attackers can flood the VoIP network equipment such as VoIP Phone with large amount of Real Time Protocol packets. This is a method to invite services from remote users to flood the network with more traffic. Such type of excess traffic on network create non availability of sufficient transmission services to the standard users, moreover, the attacker may knock about gateway connection. Electrical power supply and cooling services can be affected by attackers by disturbing VoIP network and services. Such type of attack includes equipment failure, terrorism, accident, etc. Firewall implementation at hardware and software level for detecting false stop message to media devices may be considered to avoid such attack [23, 25].

4. Challenges in VoIP Communication System

VoIP is not secure as compare to the traditional telephone systems. Before Implementation of this technology successfully, it is necessary to understand challenges associated with it [8]. Some of the challenges in VoIP communication system are shown in Table 3 along with problems associated and possible solutions.

4.1. Data Manipulation:

Call records are maintained in the call database know as Call Data Records (CDR). This data base can be manipulated by the attackers. For example, unauthorised access to phone calls, deleting the phone numbers from database, modification of calls etc. Data manipulation is serious matter in VoIP systems and may encourage the criminal activities. Encryption and decryption of communicating data only by authorized parties may be implemented [23].

4.2. *Address Resolution Protocol (ARP)*

By forwarding false ARP data to VoIP system, an attacker may combine his Media Access Control (MAC) number and a different Internet Protocols address in the ARP cache of attacker's node. In this way attacker may gain access to VoIP system. To avoid such attacks, Dynamic ARP Inspection (DAI) may be useful which intercepts every ARP packet [24].

4.3. Substitution attack

In case of Substitution attack the hacker uses to capture conversations between parties and manage the alteration of some portion of speech with meaning. Such type of fraud may be happening in banking or other financial institutions to draw cash by knowing user name and passwords. Biometrics implementation may help in protecting the system from attack [26].

Table 3. Challenges in VoIP Communication System

Types of challenges	Problems	Possible solutions
o Data Manipulation	Deleting/modification of VoIP calls database	Encrypting/decrypting only by authorized party
o Address Resolution Protocol (ARP)	Gaining access to VoIP system	Dynamic ARP Inspection
o Substitution attack	voice conversations capture and alteration	Biometrics implementation
o Overflow in Buffer	Move of data to another buffers	Updating of OS time to time
o (DHCP) attack	Alteration of IP address data by sending DHCP request	IEEE 802.1x specification provides the mechanism
o System Software Attack	Installing of different OS	Control gateways with latest patches of OS
o Backup Equipment and Power	Power failure and no backup	Parallel system planning

4.4. Overflow in Buffer

In this type of attack system buffer is manipulated to hold more data than its actual size, so that data can move to the another buffers. Overflow in buffer attack may be used by the attackers for phone devices and nodes in a VoIP network. In such cases to avoid attack, operating systems like Windows, Linux must be updated time to time [27].

4.5. Dynamic Host Configuration Protocol (DHCP) attack
The attacker send DHCP request for MAC addresses and alter the data of IP address in the system. This may prevent the nodes for getting IP address and bound the server to obtain IP address from another VoIP server. The IEEE 802.1x specification provides a mechanism to prevent such attack [27, 28].

4.6. System Software Attack
System Software Attack is attack on operating system software like windows, Linux, Mac. etc. Any software for softphone may be installed on the operating system of any VoIP system. This may give chance to attacker to access important call data from the VoIP server. To avoid such attack control gateways should be implemented and latest patches of operating system must be updated [5].

4.7. Backup Equipment and Power
Backup power requirement is estimated based on the number of VoIP phones attached with the equipment. Emergency power requirement is estimated based on the fact that for how much long time the equipment needs to remain in working condition. VoIP network can be planned with parallel systems so that in case of failure of one equipment other equipment should start working and network traffic is not affected. Backup phones may be considered as safety measure in case of failure of IP phone for making emergency call [25].

5. Problems in VoIP communication system
VoIP communication can experience problem of poor call quality of voice transmission. Some of the major problems in VoIP communication system are mentioned below.

5.1. Jitter and Latency
Jitter is related to arrival data packets at destination not in its original sequence. In packet switched networks data packets travels in different paths from source to the destination and vice versa. Assembling and disassembling of data packets is a continues process in VoIP system during communication between two or more devices. This process degrades the voice and video quality due to data packets lost in between communication path. Jitter affect the voice and video quality of VoIP system. Latency is the time taken by a data packet to reach from its source to its destination. Latency in VoIP system is related to the time taken by voice from speaker's mouth to listener's year [29].

5.2. Delay
Three types of delays are possible in VoIP communication, the propagation Delay, handling delay and queuing delay. Time taken by light to pass through vacuum medium is 1,86,000 mps (mile/sec) and by electrons via metal like copper is about 125,000 miles per second. Fiber network induces a delay about 70 ms (milliseconds). This type of delay known as propagation delay causes in VoIP system speech quality degradation. Secondly, the handling delay, is a big issue in VoIP system as it is concerned with forwarding of frames through packet switched network. Lastly, the queuing delay where packets in data network arrived at destination in que and because of congestion, packets may not arrive at given time. Queuing delay also degrade the Voice and video quality in VoIP system [29].

5.3. Internet Connectivity
Generally, network is configured for data transfer, if network is not properly configured for audio and video transfer the audio and video quality may degrade. Internet Service Provider (ISP) design their networking based on web surfing, they need extra protocols for video handling. Low internet connectivity. inadequate router and internal network not configured properly are the problem need to be solved for improvement of voice quality. VoIP system needs specialised routers and their proper configuration for audio and video transfer [19].

6. Conclusion

Many organizations are adopting VoIP network as compared to traditional telephone network based on switching circuit. The reason for the choice is the VoIP implementing cost and easy to install. But security of the system is big challenge to secure confidential data from attackers. We surveyed on VoIP security and found that the security issues like as eavesdropping, rogue sets and spoofing attack etc. are big challenges against the confidentiality and integrity of data on VoIP network. We presented securities issues in VoIP system along with problems and possible solutions in the form of table. We continue with survey on possible problems and challenges that have big effect on VoIP security such as data manipulation, substitution attack, power backups, jitter, latency and delay. Moreover, we presented types of challenges along with problems and possible solutions in the form of table. It is concluded that lots of efforts are made by researcher since the inception of Voice over Internet Protocols technology to improve its voice and video quality but problems still remain. Further, researchers need to work on the security issues and challenges to provide solutions against the security attacks which are still active in present days VoIP communication systems.

References

[1] Gupta V and Dharmaraja S 2013 Reliability and performance modelling of VoIP system with multiple component failures *International Journal of Reliability and Safety* **7(1)** 58-74

[2] Naeem M, Namboodiri V and Pendse R 2010 Energy implication of various VoIP codecs in portable devices *IEEE Local Computer Network Conference* 196-199

[3] Chhabra A and Singh D 2015 Assessment of VoIP E-model over 802.11 wireless Mesh network *IEEE International Conference on Advances in Computer Engineering and Applications* 856-860

[4] Raj R, Gagneja A and Singal R 2012 Voice over Internet Protocol–The Technology and Its Application *International Journal of Scientific & Engineering Research* **3(5)** 1-6

[5] Al Saidat, M R S, Al Kasasbeh B and Al Adwan F 2015 Develop a secure SIP registration mechanism to avoid VoIP threats *IEEE International Conference on Open Source Software Computing* 1-6

[6] Jalendry S and Verma S 2015 A detail review on voice over internet protocol (voip) *Int. J. Eng. Trends Technol* **23(4)** 161-166

[7] Lam T, Galkandage C, Dagiuklas T and Iqbal M 2020 Much needed Group-based Quality of Experience on Unified Communications to support post COVID-19 Disruptive Business Models *State-of the-Art and Research Challenges* 1-17

[8] Okereafor K and Marcelo A 2020 Addressing Cybersecurity Challenges of Health Data in The Covid-19 Pandemic *International journal in IT & Engineering* **8(6)** 1-12

[9] Castanon G, Sarmiento, A M, Ramirez-Velarde R and Aragón-Zavala A 2009 Software tool for network reliability and availability analysis *Wire Journal International* **74** 1-12

[10] Raja A, Azad R M A, Flanagan C and Ryan C 2008 VoIP speech quality estimation in a mixed context with genetic programming *Proceedings of the 10th annual conference on Genetic and evolutionary computation* (ACM) 1627-1634

[11] Myakotnykh E S and Thompson R A 2009 Computational quality model for wideband voice-over-IP communications *IEEE International Workshop Technical Committee on Communications Quality and Reliability* 1-6

[12] Werapun W, Abou El Kalam A, Paillassa B and Fasson J 2009 Solution analysis for SIP security threats *IEEE International Conference on Multimedia Computing and Systems* 174-180

[13] Maher A, Hegazy A F, Hasan B and Hashim M 2010 Beyond steganography for Voice Over IP (VOIP) *IEEE International Conference on Computer Engineering & ystems* 46-51

[14] Khan L A, Baig M S and Youssef A M 2010 Speaker recognition from encrypted VoIP communications *Digital Investigation* **7(1-2)** 65-73

[15] Cattaneo G, Catuogno L, Petagna F and Roscigno G 2015 Reliable voice-based transactions over VoIP communications *IEEE 9th International Conference on Innovative Mobile and Internet Services in Ubiquitous Computing* 101-108
[16] Golait D and Hubballi N 2016 Detecting anomalous behavior in VoIP systems: A discreteevent system modeling *IEEE Transactions on Information Forensics and Security* **12(3)** 730-745
[17] Chen M J, Wen C C, Lin H C and Chu Y S 2016 ASIC design and implementation for VoIP intrusion prevention system *IEEE International Conference on Applied System Innovation* 1-4
[18] Gupchup J, Hosseinkashi Y, Ellis M Johnson S and Cutler R 2017 Analysis of problem tokens to rank factors impacting quality in VoIP applications *IEEE Ninth International Conference on Quality of Multimedia Experience* 1-6
[19] Abbass W, Bakraouy Z, Baina A and Bella M 2019 Assessing the Internet of Things Security Risks *Journal of Communications* **14(10)** 958-964
[20] Mahini H and Mousavirad S M 2020 WiFi Intrusion Detection and Prevention Systems Analyzing A Game Theoretical Perspective *International Journal of Wireless Information Networks* **27(1)** 77-88
[21] Luh R, Temper M, Tjoa S, Schrittwieser S and Janicke H 2020 PenQuest: a gamified attacker/defender meta model for cyber security assessment and education *Journal of Computer Virology and Hacking Techniques* **16(1)** 19-61
[22] Waraga O A, Bettayeb M, Nasir Q and Talib M A 2020 Design and implementation of automated IoT security testbed *Computers and Security* **88** 101648
[23] Almuairfi S J and Alenezi M 2020 Anonymous Authentication for Privacy Preserving of Multimedia Data in the Cloud *Handbook of Research on Multimedia Cyber Security* (IGI Global) 48-72
[24] Rippon W J 2006 Threat assessment of IP based voice systems *1st IEEE Workshop on VoIP Management and Security* 19-28
[25] Chong H M and Matthews H S 2004 Comparative analysis of traditional telephone and Voice-over-Internet Protocol (VoIP) systems *IEEE International Symposium on Electronics and the Environment* 106-111
[26] Liu Z H and Wang H X 2014 Pseudo-zernike moments-based audio content authentication algorithm robust against feature-analysed substitution attack *Multimedia tools and applications* **70(3)** 2271-2291
[27] Butcher D, Li X and Guo J 2007 Security challenge and defense in VoIP infrastructures *IEEE Transactions on Systems, Man, and Cybernetics, Part C (Applications and Reviews)* **37(6)** 1152-1162
[28] Hubballi N and Tripathi N 2017 A closer look into DHCP starvation attack in wireless networks *Computers and Security* **65** 387-404
[29] Meddahi A and Afifi H 2006 Packet-E-Model: E-Model for VoIP quality evaluation *Computer Networks* **50(15)** 2659-2675

Significance of Renewable Energy for Empowering Cooperative-Farming Sector in Majuli Island of India

Athira R[a], A. Parida[b], S. Choudhury[c]

[a,b,c] North Eastern Regional Institution of Science and Technology, Nirjuli, Arunachal Pradesh, India

Email: athuammu@hotmail.com

Abstract: Agriculture is the backbone of the Indian economy and plays a vital role in the overall development of the country. Many studies and practices have proved the importance and benefits of electricity in agriculture throughout the world. However, the fossil fuel-based conventional energy supply to agriculture is always associated with environmental issues. In an attempt to study the significance of renewable energy in the cooperative farming sector particularly in the biodiversity regions, Majuli Island of Assam, India has been selected for the case study. Even though the major income source in the said locality is agricultural farming, they are deprived of getting the optimum benefit from this indigenous income source due to lack of technological intervention. Technological intervention in this sector in the biodiversity region is not imaginable without exploring the feasible intervention of renewable energy at an affordable energy cost. Therefore, in this paper, the significance of renewable energy in agriculture considering the environmental issues has been highlighted in general and in cooperative farming in particular.

Key Words: Renewable Energy, Cooperative Farming, Environmental Issues, Sustainable Livelihood.

1. Introduction

The World's biggest waterway island, Majuli, situated in the Brahmaputra stream of Assam state, is recently declared as first island district of India [1-3]. It covers a zone of 352 sq. km and a populace of around 1, 67, 304 (approx.) [2], [4]. The beautiful excellence and uniqueness of the Island make it a vacationer location for people groups, all around the globe [4]. Both abroad and homegrown sightseers visit here on a regular stretch and along these lines makes a road for the job of the Island [4-5]. Aside from Tourism, small scale cottage industries are also providing employment opportunities in this area[2]. From the primer examination by the creators, it has been discovered that the essential type of revenue here is principally from horticulture, around 75% of the all out populace totally relies on farming for their vocation. [1-2], [4].

1.1 Cooperative-Farming

"Cooperative" in different sectors is coexisting from the last few decades, farming, is one among them. However, cooperative-farming is a novel concept conceptualized appropriately in various studies [4]. Cooperative farming is a type of farming system where each farmer remains as the proprietor of their land but the farming activities are carried out collectively by a group of farmers through cooperative activities and technological intervention [4], [6], [6]. The profit and loss from these farms are distributed among the farmers based on the ratio of the land they have own [4]. With the help of certain agencies formed by the farmers and for the farmers, individual farmers performed their farming activities jointly in their holding [4], [6-8]. These agencies also help in purchasing and sending of farm inputs and outputs. The basic idea behind agricultural cooperative is to pool the resources for getting the benefit of 'Teamwork' and thereby reducing risk rate. Cooperative farming in different forms has already tried and still practicing in different countries such as the UK, Germany, France, etc. for the last few decades [8-11].

1.2 Cooperative Farming and Renewable Energy

Many studies have proved that the highest efficiency in farming can be achieved only by the means of electrical energy and hence technological intervention. Technological intervention in the area of agriculture will increase the competence of certain farming activities. Usually, farming activities are carried out by manual labor and with conventional tools such as sickle, ploughing, etc. but renewable energy-based mechanized farming will reduce the manual work and automats the farming activities thereby saving time and energy. Renewable energy can play a major role in agriculture and particularly in cooperative farming for sustainable growth. Moreover, the intervention of renewable energy in farming will not only help in rural development missions but also creates a sustainable environment for the entire world in terms of reduced carbon-cost [12]. Based on the case-study conducted at the *"Borgayan"* village of Majuli district, the following points are the contributions of this manuscript.

1. The significance of the renewable sources of energy in the corporative farming sector for the said location and its feasibility study has been discussed systematically.
2. The analysis of the existing agricultural practices and socio-economic scenario of the proposed case-study location has been detailed
3. The benefits of cooperative farming, the intervention of technology, and the introduction of renewable energy in the cooperative farming structure have been discussed.

2. Literature Review

Some of the similar reviews done by various authors in their respective journals are: Bardi, Toufic, and Lavacchi, (2013), in their study "Turning electricity into food: the role of renewable energy in the future of agriculture, says that encouraging the use of electric power in the farm is a worthy effort, even though it is not possible to replace renewable energy with fossil fuel simply, leaving everything unchanged, but it is possible only utilizing more efficient use of energy and mineral resources, as well as an approach in a more respectable way to preserve soil and local resources. R Athira, and et al(2019), in their research, "Proposed Cooperative Farming as a Catalyst to Livelihood Augmentation for the Marginal Farmers of Majuli Island, said that the cooperative system will not only increase the standard of living but also increases the feeling of friendship, cooperation and thereby achieves synergy in the agricultural sector. Keely AR(2016) in his study, "Renewable Energy in Pacific Small Island Developing States: the role of International Aid and the Enabling Environment from Donor's Perspective, studies that, High targets supported by well-structured action plans and effective regulator body is responsible for RE and also financial aspects of utilities hold a higher importance than technological aspects of utilities. Dhunny, Allam, Lobin, and Lollchund (2019) in their research," Sustainable renewable energy planning and wind farming optimization from a biodiversity perspective, proves that, in an economic analysis performed for understanding energy generation of scenarios in respect of the selected site, reveals a positive result inefficiency of the model and its potential applicability to other sites, who are looking for sustainable energy planning. Elaine Forde, (2017), in his study, "The Ethics of energy provisioning: Living off-grid in rural Wales", argues that living off-grid comprises of materially engaged ethics of usage of energy, emerges from and situated in everyday realities of off-grid energy provisioning. R, Borlase Matthews (1922), in his topic," Electro Framing or the Application of electricity to agriculture" in his experimental study suggested that the highest efficiency in farming can be only achieved through electricity. Chetan Dwarkani, (2015) in their study, "Smart Farming System Using Sensor for Agricultural Task Automation", developed a user-friendly smart farming system and suggest that it liberates agricultural productivity and mode of production can be changed. Fabio Riva and Emanuela Colombo (2020). In their research, scientific reports only show controversial results regarding the impact of electrification programs and most of the time provide attributable to inappropriate planning of off-grid capacities of the system. Bikash Das (2017) in his study on, "An overview of the economic condition of Majuli, the largest inhabited river island of the world", says that most people of River Island depend upon agriculture, but, the traditional agricultural system o Majuli results in low productivity." V.J.P.D. Martinho (2020), in his study," Relationship between agriculture energy and farming indicators", says that reviews concerning agriculture energy and pollution highlight that there is a relevant relationship between energy consumption, economic development, and environmental impacts. Kurunathilake H (2016), in this research," Renewable energy integration into community energy system: a case study of new urban residential development", reveals that the Renewable energy system depends upon local situations energy availability, etc.

3. Research Methodology Area

3.1. Case Study Area

The research was carried out in a village named Borgayan Village, situated in Majuli District, Assam. The literacy rate of the village, compared to Assam, is very low and no modernization has happened yet. People groups here totally relies on horticulture for their occupation just as for their every day needs. Rice is the principle cultivation here. Aside from this, many are additionally into kitchen cultivating. [4].

3.2. Data Collection

The research was done using both primary and secondary data. Primary data was collected using survey method, a field visit was done and directly interacted with the farmers at work, thereby first-hand data were used for the study. Other primary sources includes interview method, random sampling, personal observation method, etc. The Secondary data collection was made based on different sources such as census, various related journals, other publications, articles, e-journals, books, recent reports, etc.

An isolated island, no connected examination has been done in Majuli, till this date. Accordingly, the "Theory Building" idea has been utilized for proposing the model, of this paper. Theory building, is a broadly utilized idea in research territories, as it gives an examination structure to the investigation, additionally a proficient field improvement should be possible utilizing this idea and also it helps in further applicability for solving the problems and issues related to the practical real world. Theory, however a dreary artistic expression, practical, without theory, can be hazardous and dull. In explanatory reasonable examination, a subcategory of "theory building" idea, adds new bits of knowledge into conventional issues through logical relationship building. So in this paper, we utilized the "theory Building" concept, along with primary and secondary data, for the study [13].

4. The Present Agricultural Scenario of Majuli Island

Majuli Island is situated in the Brahmaputra River of Assam. It was one of those three subdivisions of Jorhat District, but recently in September 2016, the river island Majuli is declared as one of the districts of Assam [1]. In spite of the fact that a confined island, the picturesque excellence improved with greenery makes it a traveler center point for people groups from various pieces of the world. Farming is the essential kind of revenue here, around 75% of the absolute populace relies on horticulture for their occupation. The climatic condition and soil of Majuli are very appropriate for development. Primary cultivating here is rice, mustard, vegetables, and so on from the complete region of Majuli around 41.43 percent of the zone is utilized for development. Ranchers here are generally negligible ranchers, having little landholdings of 0.95 hectares [4]. Modernly, Majuli is a little in backward spot in the entire of Assam. Absence of modernization, innovative headway, absence of legitimate transportation facility, geological disconnection, and so forth can be the explanation behind this. Despite the fact that, couple of cottage and small scale industries are existing in Majuli, for example, ceramics industry, boat making industry, handloom, stick bamboo, and so on [2]. Aside from this sericulture is additionally a significant agro-based industry of Majuli. Majuli is additionally famous for its fishing area, around 10%-20% of clans here, fundamentally ST and SC, are occupied with fishing [4].

4.1. Current Scenario of Borgayan Village

The town we chose for the research was Borgayan. As per statistics 2011, the all out populace of the town is 1320. Rice and mustard are the fundamental development here in the town, around 90% of the populace are into horticulture. Three primary sorts of rice crops are developed here. The yield development graph is given in Table 1. Ahu-Dhan and Mustard are the fundamental harvests here. This town is additionally well known for the black gram and curd [2]. Peoples here additionally insignificantly do kitchen cultivating, just to meet their every day needs. Apart from this, the other sources of revenue are from livestock, fish ponds, and so forth some are also workings labors at nearby small-scale industries.

Table-1: Crop Cultivation of Majuli

Crops	*Harvesting*
Ahu-Dhan (rice)	*Jan-July*
Baw (rice)	*April-Dec*
Mustard	*Sep-March*
Chana Dal	*Nov-May*

It is found that farming here is done using the traditional system of agriculture. Due to a lack of modernization and technology availability, peoples here get a minimum profit. For irrigation peoples here depends upon "rain". Only 5% of farmers use pumps for irrigating their land and that too a diesel one. Based on land availability, some also do multi-cropping. Mustard is the only crop, used for trade purposes. Harvested mustards are collected by brokers directly from the farm and sell it to the mill, thereby brokers are more benefited and farmers get less benefit. For rice, farmers here, after harvesting, store the grains in a "bharl", a small room made up of bamboo and dried grains. Based on the need, these grains are taken to the mill and processed into rice, the rice husk is used as cattle feed and rice for their daily needs. Here, a conventional method of farming system is used. As every farmer, having land, does their farming on that particular lad itself. Here, the profit and loss is beard by the individual farmer himself. As a being an isolated island, Majuli is facing issues like soil erosion, flood, etc, when such calamities affect farming the farmers have to face the entire loss, and sometimes the loss can be too high for a marginal farmer to bear [1]. The other issue of this system is that the small farmlands divided with boundaries create a problem for proper irrigation as well as for proper drainage related issue. Since there was no other option available, farmers were facing the entire struggles.

5. Renewable Energy and Cooperative Farming Correlation

5.1. "Cooperative" in detail

Many studies have proved that "cooperative farming" or "collective farming" can resolve many issues of individual farming [14-15]. The basic concept of these farming is to pool together the land and resources. In this system, each farmer will remain as the owner of their respective land but farming will be carried out jointly. Here the profit and loss incurred from agriculture will be divided based on the ratio of land, the farmer holds [6]. Based on the number of days/ hour, they work, wages can be decided and distributed accordingly. In this system, the land earlier used for dividing boundaries can be used for cultivation, hence, wastage can be reduced. As working together, the feeling of brotherhood and spirit of synergy can be achieved [4], [6], [8]. It is found that in many developed countries, a large proportion of farmers rely on a cooperative system of agriculture to govern both forward and backward market transactions [6], [8]. Many producers are joining the force in agriculture cooperative by becoming its member and using the services, the collective organization provides. Existing cooperative farming can be divided into two: a) agriculture production cooperative, which helps in land pooling and using resources jointly b) agricultural service cooperative, which provides basic services required for farming such as warehousing, storing, etc. mainly the agricultural inputs are purchased from the market, at a reasonable price, but when the market fails to do so, cooperatives form [4].

5.2. Significance of Renewable Energy Based Electricity

In view of numerous ongoing examinations it is discovered that most extreme productivity in agriculture can be arrived at simply by the utilization of the electrical technique [16]. The utilization of power in cultivating is numerous, for example, in the poultry area, electric light, and etc, the details are shown in Table-2.

Table 2: Details of usage of Electricity in Farm

Use in Farm Building	• *Electric light for dairy, yards, milking, chaff, food preparation, dust extractor* • *Dairy- milk separation, cooler* • *Poultry- increasing egg production, egg tester, etc.*
Use in Farming Land	• *Manure distribution by electric machines* • *Plowing* • *Irrigation purpose* • *Electric engines for farmland works*
Treatment of Crops	• *Preparation of seed- bathing of seed, seed drying, etc.* • *Gathering groups- hay drying, pulp extraction from fruits, etc.*

Apart from this, a lot of works that ought to be substantial by electric power, is currently being done by manual labor [17]. However, these farmers are not aware that electricity can assist them to a great extent until someone shows them the possibilities. Currently, they are forced to work with all their effort with disregard of time. Though electric light is avoided or considered to be less important, bringing these electric light will allow the safety and convenience of their work apart from agriculture, livestock farming can also be done using electricity [16]. Sufficient milk from cooperative farming can be stored and used for other purposes in the food processing section can be done easily done using cooler facilities through electro farming [16],[18]. An experiment done on a foreign farm shows that certain electric light treatment can be used for egg-laying and earlier maturity of livestock [18]. In India, around 70% of the population is into agriculture, contribution to GDP is 8%. The main reason behind this deprived performance is the lack of automation in the agriculture sector. Because of the absence of appropriate advancement, farmers are not able to update themselves. They are not even able to get to know about the benefits and programs from the government. With the help of electricity, many issues can be solved. Digitalization is also an important element along with electricity [17], [19-20]. The sector agriculture plays an important role in reducing poverty, providing food, earning foreign exchange, etc. but the main input needed to sustain the agro sector and its growth is irrigation which requires access to electricity at farms [20]. Studies have proved that electrification to the farm can increase yields and consequently income from the farm through adopting electric pumps for irrigation and also other needy applications [17], [20]. This is a labor-saving technology for farmers and also reduces the expenditure on hired labor and time consumption. Adoption of electricity can also benefit farmers from the issue of getting access to information related to getting the best practice of farming through an internet connection [4], [21].

5.3. Adopting Renewable source of Energy

In many different parts of the world, electricity is being used in the agriculture sector and are getting maximum benefit too [22]. But it is found that modern technology-based agro farming is depending upon energy supply obtained from fossil fuels or we can say the conventional source of electricity. Fossil hydrocarbon is a limited source to rely on completely and also its harmful effect on the environment creating an urgent need for reducing their use [19], [23]. By this agriculture, which meant to be the highest livelihood provider, ultimately resulting in one of the major emitters of greenhouse gas [24]. This question is about the sustainability of the environment. The continued use of fossil fuel may lead to a slowdown of production in near future causing serious issues to the agricultural sector [19], [25-26]. Many researchers on their different studies have proved that a renewable source of energy is the best to rely on to avoid a negative impact on the environment created by the conventional source of energy supply. So, it is important to resolve this issue, so that we can aim towards the conservation of precious resources such as the fertility of soil and water [26-27]. Renewable energy provides a wide variety of energy production in different forms, at different cost with different efficiencies, which reduces the dependency on the conventional source of energy and ensures long term energy security for the country as well [9]. RE plays a vital role in achieving primary and secondary energy supply goals such as improved diversity, energy supply security, reducing local pollutants, etc. studies have observed that there are ample amount of opportunities available with favorable geology and geographical conditions with a huge customer base and widening gap between demand and supply. Sustainable renewable energy can be of vital link for the development of the nation [28-31].

5.4. Feasibility Study on Solar PV based Renewable Energy in the Case-Study Site

The Majuli biodiversity region is situated in the global positioning of 27.0016° N, 94.2243° E. The average solar radiation *($kWh/m^2/day$)* data has been collected for the year 2019 with a sample size of 7 samples/day basis using the Solar Radiation (Silicon Pyrometer) Smart Sensor - S-LIB-M003. The collected data has been shown in Fig.1.

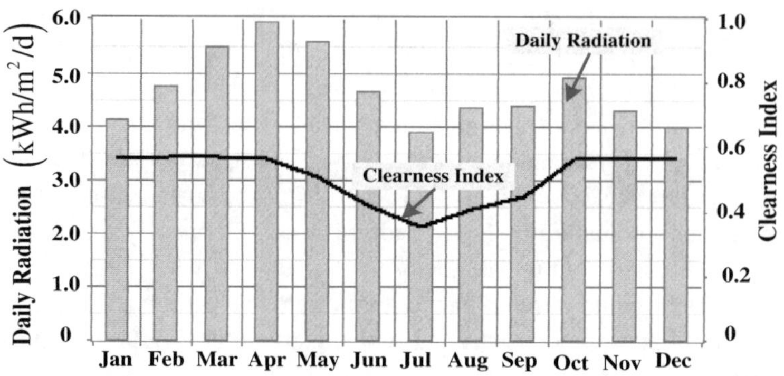

Figure-1: Average solar radiation in a year at 27.0016^{0}N, 94.2243^{0}E

It has been observed that the average solar radiation at the test site is well above the standard 0.2 (kWh/m^2/day). This is the minimum radiation level above which it is suitable for solar PV power generation site economically [32-35]. Therefore, the site is suitable for solar PV based on renewable power generation technologies. Based on a proposed 5kW solar power plant, the cost analysis has been performed in an attempt to compute the cost of energy as shown in Table-3. The estimated cost is found to be \$0.1756/- which is significantly less than the present cost of energy from conventional sources at the same location.

Table-3: Cost of energy with proposed 5kW solar PV power plant

Cost Factors	*Value*
Initial Capital Investment (\$)	*6600/-*
Cost of Energy (\$/kWh)	*0.1756/-*
Net Present Cost (\$)	*1220/-*
Operating Cost (\$/year)	*675/-*

.

6. Proposed Electro Based Cooperative Farming Model

Based on certain studies we designed a model, a model based on electricity and technology. The main idea of this proposed model is to lessen human labor and getting a clearer idea of responsibilities and thereby obtaining a productive environment through the activities of agriculture using a cooperative farming system based on electricity and technological advancement for the betterment of livelihood.

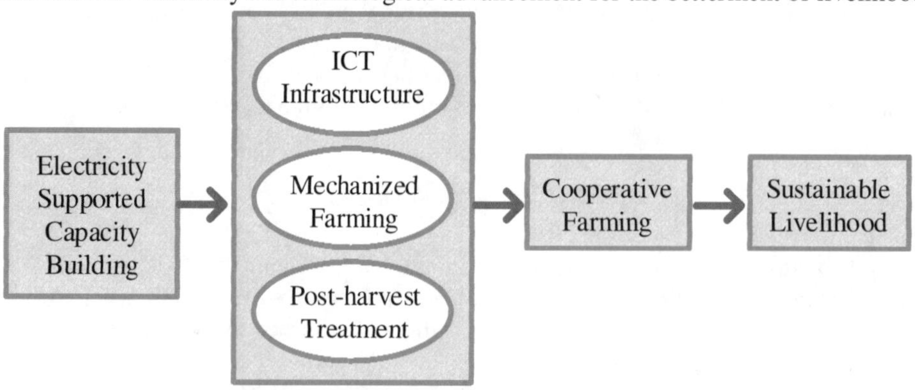

Figure-2: Proposed electro based cooperative farming model

- ***Electricity influenced Capacity Building***

Now, we all know that modernization is the basic concept behind this model. Smartly achieving maximum benefit is the concept here. To reduce those manpower labor work and optimal utilization of time can only be achieved using electricity. As it plays a very important role in all of this. Each part of the model, need electricity. We cannot deny the fact that nothing in this cooperate model can be performed without the use of electricity. We need electricity in farm machinery, food processing, use of information and technology, to get an update about all the marketing and financial stuff. The main purpose of this electricity influenced capacity building is to support the other elements of the cooperative section, through electricity, to make the working more productive in less time and also in a smart way.

a) ICT Infrastructure

This section will carry out activities related to technology and information. New farming techniques, ideas, updated information regarding the outside world, the subsidy is given by the government, and other initiatives by the government can only be acquired using this unit. Since Majuli is an isolated island, the village in which the study was carried out, found that peoples there are completely unaware of the subsidy and benefits providing by the government. The main motive behind the introduction of this unit is to build a platform for communication with the outer world using information communication technology and being updated with the technological world

b) Mechanized Farming

This unit will be dealing with the mechanical part. The usage of various farm machinery such as tractors, irrigators, other farm machines used for plowing, harvesting, etc. can be done here. This area will give training to peoples and will be carrying out the mechanical parts. Not only in the farming area but also food processing section. Plants, machinery, etc. will be dealt with by this unit. Under this unit, an idea will be given regarding the farming pattern, current technology, farming culture related to climatic condition, soil pattern, etc. this stuff can be very familiar to peoples who are into farming there. But the newcomers or new generations who are new to this field will not have many ideas about the land and soil. The main purpose of this unit is to give needy ideas and knowledge to upcoming people about the fertilizer, its amount to be used, its impact on soil, making them aware how to deal with manure, identification of soil fertility, cropping pattern, smart irrigation of land using modern technologies, etc.

c) Post-Harvest Treatment

The agricultural output, need to be processed in mills and products like grains which are usually taken to mill to separate the husk, can be done here in a cooperative unit. Apart from this, the excess products that can be processed into new brands for trade purposes can also be done here. Hence, all the activities related to food, food processing, agricultural output, etc. can be carried out in this "post-harvest treatment" unit. The processed food which needs to be sent for trade purpose can be done by this section. Also, the decisions regarding linkage with the outside market such as purchasing agricultural inputs like seed, fertilizers, etc. can be made by this unit. The overall marketing, advertising, selling related kinds of stuff can be performed by this group. Therefore, these three renewable energy-based sections will ultimately contribute to cooperative farming. Where peoples will be carrying out their farming activities in a cooperative way as discussed earlier in this article, and finally contributing to sustainable livelihood development. As studies have declared that sustainable development is a paramount issue that requires urgent action and changes from the side of government and society as a whole [36]. In this model, we focused on renewable sources of energy which will switch from the demand need for fossil fuel in a farming area and thereby enjoying a sustainable cooperative development of the area.

7. Conclusion

Hence, the basic need for electricity and its importance in the agricultural sector is discussed. As many actives, of machine nature, are currently being done manually. This leads to high cost, time-consuming and resulting in low productivity, therefore, the paper focus on achieving a cooperative sustainable farming system by replacing green energy sources with conventional one. An electro based cooperative model was developed and proposed in the paper, the technical based cooperative model, if implement properly can result in various benefits such as providing a platform for communication with the outer world, systematic and structured working pattern where each one gets a clear idea about the work, increased production, economic development of farmers, market linkage, optimal utilization of land, organized source of employment, bulk surplus, being a cooperative, high valued machinery can be purchased, etc. the benefits are enormous. With this cooperative model, since people working jointly will bring the synergy and love of brotherhood can also be achieved. If implemented properly can also provide employment opportunities within the village as well.

Acknowledgments

The authors are thankful to the funding agency: National Mission on Himalayan Studies (NMHS) for their support through project no.GBPNI/NMHS-2017-18/SG15. The authors would also like to acknowledge Technology Intervention for Tribal Empowerment (TITE) scheme of Department of Science and Technology, Government of India (Project no: SEED/TITE/001/2017(C).

References

[1] Goswami, R., et al., (2013) Arsenic in the groundwater of Majuli- The largest river Island of the Brahmaputra: Magnitude of occurrence and human exposure, *Journal of Hydrology*, vol. 518, pp. 354-362. http://dx.doi.org/10.1016/j.jhydrol.2013.09.022

[2] Bikash Dash, (207) An overview of the economic condition of Majuli, the largest inhabitant river Island of the world, *International Journal of Interdisciplinary and Multidisciplinary Studies (IJIMS)*, vol 4, pp. 91-96.

[3] Kmar, M Goswami, R., Awasthi, N., Das, R., (2019) Provenance and fate of trace and rare earth elements in the sediment-aquifers system of Majuli River Island, India, *Chemosphere,* vol. 237. doi:http://doi.org/10.1016/j.chemosphere.2019.124477

[4] Athira R, A Parida, S Choudhury, (2019) Proposed Cooperative Farming as a catalyst to livelihood augmentation for Marginal farmers in Majuli Island, *2nd International Conference*, ISMS 2019.

[5] A. Borborah A., M.G. Borborah, S. Choudhury (2016) Prospects of Ecotourism in Majuli. In: Purkayastha J. (eds) Bio prospecting of Indigenous Bio resources of North-East India. *Springer, Singapore*. https://doi.org/10.1007/978-981-10-0620-3_20

[6] Vladimir Milovanovic, Lubos Smukta, (2018) Cooperative rice farming within rural Bangladesh, *Journal of Cooperative Organization and Management*, vol. 6, pp. 11-19. https://doi.org/10.1016/j.jcom.2018.03.002

[7] Gong T (Charles), Battese GE, VIllano RA, (2019) Family Farms Plus Cooperatives in China:

Technical Efficiency in Crop Production, *Journal of Asian Economics*, vol. 64, pp. 101129. http://doi.org/10.1016/j.asieco.2019.07.002

[8] Hazoor HM, Safdar HT, Salman A, Saad BN, (2012) Future of Cooperative Farming in Pakistan, *Journal of Biology, Agriculture and Healthcare*, ISSN 2224-3208, vol 2, No.6.

[9] Huang, Z,.-h., et al., (2015) Situational features and governance structure of farmer cooperatives in China: Does initial situation matter?", *The Social Science Journal*, vol. 53, pp. 100-110. http://dx.doi.org/10.1016/j.soscij.2015.09.003

[10] Suh Jungho, (2015) Communitarian Cooperative organic rice farming in Hongdong District, South Korea", *Journal of Rural Studies*, vol. 37, pp. 29-37. https://doi.org/10.1016/j.jrurstud.2014.11.009

[11] Abate T.G., (2018) Drivers of agricultural cooperative formation and farmers' membership and patronage decisions in Ethiopia", *Journal of Cooperative Organization and Management*, vol. 6, pp. 53-63. http://doi.org/10.1016/j.jcom.2018.06.002

[12] Ajates Gonzalez, R., (2017) Going back to go forwards? From multi stakeholders cooperative to open cooperatives in food and farming", *Journal of Rural Studies*, vol. 53, pp. 278-290. http://dx.doi.org/10.1016/j.jrurstud.2017.02.018

[13] Wacker G John, (1998) A definition of theory: research guidelines for different theory-building research methods in operations management, *Journal of Operations Management*, vol. 16, pp. 361-385.

[14] Adefila James, Madaki Joshua, (2014) Roles of Farmers cooperative in Agriculture development in Sabuwa Local government area of Katsina State Nigeria, *Journal of Economics and Sustainable Development*, ISSN 2222-1700, vol.5, No. 12.

[15] Shen Mingrui, Shen Jianfa, (2018) Evaluating the cooperative and family farm programs in China: A rural governance perspective, *Land Use Policy*, vol. 79, pp. 240-250. https://doi.org/10.1016/j.landusepol.2018.08.006

[16] Matthews B. R., (1922) Electro Farming, or the application of electricity to agriculture, *I.E.E. Journal*,Vol. 60, No. 311.

[17] Chetan DM, Ganesh RR, Jagannnathan S, R Priyatharshini, (2015) Smart Farming System using sensor for agriculture task automation, *IEEE International conference* (TIAR 2015)

[18] Alice J. Norman and R. C. Bowes, (1984) Electricity in the farm dairy: requirements, use and conservation", *Journal of Society of Dairy Technology*, Vol.37, No. 2. https://doi.org/10.1111/j.1471-0307.1984.tb01233.x

[19] Ugo Bardi, Toufic El Asmar, Alessandro Lavacchi, (2013) Turning electricity into food: the role of renewable energy in the future of agriculture, *Journal of Cleaner Production*, vol. 53, pp. 224-23. https://doi.org/10.1016/j.jclepro.2013.04.014

[20] Namrata Chindarkar, Yvonne Jie Chen and Shilpa Sathe, (2017) Impact of Farm electricity supply management on farm households in Gujarat, India, 28 June 2017

[21] Fabio Riva, Emanuela Colombo, (2020) System-dynamics modeling of the electricity-development nexus in rural electrification based on Tanzanian case Study, *Energy for Sustainable Development*, vol. 56, pp. 128-143. https://doi.org/10.1016/j.esd.2020.04.001

[22] V.J.P.D. Martinho, (2020) Relationships between agriculture energy and farming indicators, *Renewable and Sustainable Energy Reviews,* vol. 132, pp. 110096. https://doi.org/10.1016/j.rser.2020.110096

[23] B. D. Simonevan, Bogaardt Marc- Jeroen and Wolfert Sjaak, (2019) Ethics of smart farming: Current questions and directions for responsible innovation towards the future, *NJAS – Wageningen Journal of Life Science*, vol. 90-91, pp. 100289. http://doi.org/10.1016/j.njas.2019.01.001

[24] Sen, S., Ganguly, S., Das, A., Sen, J., Dey, S., (2015) Renewable Energy Scenario in India: Opportunities and Challenges, *African Earth Science*, vol. 122, pp. 25-31, 2015. doi: http://dx.doi.org/10.1016/j.jafrearsci.2015.06.002

[25] Lee Song Ki, Choe Chan Young, (2019) Environmental performance of organic farming:

Evidence from Korean Small-holder soybean production, *Journal of Cleaner Production*, vol. 211, pp. 742-748. https://doi.org/10.1016/j.jclepro.2018.11.075

[26] A.Z. Dhunny, Z Allam, D. Lobine, M.R. Lollchund, (2019) Sustainable renewable energy planning and wind farming optimization from a biodiversity perspective, *Energy*, vol.185 pp.1282-1297, 2019. https://doi.org/10.1016/j.energy.2019.07.147

[27] Forde E., (2017) The ethics of energy provisioning: Living off-grid in rural Wales, *Energy Research & Social Science,* vol. 30, pp. 82-93. http://dx.doi.org/10.1016/j.erss.2017.06.018

[28] Bruno S S, Diana M T, (2019) Innovations for sustainable development: moving towards a sustainable future, *Journal of Cleaner Production,* vol. 208, pp. 325-332. https://doi.org/10.1016/j.jclepro.2018.09.244

[29] Keeley AR, (2016) Renewable Energy in Pacific Small Island Development States: The Role of International Aid and the Enabling Environment from Donor's perspectives, *Journal of Cleaner Production*, vol. 146, pp. 29-36. https://doi.org/10.1016/j.jclepro.2016.05.011

[30] Kaunathilake H, Perera P, Ruparathna R, Hewage K, Sadiq R, (2016) Renewable energy integration into community energy systems: A case study of new urban residential development, *Journal of Cleaner Production*, vol. 173, pp. 292-307. doi: 10.1016/j.jclepro.2016.10.1067

[31] Ian H. Rowlands, (2007) The Development of Renewable Electricity Policy in the province of Ontario: the influence of idea and timing, *Review of Policy Research,* Volume 24, pp. 185-207.

[32] S. Choudhury, A. K. Patra, A. Parida, S. Chatterjee, (2018) Renewable energy capacity estimation for Indian energy sector using energy demand forecasting through Fuzzy Time Series, Advances in Smart Grid and Renewable Energy, *Springer*, pp-551-558. doi.org/10.1007/978-981-10-4286-7_54

[33] A. Parida, D. Chatterjee, (2016) A solar PV augmented hybrid scheme for enhanced wind power generation through improved control strategy for grid connected doubly fed induction generator, *Cogent Engineering,* 3: 1147337. doi.org/10.1080/23311916.2016.1147337

[34] A. Parida, S. Choudhury, S. Chatterjee, D. Chatterjee, (2016) A cost effective trade-off based renewable power augmented energy efficient load model for manufacturing industries for demand side management", *Cogent Engineering*, 3: 1239299. doi.org/10.1080/23311916.2016.1239299

[35] A. Parida, S. Choudhury, S. Chatterjee, D. Chatterjee, (2016) A rule based electrical energy distribution model for energy efficiency in technical institutions: A case study, *Energy and Buildings,* 126, pp-508-516. doi.org/10.1016/j.enbuild.2016.05.069

[36] S. Choudhury, A. Parida, R. M. Pant, S. Chatterjee, (2016) GIS augmented computational intelligence technique for rural cluster electrification through prioritized site selection of micro-hydro power generation system, *Renewable Energy,* 142, pp-487-496. https://doi.org/10.1016/j.renene.2019.04.125

IOP Conf. Series: Materials Science and Engineering 1020 (2021) 012022 doi:10.1088/1757-899X/1020/1/012022

Comparison of NMOS and PMOS Input Driving Dynamic Comparator in $45nm$ Technology

S Hussain[1*], R Kumar[1] and G Trivedi[2]

[1]Department of Electronics and Communication Engineering (ECE), NERIST, Nirjuli, Arunachal Pradesh, India

[2]Department of Electronics and Electrical Engineering (EEE), IITG, Guwahati, Assam, India

E-mail: [1*]`s.hussainec@gmail.com`, [1]`rk@nerist.ac.in`,[2]`trivedi@iitg.ac.in`

Abstract. A comparison of NMOS and PMOS input driving dynamic comparator is carried out in $45nm$ technology. NMOS driving dynamic comparator shows lesser delay and consumes less power as compared to PMOS driving comparator. The advantage of the later is the noise immunity at higher frequencies. The *normalized* energy-noise-delay-product (ENDP) for NMOS dynamic comparator and PMOS dynamic comparator is *2.08* and *4.48*, respectively.

1. Introduction

Analog-to-digital converter (ADC) is the channel between the analog and digital world. ADCs are classified into two groups based on the sampling rate: Nyquist rate and Oversampling. Flash ADC is a Nyquist rate ADC which is also the fastest ADC. It is comprised of comparator and an encoder. Comparator is the backbone of an ADC. Several comparators have been designed in the recent past which includes TIQ based comparator [1], OPAMP based comparator, and dynamic comparators [3, 4, 5, 6, 7]. A TIQ based comparator sets a threshold point by changing the dimension of NMOS and PMOS transistors. Other comparator designs require an additional resistor ladder circuit. Two conventional dynamic comparator designs are compared in this paper. The first one is NMOS driven dynamic comparator and the other is PMOS driving dynamic comparator. The advantage of using a dynamic comparator is that it is efficient as compared to other comparators in terms of power dissipation and delay. The dynamic comparator works in three stages: reset stage, comparison stage and regeneration stage. The output code generated by the comparator bank is termed as *thermometer code*. The number of comparators depends upon the resolution of flash ADC. The required number of parallel computing comparators is $2^n - 1$ for n number of bits.

The characteristics of NMOS and PMOS driving dynamic comparators in $45nm$ technology is laid down in this paper. The *energy-noise-delay product* (ENDP) is calculated to observe the overall performance of the two comparators. A strong ARM latch [6] is used as a comparison circuit to observe the outputs characteristics. Several other dynamic comparators are designed in the recent advancements that are based on single tail [2] or dual tail [5] architectures to provide better results. In this paper, only a single tail architecture is used for comparison because it is the base for other architectures. Based on the comparison matrix, choice of input driving transistors can be made for a particular application.

The paper is divided into four sections to depict the granularity of the comparison. *Section 1* provides a brief introduction of the selected topic. *Section 2* demonstrates an overview of both the dynamic comparators. *Section 3* showcases the simulated results and comparison matrix of both the designs. And, finally, *Section 4* inferences the idea of comparison and its role in ADC design. The simulations are carried out in 45nm technology using *LTSpice* simulator.

Table 1. Status of transistors of NMOS and PMOS dynamic comparator

Stage	Transistor	Status	Region of Operation
Reset Phase	M1	OFF	Cutoff
	M2, M3	ON	Saturation
	M4, M5	ON	Linear
	M6, M7	OFF	Cutoff
	M8, M9	ON	Linear
Comparison Phase	M1	ON	Linear
	M2, M3	ON	Saturation
	M4, M5	ON	Linear
	M6, M7	OFF	Cutoff
	M8, M9	OFF	Cutoff
Regeneration Phase	M1	ON	Linear
	M2, M3	ON	Saturation
	M4, M5	ON	Linear
	M6, M7	ON	Linear
	M8, M9	OFF	Cutoff

Table 2. Comparison matrix of NMOS and PMOS dynamic comparator

Parameters	NMOS Comparator	PMOS Comparator
Delay (ns)	35.1	55.5
Power Dissipation (W)	4.328	4.44
Noise (nV/Hz1/2)	3.9	3.28
Normalized PDP (J)	1.51	2.46
Normalized EDP (Js)	5.332	13.76
Normalized ENDP (JsV/Hz1/2)	2.08	4.48

2. Dynamic Comparator

Dynamic comparator has three stages: *reset*, *comparison* and *regeneration*. During the reset phase, the differential output is set to logic '0' or logic '1'. After the *reset phase*, the comparator enters *comparison phase* and compares the differential inputs. When comparator enters *regeneration phase*, the outputs are regenerated to reach HIGH or LOW voltage levels

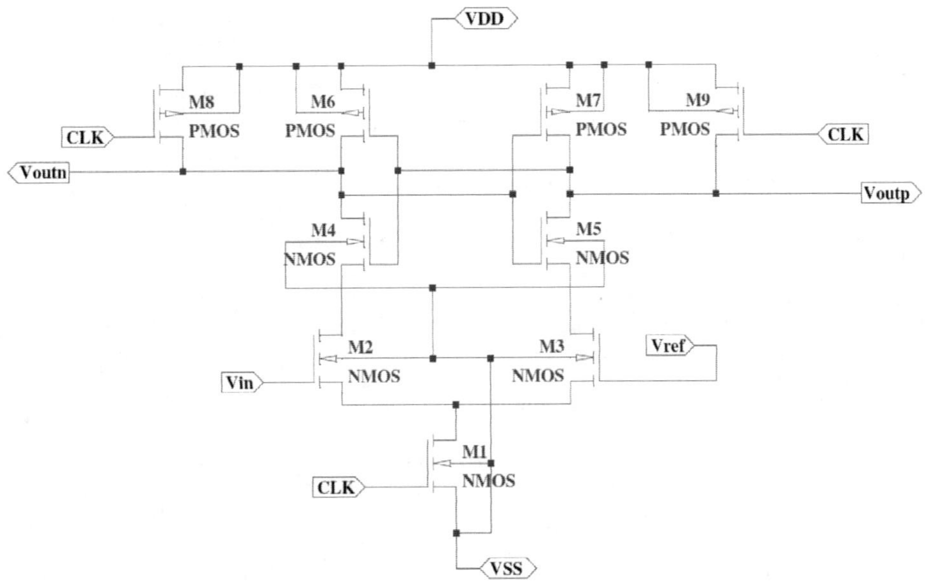

Figure 1. Circuit diagram of NMOS dynamic comparator

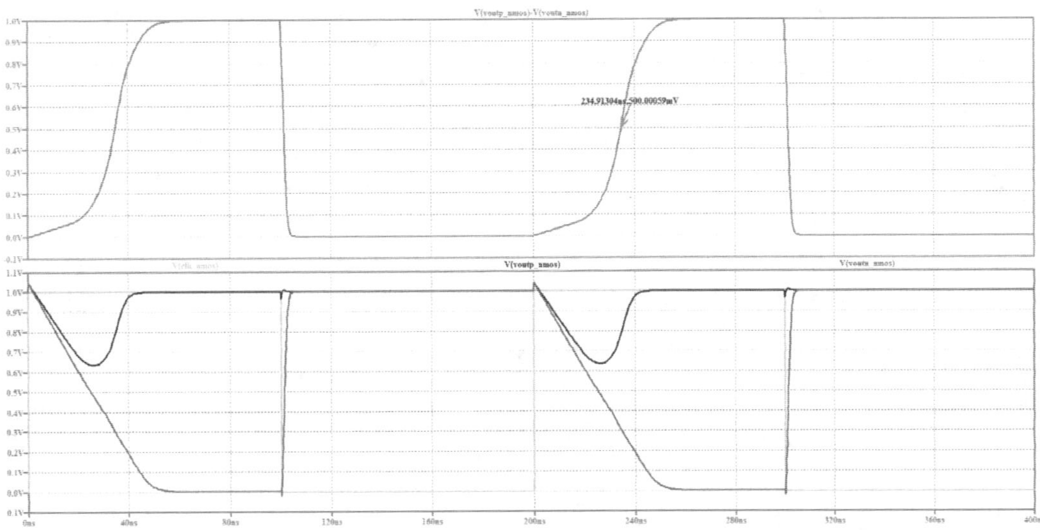

Figure 2. Transient analysis of NMOS dynamic comparator

based on differential inputs. Strong ARM latch based dynamic comparator is used to compare both the designs. NMOS and PMOS driving comparator circuit is depicted in figure 1 and figure 3, respectively. In both the designs, the tail transistor is either in active region or cutoff region. The input transistors are in saturation region and latch is in linear region. The clock pulse given to the NMOS comparator is the reciprocal to the pulse given to the PMOS comparator. The PMOS comparator pulls up the output faster than the NMOS comparator and the NMOS comparator pulls down the output faster than the PMOS comparator. A lower technology node is considered for simulation in order to consider the effect of small channel length devices. A comparison of NMOS input driving comparators is carried out earlier in [3],

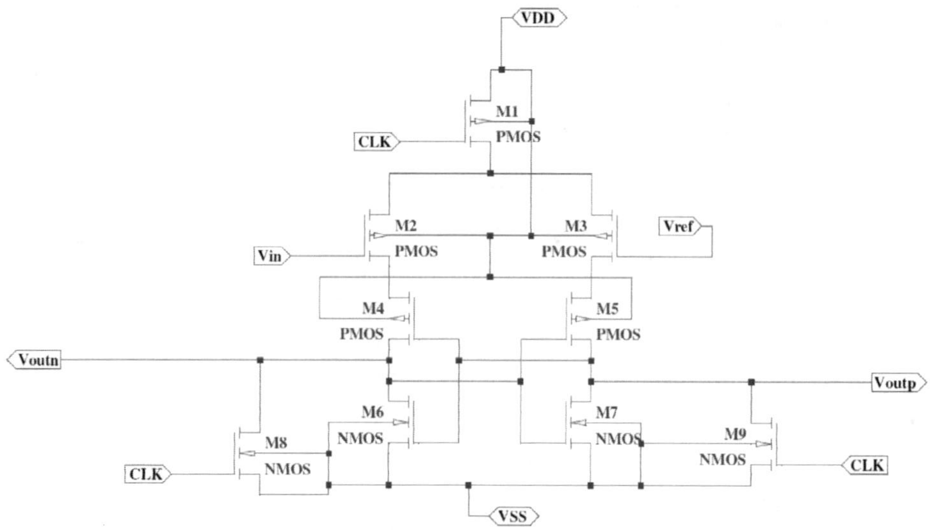

Figure 3. Circuit diagram of PMOS dynamic comparator

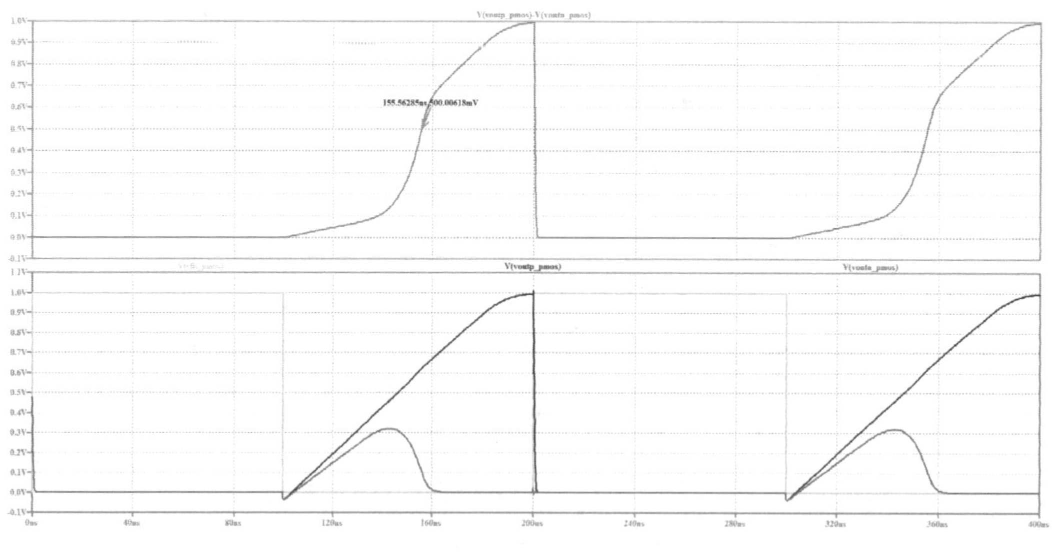

Figure 4. Transient analysis of PMOS dynamic comparator

[11]. In order to explore the possibilities of PMOS driving comparators a comparison is made for application specific designs. PMOS input transistors based amplifiers provide lower noise in high frequencies. For fast amplifiers or comparators NMOS input transistors are preferred.

NMOS input driving dynamic comparator consists of an NMOS input amplifier and a latch circuit. The circuit diagram is depicted in figure 1. During the *reset phase*, CLK pulse is set to GND (or $0V$) and the tail transistor M1 is turned OFF whereas the PMOS load transistors M8 & M9 turns ON pulling up both the output nodes to VDD (or $1V$). During the comparison phase, CLK pulse is set to VDD and M1 turns ON. The input pairs M2 & M3 amplifies the differential input. During the regeneration phase, the latches pulls up or down the output node based on the differential input provided by the input differential NMOS pairs. The status of the transistors is shown in table 1. The transient analysis of NMOS transistor is shown in figure 1.

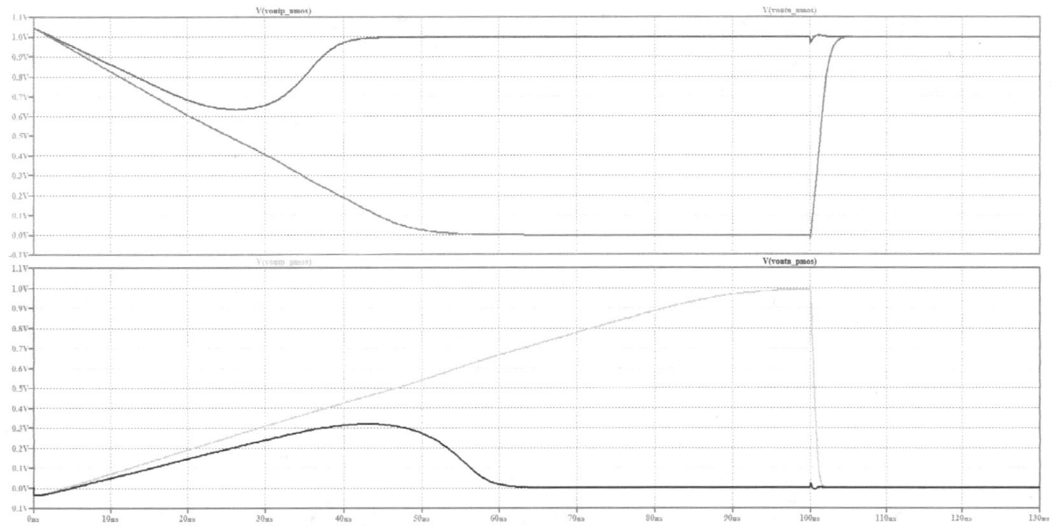

Figure 5. Comparison of NMOS and PMOS dynamic comparator: Transient Analysis

The voltage gain equation of the NMOS comparator is given by *equation* 1.

$$Voltage\ Gain,\ A_{v,n} = (g_{mn2,3}V_{th,n})/I1 \quad (1)$$

where, $g_{mn2,3}$ is the transconductance of the input NMOS transistor pairs, $I1$ is the tail current and $V_{th,n}$ is the threshold voltage of an NMOS transistor.

Let, t_{amp} be the amplifier delay and t_{latch} be the latch delay, then total delay of the NMOS comparator is given by *equation* 2.

$$\begin{aligned} Delay,\ t_{delay,n} &= t_{amp} + t_{latch} \\ &= 2.\frac{(Cl|V_{thp}|)}{I1} + \frac{Cl}{(g_{m,eff})}.\ln\left(\frac{VDD}{(4.|V_{thp}|\Delta Vin}\right)\sqrt{\frac{I1}{\beta_{2,3}}} \end{aligned} \quad (2)$$

where, Cl is the load capacitance, $I1$ is the tail current, $g_{m,eff}$ is the effective transconductance of the latch, V_{thp} is the threshold voltage of the PMOS transistor, VDD is the supply voltage, ΔV_{in} is the differential input, and $\beta_{2,3}$ is the current coefficient of input NMOS transistor pair.

The regeneration time is given by the *equation* 3.

$$Regeneration\ time,\ t_{reg,n} = \frac{C_{x,y}}{g_{m4,5}\left(1 - \frac{C_{x,y}}{Cl}\right)} \quad (3)$$

where $g_{m4,5}$ is the transconductance of the latch NMOS pair, $C_{x,y}$ is the capacitance across output nodes of the amplifier and Cl is the load capacitance.

Similarly, PMOS input driving dynamic comparator consists of a PMOS input amplifier and a latch circuit. The circuit diagram is depicted in figure 3. During the *reset phase*, CLK pulse is set to VDD (or $1V$) and the tail transistor M1 is turned OFF whereas the NMOS load transistors M8 & M9 turns ON pulling down both the output nodes to GND (or $0V$). During the comparison phase, CLK pulse is set to GND and M1 turns ON. The input pairs M2 & M3 amplifies the differential input. After comparison, the comparator enter into the *regeneration phase*, the latches pulls up or down the output node based on the differential input provided by the input differential PMOS pairs. The status of the transistors is shown in table 1. The transient analysis is given in figure 4. The gain and delay equations of PMOS comparator is similar to that of the NMOS and is given by the following *equations 4,5,6*.

$$Voltage\ Gain,\ A_{v,p} = \frac{g_{mp2,3}.V_{th,p}}{I1} \tag{4}$$

$$\begin{aligned} Delay,\ t_{delay,p} &= t_{amp} + t_{latch} \\ &= 2.\frac{Cl.|V_{th,n}|}{I1} + \frac{Cl}{g_{m,eff}}.\ln(\frac{VDD}{4.|V_{th,n}|\Delta V_{in}}\sqrt{\frac{I1}{\beta_{2,3}}} \end{aligned} \tag{5}$$

$$Regeneration\ time,\ t_{reg,p} = \frac{C_{x,y}}{g_{m4,5}\left(1 - \frac{C_{x,y}}{Cl}\right)} \tag{6}$$

where, $g_{mp2,3}$ is the transconductance of the PMOS input transistor pairs, $V_{th,p}$ is the threshold voltage of the PMOS transistor, $I1$ is the tail current through $M1$, Cl is the load capacitance, $V_{th,n}$ is the threshold voltage of the NMOS transistor, ΔVin is the differential input, VDD is the supply voltage, $g_{m,eff}$ is the effective transconductance of the latch, and $\beta_{2,3}$ is the current coefficient of input PMOS transistor pair.

3. Observation and Results

The simulations are done in $45nm$ technology using *LTSpice* simulator. The supply voltage *VDD* is set to $1V$ because of the maximum voltage supported by the technology node. For a differential input of $10mV$ and clock pulse of period $200ns$, the comparison matrix is tabulated in figure 2. Figure 5 compares the transient behaviour of both the comparators. It is observed that if the supply voltage is increased, the difference in the delay for voltage change is more for PMOS comparator and minimal for NMOS comparator. The delay change in NMOS varies from $235.1ns$ to $233.9ns$ whereas, for PMOS it varies from $155ns$ to $117.17ns$. For input frequency of $100MHz$, the calculated normalized ENDP (*JVs/Hz1/2*) for NMOS and PMOS comparator are *2.08* and *4.48*, respectively.

4. Conclusion

It can be inferred from the above results that the NMOS dynamic comparator provides better results. However, PMOS comparator at higher frequencies provides better noise immunity. The PMOS dynamic comparators can be used in applications where noise is an important factor. The *normalized* power-delay product (PDP) for NMOS and PMOS dynamic comparator are calculated to be *1.51* and *2.46*, respectively. The *normalized* energy-delay product (EDP) for NMOS and PMOS dynamic comparator are calculated to be *5.332* and *13.76*, respectively. The *normalized* energy-noise-delay product (ENDP) for NMOS and PMOS dynamic comparator are calculated to be *2.08* and *4.48*, respectively. The overall result favours the wide use of NMOS input driving dynamic comparator. The study provides a platform for use of application based comparators.

4.1. Acknowledgments

Authors would like to thank Meity, Govt of India for providing research grant under Visvesvarya PhD Scheme. Authors would also like to thank IITG and NERIST for supporting in the execution of the research.

References

[1] Yoo J 2003 A TIQ based CMOS FLASH A/D Converter for System On-Chip Applications *Ph.D. dissertation*

[2] Abbas M, Furukawa Y, Komatsu S, Yamaguchi T J, and Asada K 2010 Clocked comparator for high-speed applications in 65*nm* technology *IEEE Asian Solid-State Circuits Conference*, 1-4

[3] Hussain S, Kumar R and Trivedi G 2017 Comparison and Design of Dynamic Comparator in 180*nm* SCL Technology for Low Power and High Speed Flash ADC *IEEE International Symposium on Nanoelectronic and Information Systems (iNIS)*, 139-144.

[4] Chan C H, Zhu Y, Chio U F, Sin S W, Seng-Pan U, and Martins R P 2011 A reconfigurable low-noise dynamic comparator with offset calibration in 90nm CMOS *IEEE Asian Solid-State Circuits Conference* 233-236.

[5] Mashhadi S B and Lotfi R, 2013 Analysis and Design of a Low-Voltage Low-Power Double-Tail Comparator *IEEE Transactions on Very Large Scale Integration (VLSI) Systems* **22** 343-352.

[6] Razavi B 2015 The Strong ARM Latch A Circuit for all Seasons, PF, *EEE Solid-State Circuits Magazine* **7** 2, 12-17.

[7] Gao J, Li G, and Li Q. 2015 High-speed low-power common-mode insensitive dynamic comparator *Electronics Letters* **5**, 2, 134-136

[8] Hao Xu, Asad A. Abidi, 2019 Analysis and Design of Regenerative Comparators for Low Offset and Noise *IEEE Transactions on Circuits and Systems-I: Response*, **66**

[9] Tang X, Kasap B, ShenL, Yang X, Shi W and Sun N 2020 An energy efficient comparator with dynamic floating inverter pre-amplifier *Proc. IEEE Symp. VLSI Circuits (VLSI)* C14-C141.

[10] Bindra H S, Lokin, Annema A J, and Nauta B 2017 A 30fJ per comparison dynamic bias comparator *Proc. 43rd IEEE Eur.Solid State Circuits Conf. (ESSCIRC)* 71-74.

[11] Halim S. L. M., Hassan A, N D B M and Rahim A. A. A. 2012 Comparative study of comparator and encoder in a 4-bit Flash ADC using 0.18μm CMOS technology *International Symposium on Computer Applications and Industrial Electronics* 35-38.

Sentiment analysis and prediction of Indian stock market amid Covid-19 pandemic

Chetan Gondaliya*[1], Ajay Patel[2], Tirthank Shah[3]

[1]*AMPICS, Ganpat University, Mehsana, Gujarat*

[2]*AMPICS, Ganpat University, Mehsana, Gujarat*

[3]*Institute of Management, Nirma University, Ahmedabad, Gujarat*

[1]cpg01@ganpatuniversity.ac.in,[2]ajaykumar.patel@ganpatuniversity.ac.in,[3]tirthanks@gmail.com

Abstract. Outbreak and spread of the Covid-19 pandemic have touched to the core of our sentiments. Indian stock market has seen a roller coaster ride so far this year amid the Covid-19 pandemic. Sentiments have turned out to be a significant influence on the movement of the Indian stock market and pandemic has only added more steam. This study with the limelight on the Covid-19 pandemic is an endeavour to investigate the classification accuracy of selected ML algorithms under natural language processing for sentiment analysis and prediction for the Indian stock market. The study proposes the framework for sentiment analysis and prediction for the Indian stock market where six ML algorithms are put to test. Consequently, the study highlights the superior algorithms based on accuracy results. These superior algorithms can be potent input to build robust prediction models as a logical next step.

Keywords: ML algorithms, Natural language processing, Sentiment analysis, Sentiment prediction, Indian stock market

1. Introduction

Voicing out opinions and expressing reactions have been central to human development. The advent of social media sites such as Twitter, Myspace, Facebook, YouTube etc. and networking sites such as LinkedIn, Xing, Meetup etc. have provided an instant digital platform to express the opinions as well as reaction to any news or announcements in textual format. Such enormous textual data has turned out to be a gold mine to researchers. The Covid-19 pandemic has been centered on generating a huge number of text messages as an expression of various sentiments. These text messages are invariably the writer's expression of opinion or reaction towards pandemic, product, service, people, news, events and other such instances. Analyzing such texts through appropriate techniques to unearth the sentiments of the writer behind it, can lay a strong foundation for predictive models.

'Sentiment analysis' as a term was first used by [1], which has gained significant traction in the research world. Sentiment analysis is widely applied to diverse fields such as movie reviews [2], buying behavior in financial products [3], towards improvement in products and services [4], improving services of government agencies by analyzing customers' review [5]. Stock market movements in general and price movements of the stocks, in particular, have been an immense source of opinion and reaction generator from several viewers, traders and investors. It is understood that in short-run stock market movements are majorly influenced by sentiments of participants. Many studies have focused on sentiment analysis and prediction based on models in context with stock market movement. Investors' sentiment on the company's earnings announcements has an impact on the price movement of the US stocks [6]. Firm specific sentiments and general market sentiments have been studied to understand and to predict the influence of sentiments on the pricing of selected sample US companies [7]. Analysis of sentiment signals from experts based on Twitter feeds has shown better predictive power for stock market price movements [8].

However, it is to be noted that building a prediction model based on particular sentiment analysis without identifying the optimal algorithm technique to perform such sentiment analysis may lead to misleading inferences. Hence, this paper is an attempt to conduct and compare sentiment analysis and prediction of sentiment on Indian stock market news by applying six algorithms techniques namely; Decision Tree method, Random Forest method, Logistic Regression method, Naïve Bayes method, Support Vector Machine method and KNN method with the aim to identify the most suitable technique for sentiment prediction based on accuracy with special focus on the time period which covers the outbreak and spread of Covid-19 pandemic. This study contributes towards the identification of optimal algorithm techniques to build and test a superior predictive model as the next logical step of research. The sources for text messages to undertake sentiment analysis are identified as forums discussion, financial news, stock market tweets and RSS feed related to the Indian stock market. Different approaches have been used by researchers for sentiment analysis. Machine learning techniques are used for sentiment analysis and prediction based on Twitter feed [9]. The vector space model approach is applied to measure sentiment orientation [10]. While the unsupervised approach is used to conduct sentiment analysis on financial news [11].

This study is divided into sub-sections where the second section highlights the literature review, the third section represents data, methodology and proposed framework, findings and discussion which is heart of any empirical testing is elaborated in the fourth section and concluding remarks and future scope of research is deliberated in the fifth section.

2. Literature review

Sentiment analysis has increasingly received significant research attention. Opinion mining has emerged as a key tool to comprehend the sentiments of the target audience in order to build superior predictive models. An elaborative review of the evolution of sentiment analysis was conducted by [12] with special context to research topics and highly cited papers. Research work in the domain of computer-based sentiment analysis has seen an upsurge since 2004 and application of sentiment analysis in divergent fields like cyberbullying, elections, stock markets, medicine, disasters etc. has been witnessed [12]. As the prediction of stock market movement has always lured many researchers as well as practitioners, it has attracted huge attention from researchers.

[13] has conducted sentiment analysis by using python script language on Indian stock market news concerning predicting the movement of Sensex and Nifty as the stock market index. Like-wise supervised sentiment analysis was used as a tool to predict the buying behavior of participants in the Indian Futures market which is a type of financial derivative [3]. How social media sentiment influences the stock price movement with special reference to Apple Inc. was analyzed and accuracy was estimated [14]. The part-of-speech graphical model was used under model-based opinion mining to test the prediction power in the Iranian stock market based on the text of the opinion of users [15]. A study by [16] suggested the utility of data from microblogging sites for predicting behavior and movement of the stock market where variables such as trading volume, volatility and returns of stock were taken into consideration.

Deep learning approaches or techniques have been adopted by studies such as [2], [4], [16]. The study [2] reported exploration and comparison of various methods under deep learning. As per [4], the deep learning model was used to construct a classifier for detection of sentiment which has reported a higher range of precision in the domain of satisfaction of customer and opportunities identification for products and services improvement. Five different methods of regression; Ensemble Averaging method, Neural Network, Random Forest, Multiple Regression and Support Vector Machine method is used to perform sentiment analysis for assessing the prediction power in the context of the S&P 500 index a stock market index of US [16].To classify sentiment, the machine learning technique of Naive Bayes classification was applied and the study has found that there was an influence of sentiments represented on various platforms on the price of Apple Inc.[14]. Enhanced learning-based method (enhanced NN system) was used by [17] which has reported that the performance of this method can be increased by having selected proper window size. Feature extraction technique based on N-gram was applied to categorize tweets in regards to feature vector while tweet class prediction was conducted based on SVM and Naïve Bayes classifiers for prediction analysis in context with the influence of stock market news on the future movement of stock prices by [18]. A study by [19] has suggested that the SVM model with the segment index has shown accuracy of prediction on the higher side as compared to the SVM model without a combined segment index in regards to using social media text to predict the movement of the stock market.

3. Data set and methodology

3.1. Data

As the data sources for the collection of text messages have become abundant, it is crucial to identify appropriate text sources or platforms from which text messages are selected for sentiment analysis. Four key sources namely; RSS Feed, Forum discussion, Twitter and News portals were selected for extracting opinions and reactions to stock market movement in text format. The data has been collected for the very recent time period from 1st-January-2020 till 24th-August-2020. The Figure-1 describes the data sources which is given below;

Table 1. Data collection sources.

RSS Feed	Forum discussion	Twitter	News portals
Money Control	Money control	Hashtag	Marketmojo
Economics times	Yahoo conversations		Outlook India
United news of India	Traderji		Money Control
Business standard	Valuepickr		Economic times
			United news of India
			Millenniumpost
			Htsyndication
			The Himalayan
			Business standard
			Free press
			Equity bulls
			Business line
			The statesman
			Investing
			Stock adda
			Bse2nse
			Reuters

Labeling of textual data has been done under three categories; Positive, Negative and Neutral. Word association for positive and negative words is done as per Table-2 which is represented as below.

.

Table 2. Word association for labelling under positive and negative.

Sr. No.	Positive phrases	Negative phrases
1	Bullish	Bearish
2	High	Low
3	Outperform	Underperform
4	Surge	Decline

3.2. Methodology

Natural language processing (NLP) is considered to be a powerful tool for understanding and interpreting human languages like speech and text. Natural language processing is useful in text analysis and text mining. Frequently, used three techniques for natural language processing are; 1) Bag of words 2) N-grams 3) TF-IDF (Term Frequency — Inverse Data Frequency)

3.2.1 Bag of words method

Under natural language processing, the bag of words model is used for the representation of text data in a simplified way. Bag of words method works as the name suggests by creating a bag that is represented as a collection of words for particular text such as a document or a sentence. This model is hugely used for document classification.

3.2.2 N-grams method

The N-grams method is quite useful in natural language processing. N-grams model generates document term matrix in which counts are there a part of cell but not as single terms represented in columns but as a representation of adjacent words' combination of particular length 'N' in specific text. The example of how N-grams model works is as under;

Example: "NLP is an interesting topic"

Table 3. N-grams.

n Name	Tokens
2 bigram	["nlp is", "is an", "an interesting", "interesting topic"]
3 trigram	["nlp is an", "is an interesting", "an interesting topic"]
4 four-gram	["nlp is an", "is an interesting", "an interesting topic"]

Formula:

The underlying formula for N-grams model is as under;

If we calculate the probability of 'w1' word coming right after 'w2' word, then the formulation of equation is represented as follows;

count (w2 w1) / count(w2)

it means in a given sequence the number of times word occurs, divided by number of times the word occurs before the expected word in the corpus.

3.2.3 TF-IDF model

TF-IDF model stands for 'Term Frequency — Inverse Data Frequency'.

- Term frequency (TF): Term frequency provides a frequency of the word in each given document in a corpus or collection. Term frequency is represented in ratio format where the number of times the word appears in a given document is numerator while the total sum of words in the document is the denominator. As the occurrences of the word increases, the ratio tends to go up. Every document can have the term frequency of it.

- Inverse Data Frequency (IDF): Inverse Data Frequency is used to measure the rare words' weight in all the documents which are there in the corpus or collection. High Inverse Data Frequency (IDF) score will be represented for the words which have rare occurrences.

Formula:

$$w_{i,j} = tf_{i,j} \times \log(\frac{N}{df_i}) \quad (1)$$

Where,

$tf_{i,j}$ = Number of occurrences of i in j

df_i = Number of documents containing i

N = Total number of documents

Tools that are applied for the analysis purpose are namely; Python 3.7, NLTK toolkit for python and NLP.

3.3. Proposed framework

Data feed from four sources; Forum discussion, Twitter, News and RSS feed are used as input. On these text data pre-processing is done where removal of punctuation and stopwords is done also tokenization and lemmatization is done. Henceforth, pre-processed data is used on which three NLP techniques namely; Bag of words, TF-IDF and N-grams are applied. The feature set is created under every technique on which six ML algorithms are applied to test the sentiment prediction capability of each algorithm if it is correctly classifying sentiment under any one of the categories which are; Positive, Negative and Neutral. The flow chart of the proposed framework of sentiment analysis and prediction is depicted in Figure-3 as below.

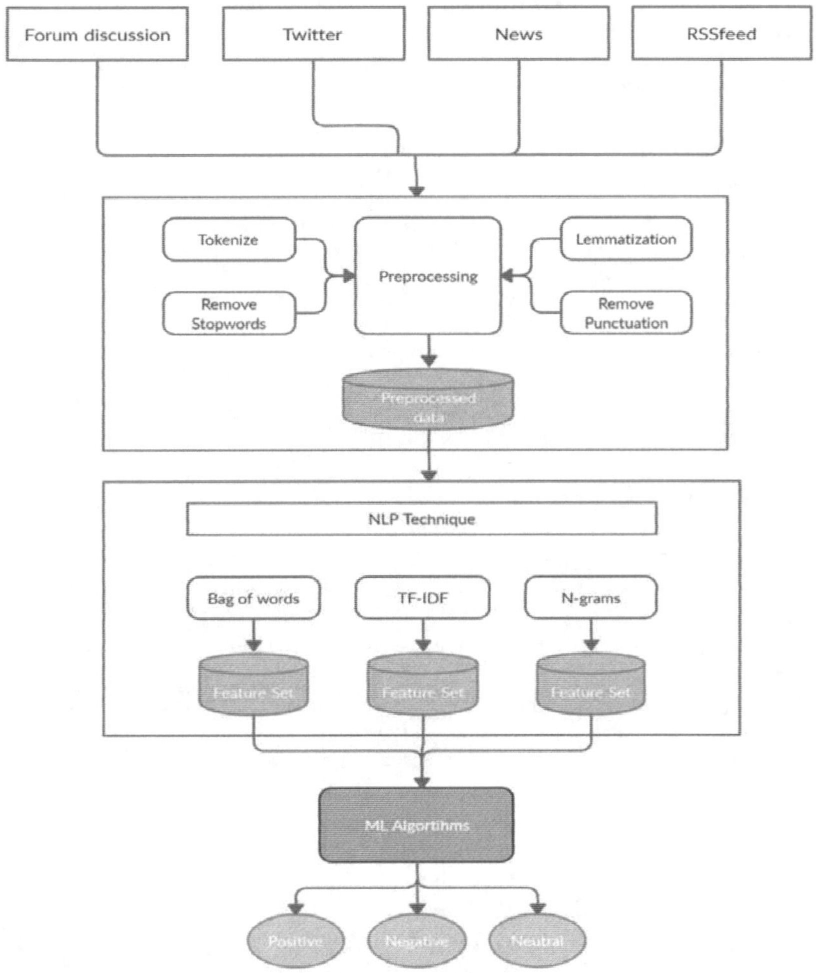

Figure 1. Proposed framework of sentiment analysis and prediction

4. Result and discussion

This study is an endeavour to identify the optimal method of sentiment analysis and sentiment prediction in context to Indian stock market news by making a comparison among various methods with specific reference to time period which covers the Covid-19 pandemic impact. For sentiment analysis and to test prediction power of algorithms for sentiment, the text is labeled under any of three categories; positive, negative and neutral. This study has tested how effectively these six algorithms with two techniques predict the sentiments by accurately labeling the text under the correct category. Figure-4 represents the accuracy level of six algorithms for sentiment prediction applied with the Bag-of-words technique as well as the TF-IDF technique.

Table 4. Classification accuracy for six algorithms.

Sr. No.	Algorithm	Technique	Classification Accuracy (%)
1	Decision Tree	Bag-of-words	77
		TF-IDF	75
2	KNN	Bag-of-words	53
		TF-IDF	68
3	Logistic Regression	Bag-of-words	78
		TF-IDF	76
4	Naïve Bayes	Bag-of-words	62
		TF-IDF	61
5	Random Forest	Bag-of-words	76
		TF-IDF	74
6	Support vector machine	Bag-of-words	78
		TF-IDF	76

While comparing Bag-of words technique with TF-IDF technique for the given data, it was observed that the Bag-of-words technique has shown higher classification accuracy in terms of sentiment prediction for five of six algorithms such as Decision Tree, Logistic Regression, Naïve Bayes, Random Forest and Support vector machine. The TF-IDF technique has shown higher accuracy only in the case of the KNN algorithm. When a comparison is made among six algorithms where Bag-of words technique is applied, both Logistic Regression and Support vector machine algorithms have indicated the highest classification accuracy of 78% for sentiment prediction. Naïve Bayes algorithm has generated the lowest classification accuracy of 62% for sentiment prediction. Below Figure-5, indicates graphically the accuracy level of all six algorithms with Bag-of words and TF-IDF technique.

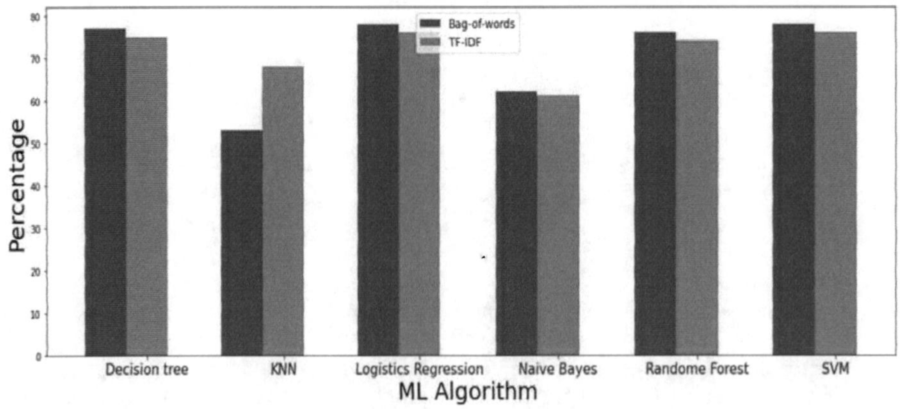

Figure 2. Bar chart representation of accuracy of six algorithms with two techniques

The recorded sentiment of the duration from 1st-January-2020 till 24th-August-2020 is depicted in Figure-6. It can be observed from the Figure-6 how the sentiment movements are gyrating.

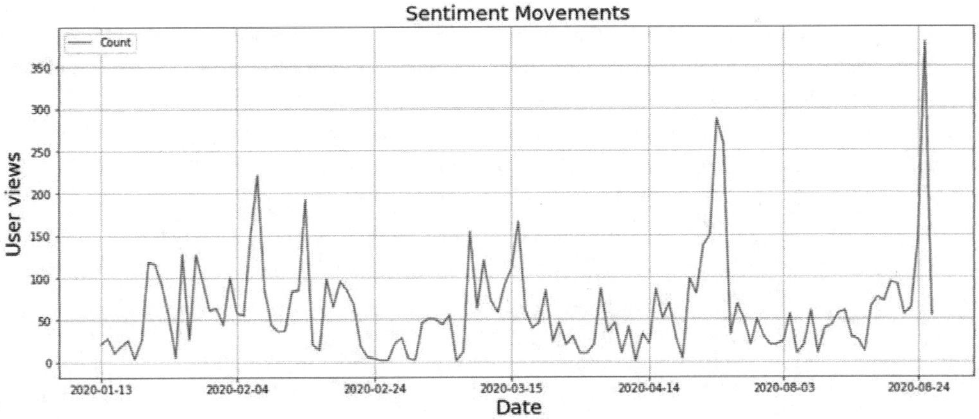

Figure 3. Sentiment graph

5. Concluding remarks and future scope

Based on the data which represents very recent time period and crucially includes the outbreak and spread of Covid-19 pandemic in India for performing the sentiment analysis to identify the optimal algorithm, Logistic Regression and Support vector machine algorithms have presented better results – making these two algorithms superior in sentiment prediction in comparison with other algorithms when Bag-of words technique is applied. The study contributes significantly towards the identification of superior algorithms in terms of accuracy for sentiment prediction. As natural language processing is increasingly becoming a potent tool for understanding text messages, this article provides a solid platform to test six selected algorithms with special reference to Indian stock markets news and covering the most recent time period of Covid-19 pandemic impact.

In the future, such research can be expanded to include other methods of sentiment analysis with more text data. Building the prediction model for stock market price movement based on identified superior algorithms in this study will throw light on if such predictive models generate better results or not. Conducting research with wider data sources and for a longer time period may also provide insight into if results generated in this study get substantiated in other such studies or not.

6. References

[1] Nasukawa, T., & Yi, J. (2003). Sentiment analysis. Proceedings of the International Conference on Knowledge Capture - K-CAP '03. doi:10.1145/945645.945658

[2] Chakraborty, K., Bhattacharyya, S., Bag, R., & Hassanien, A. A. (2019). Sentiment Analysis on a Set of Movie Reviews Using Deep Learning Techniques. Social Network Analytics, 127-147. doi:10.1016/b978-0-12-815458-8.00007-4

[3] Yadav, R., Kumar, A. V., & Kumar, A. (2019). News-based supervised sentiment analysis for prediction of futures buying behaviour. IIMB Management Review, 31(2), 157-166. doi:10.1016/j.iimb.2019.03.006

[4] Paredes-Valverde, M. A., Colomo-Palacios, R., Salas-Zárate, M. D., & Valencia-García, R. (2017). Sentiment Analysis in Spanish for Improvement of Products and Services: A Deep Learning Approach. Scientific Programming, 2017, 1-6. doi:10.1155/2017/1329281

[5] Alqaryouti, O., Siyam, N., Monem, A. A., & Shaalan, K. (2020). Aspect-based sentiment analysis using smart government review data. Applied Computing and Informatics, Ahead-of-print(Ahead-of-print). doi:10.1016/j.aci.2019.11.003

[6] Bouteska, A. (2019). The effect of investor sentiment on market reactions to financial earnings restatements: Lessons from the United States. Journal of Behavioral and Experimental Finance, 24, 100241. doi:10.1016/j.jbef.2019.100241

[7] Broadstock, D. C., & Zhang, D. (2019). Social-media and intraday stock returns: The pricing power of sentiment. Finance Research Letters, 30, 116-123. doi:10.1016/j.frl.2019.03.030

[8] Groß-Klußmann, A., König, S., & Ebner, M. (2019). Buzzwords build momentum: Global financial Twitter sentiment and the aggregate stock market. Expert Systems with Applications, 136, 171-186. doi:10.1016/j.eswa.2019.06.027

[9] S., S., & K.v., P. (2020). Sentiment analysis of malayalam tweets using machine learning techniques. ICT Express. doi:10.1016/j.icte.2020.04.003

[10] Kim, Y., & Shin, H. (2018). A New Approach for Measuring Sentiment Orientation based on Multi-Dimensional Vector Space. ArXiv, abs/1801.00254.

[11] Yadav, A., Jha, C. K., Sharan, A., & Vaish, V. (2020). Sentiment analysis of financial news using unsupervised approach. Procedia Computer Science, 167, 589-598. doi:10.1016/j.procs.2020.03.325

[12] Mäntylä, M. V., Graziotin, D., & Kuutila, M. (2018). The evolution of sentiment analysis—A review of research topics, venues, and top cited papers. Computer Science Review, 27, 16-32. doi:10.1016/j.cosrev.2017.10.002

[13] Bhardwaj, A., Narayan, Y., Vanraj, Pawan, & Dutta, M. (2015). Sentiment Analysis for Indian Stock Market Prediction Using Sensex and Nifty. Procedia Computer Science, 70, 85-91. doi:10.1016/j.procs.2015.10.043

[14] Suman, N., Gupta, P. K., & Sharma, P. (2017). Analysis of Stock Price Flow Based on Social Media Sentiments. 2017 International Conference on Next Generation Computing and Information Systems (ICNGCIS). doi:10.1109/icngcis.2017.34

[15] Derakhshan, A., & Beigy, H. (2019). Sentiment analysis on stock social media for stock price movement prediction. Engineering Applications of Artificial Intelligence, 85, 569-578. doi:10.1016/j.engappai.2019.07.002

[16] Oliveira, N., Cortez, P., & Areal, N. (2017). The impact of microblogging data for stock market prediction: Using Twitter to predict returns, volatility, trading volume and survey sentiment indices. Expert Systems with Applications, 73, 125-144. doi:10.1016/j.eswa.2016.12.036

[17] Wang, Z., Ho, S., & Lin, Z. (2018). Stock Market Prediction Analysis by Incorporating Social and News Opinion and Sentiment. 2018 IEEE International Conference on Data Mining Workshops (ICDMW). doi:10.1109/icdmw.2018.00195

[18] Urolagin, S. (2017). Text Mining of Tweet for Sentiment Classification and Association with Stock Prices. 2017 International Conference on Computer and Applications (ICCA). doi:10.1109/comapp.2017.8079788

[19] Wang, Y., & Wang, Y. (2016). Using social media mining technology to assist in price prediction of stock market. 2016 IEEE International Conference on Big Data Analysis (ICBDA). doi:10.1109/icbda.2016.7509794

IOP Conf. Series: Materials Science and Engineering 1020 (2021) 012024 doi:10.1088/1757-899X/1020/1/012024

On some generalizations of reduced and rigid modules

M Dutta[1] **and K H Singh**[2]

[1]Department of Mathematics, North Eastern Regional Institute of Science and Technology, Arunachal Pradesh, India.

[2]Department of Mathematics, Manipur University, Manipur, India

E-mail: mdutta499@gmail.com and heranerist@gmail.com

Abstract. In this paper we have introduced the notion of weak reduced module and weak rigid module as the generalizations of reduced module by working on the context of nilpotent elements of module and investigated some of its properties. Various results are extended for this generalization and some counter examples are constructed to validate the generalizations.

1. Introduction

Throughout this paper R denotes an associative ring with identity and M is unitary left $R-$module (denoted as ${}_RM$). Some terms like $Tor(M)$, $ann(M)$, $ann(m)$, $M_n(R)$ and $T_n(R)$ are used here. $Tor(M)$ is defined as $Tor(M) = \{m \in M : rm = 0, if \exists r(\neq 0) \in R\}$ and it forms a submodule of M if the ring R is commutative domain. The set $ann(M) = \{r \in R : rm = 0, \forall m \in M\}$ forms a left ideal and $ann(m) = \{r \in R : rm = 0\}$. $M_n(R)$ is the set of all matrices of order n over R, where $T_n(R)$ is the set of all upper triangular matrices of order n over R and it has ring structure with respect to the usual matrix addition and multiplication.

A ring R is reduced if the set of nilpotent elements of R (i.e. $nil(R)$) contains only the zero element or whenever $a^2 = 0$ implies $a = 0$ for $a \in R$. A ring R is called rigid whenever $a^2b = 0$ implies that $ab = 0$ for all a,$b \in R$. Reduced rings are always rigid. Both are same in the situation when the ring is with identity. In [1],[2] and [3] they introduced new classification of reduced ring and rigid rings on the basis of set of central elements $C(R)$ of R, the set of nilpotent elements $nil(R)$ of R and rigid endomorphism (in [4]). Again the concept of reduced ring was extended to R-module by T.Lee and Y.Zhou in [5]. They called a left R-module a reduced module if $a \in R$ and $m \in M$ satisfying $a^2m = 0$ implies $aRm = 0$. Easily we can check that R is reduced ring iff ${}_RR$ is reduced module. Similarly a left R-module M is rigid if for $a \in R$ and $m \in M$, $a^2m = 0$ implies $am = 0$. Therefore rigid module is a generalization of reduced module itself. In [6] they constructed counter examples (Example 2.20 and Example 2.21) of rigid modules which are not reduced modules and concluded that a module is reduced if and only if it is rigid and semicommutative. Where a left R-module M is called semicommutative module if $a \in R$ and $m \in M$ satisfying $am = 0$ implies $aRm = 0$ and this definition is introduced by A.M.Buhphang and M.B. Rege in [7]. Recently Ssevviiri and Groenewald introduced the concept of nilpotent element for module in [8]. In this paper the term "weak" is used to generalize the reduced module and rigid module on the basis of the definition of nilpotent element of a module as defined in [8]. The properties of reduced and rigid modules are extended using this new generalizations and constructed counter examples in some cases.

2. The nilpotent elements in a module

In [8], the definition of a nilpotent element of a module ${}_RM$ is given as :

A non zero element m of M is called a nilpotent element if $\exists\ r \in R$ and a natural number $k > 1$ such that $r^k m = 0$ but $rm \neq 0$. Here k is said to be the degree of nilpotency of m. The zero element of M is always considered as a nilpotent element. The set of nilpotent elements of M is denoted as $nil_R(M)$.

2.1. Example
For $M = \mathbb{Z}_4$ and $R = \mathbb{Z}_4$. Here

$$\bar{2}^2.\bar{1} = 0 \text{ but } \bar{2}.\bar{1} \neq 0.$$
$$\bar{2}^3.\bar{1} = 0 \text{ but } \bar{2}.\bar{1} \neq 0$$
$$\bar{2}^2.\bar{3} = 0 \text{ but } \bar{2}.\bar{3} \neq 0$$
$$\bar{2}^3.\bar{3} = 0 \text{ but } \bar{2}.\bar{3} \neq 0$$

Clearly $nil_{\mathbb{Z}_4}\mathbb{Z}_4 = \{\bar{0}, \bar{1}, \bar{3}\}$. In case of degree of nilpotency of an element we will consider the smallest positive integer (e.g. the degree of nilpotency of $\bar{1}$ is 2).

2.2. Lemma
If ${}_RN$ is a submodule of ${}_RM$ then $nil_R(N) \subseteq nil_R(M)$.

The above lemma can be proved very easily.

2.3. Lemma
Let $\phi : {}_R M \longrightarrow {}_R N$ be a homomorphism. Then we have :
(1) $m \in nil_R(M) \Rightarrow \phi(m) \in nil_R(N)$ whenever ϕ is 1-1.
(2) $n(\in \phi(M)) \in nil_R(N) \Rightarrow \phi^{-1}(n) \in nil_R(M)$ whenever ϕ is 1-1. ($\phi^{-1}(n)$ contains only one element as ϕ is 1-1.)

Proof. (1) Let $m \in nil_R(M)$. So by definition $\exists r \in R$ and a natural number $k > 1$ such that $r^k m = 0$ but $rm \neq 0$. Therefore $\phi(rm) \neq 0$ and $\phi(r^k m) = 0$ by using the condition that ϕ is 1-1. It implies $r\phi(m) \neq 0$ and $r^k\phi(m) = 0$. Therefore $\phi(m)$ is a nilpotent element of N.
(2) Let the pre image of n, $\phi^{-1}(n) = \{m\}$. Since $n \in nil_R(N)$. So by definition $\exists t \in R$ and a natural number $k > 1$ such that $t^k n = 0$ but $tn \neq 0$. This implies $t^k\phi(m) = 0$ but $t\phi(m) \neq 0$ as $\phi(m) = n$. By the property of R-module homomorphism, $\phi(t^k m) = 0$ but $\phi(tm) \neq 0$. As ϕ is 1-1, $\phi(t^k m) = 0$ implies $t^k m = 0$. If $tm = 0$, then $\phi(tm) = 0$, which is a contradiction. Therefore $tm \neq 0$. Hence $\phi^{-1}(n) = m \in nil_R(M)$.

2.4. Lemma
If $f(x) \in nil_R(M[x])$ where $f(x) = \sum_{i=0}^{n} m_i x^i$ and $m_i \in M$ then at least one of m_i is in $nil_R(M)$.

Proof. Since $f(x) \in nil_R(M[x])$, by defintion $\exists t \in R$ and a natural number $k > 1$ such that $t^k f(x) = 0$ but $tf(x) \neq 0$. This implies $t^k(\sum_{i=0}^{n} m_i x^i) = 0$ but $t\sum_{i=0}^{n} m_i x^i \neq 0$. So $\sum_{i=0}^{n}(t^k m_i)x^i = 0$ but $\sum_{i=0}^{n}(tm_i)x^i \neq 0 \Rightarrow t^k m_i = 0$ for all i but $tm_j \neq 0$ for some j, where $0 \leq j \leq i$. Therefore m_j is a nilpotent element of M.

3. Weak reduced module and weak rigid module

We start with the following definitions and some of their basic properties:

IOP Conf. Series: Materials Science and Engineering 1020 (2021) 012024 doi:10.1088/1757-899X/1020/1/012024

3.1. Definition
A module $_RM$ is weak reduced module if whenever $a^2m = 0$ for a in R and m in M implies $aRm \subseteq nil_R(M)$.

3.2. Definition
A module $_RM$ is weak rigid module if whenever $a^2m = 0$ for a in R and m in M implies $am \in nil_R(M)$.

The direct results from the above definitions:

(i) Every reduced module (respectively rigid module) is a weak reduced module (respectively weak rigid module).

(ii) Every weak reduced module is weak rigid module.

(iii) Every reduced module is weak rigid module.

(iv) Submodules of a weak reduced module (respectively weak rigid module) are weak reduced modules (respectively weak rigid modules).

3.3. Lemma
If R is a reduced ring with $Tor(M) = \{0\}$, then (1) and (2) are equivalent:
(1) M is reduced (respectively rigid) R-module.
(2) M is weak reduced (respectively weak rigid) R-module.

Proof. (1)$\Rightarrow$(2) is obvious. For (2)$\Rightarrow$(1), if M is weak reduced (respectively weak rigid) module over R then whenever $a^2m = 0$ for a in R and m in M implies $aRm \subseteq nil_R(M)$ (respectively $am \in nil_R(M)$). Now let $n \neq 0$ be a nilpotent element of M. By using the definition of nilpotent element $\exists t \in R$ and a natural number $k > 1$ such that $t^kn = 0$ but $tn \neq 0$. Since $nil(R) = 0$ as R is R is reduced. Thus $t^k \neq 0$. By using the definition of torsion element, $t^kn = 0 \Rightarrow n \in Tor(M)$. But M is torsion free module, therefore $n \in Tor(M) = 0 \Rightarrow n = 0$, which is a contradiction. So M does not contain any non zero nilpotent element. So $aRm = 0$ (respectively $am = 0$). Hence $_RM$ is reduced (respectively rigid) module.

3.4. Lemma
$_RM$ is weak reduced (respectively weak rigid) module if and only if every cyclic submodules of $_RM$ are weak reduced (respectively weak rigid).

Proof. If $_RM$ be weak reduced (weak rigid) module and N be a cyclic submodule of $_RM$. Since every submodule of weak reduced (weak rigid) module is weak reduced (weak rigid). Therefore $_RN$ is weak reduced (weak rigid) module.
If a^2m=0 for a in R and m in M. Since $Rm = \langle m \rangle$ be a cyclic submodule of $_RM$ and Rm is weak reduced (respectively weak rigid) by assumption of every cyclic submodule of M is weak reduced (respectively weak rigid). Hence $a^2m = 0 \Rightarrow a^2(1.m) = 0 \Rightarrow aR(1.m) \subseteq nil_R(Rm) \Rightarrow aRm \subseteq nil_R(Rm) \Rightarrow aRm \subseteq nil_R(M)$ (respectively $a^2m = 0 \Rightarrow a^2(1.m) = 0 \Rightarrow a(1.m) \in nil_R(Rm) \Rightarrow am \in nil_R(Rm) \Rightarrow am \in nil_R(M)$) by using Lemma 2.2.

3.5. Proposition
Every left R-module is weak reduced (respectively weak rigid) if and only if every cyclic left R-module is weak reduced (respectively weak rigid).

Proof. It can be easily proved by using Lemma 3.3 and Lemma 2.2.

A natural question arises that under what conditions a weak reduced module M over a ring S will be weak reduced over another ring R. The proposition 2.2 given in [9] (i.e. change of rings) stated in which situation it is true for reduced module. Here in the next theorem we will prove that it can be extended to weak reduced module also. But the ring homomorphism should be onto in case of weak reduced module to show both (i) and (ii).

3.6. Proposition
Let ${}_SM$ be a module. By defining $rm = \psi(r)m$, M is a left R-module where $\psi : R \longrightarrow S$ is a ring homomorphism. If ψ is onto, Then we have
(i) If ${}_SM$ is weak reduced module, then ${}_RM$ is weak reduced module.
(ii) If ${}_RM$ is weak reduced module, then ${}_SM$ is weak reduced module.
(iii) If S is reduced ring, then ${}_RS$ is weak reduced module.

Proof. (i) Let us consider M is weak reduced S-module and $a^2m = 0$ for any a in R and m in M. Thus $\psi(a^2)m = 0 \Rightarrow [\psi(a)]^2m = 0$. Therefore $\psi(a)sm \in nil_S(M)$ for all $s \in S$ as M is weak reduced S-module. If $\psi(a)sm = 0$, then $\psi(a)\psi(r)m = 0$ as ψ is onto and $\psi(r) = s$. Thus $\psi(ar)m = 0 \Rightarrow arm = 0 \in nil_R(M)$ for all $r \in R$. If $\psi(a)sm \neq 0$, then by definition $\exists t_s \in S$ and a natural number $k > 1$ such that ${t_s}^{k_s}\psi(a)sm = 0$ but $t_s\psi(a)sm \neq 0$. But ψ is onto. So $\exists t_r \in R$ and $r \in R$ such that $\psi(t_r) = t_s$ and $\psi(r) = s$. Thus $\psi({t_r}^{k_s})\psi(a)\psi(r)m = 0$ but $\psi(t_r)\psi(a)\psi(r)m \neq 0$. Now by using the condition ψ is ring homomorphism, therefore $\psi({t_r}^{k_s}ar)m = 0$ but $\psi(t_rar)m \neq 0$. So ${t_r}^{k_s}arm = 0$ but $t_rarm \neq 0$ for all $r \in R$. Therefore by definition $aRm \subseteq nil_R(M)$.
(ii) Let us consider M is weak reduced R-module and $s^2m = 0$ for s in S and m in M. Since ψ is onto implies $\exists r_s \in R$ such that $\psi(r_s) = s$. Thus $[\psi(r_s)]^2m = 0 \Rightarrow \psi({r_s}^2)m = 0 \Rightarrow {r_s}^2m = 0$. Since M is weak reduced $R-$ module, therefore $r_srm \in nil_R(M)$ for all $r \in R$. If $r_srm = 0$, then $\psi(r_sr)m = 0 \Rightarrow \psi(r_s)\psi(r)m = 0 \Rightarrow s\psi(r)m = 0$ where $\psi(r) \in R$ for all $r \in R$. Therefore $s\psi(R)m = 0$. But $\psi(R) = S$ as ψ is onto. Thus $sSm = 0$ implies $sSm \subseteq nil_S(M)$. On the other hand if $r_srm \neq 0$, by definition $\exists t_r \in R$ and $k_r \in \mathbb{N}$ such that ${t_r}^{k_r}r_srm = 0$ but $t_rr_srm \neq 0$. It implies $\psi({t_r}^{k_r}r_sr)m = 0$ but $\psi(t_rr_sr)m \neq 0$. So $\psi(t_r)^{k_r}\psi(r_s)\psi(r)m = 0$ but $\psi(t_r)\psi(r_s)\psi(r)m \neq 0$. Let $\psi(t_r) = t_s$. Then ${t_s}^{k_r}s\psi(r)m = 0$ but $t_ss\psi(r)m \neq 0$. It implies $s\psi(R)m \subseteq nil_S(M)$. But $\psi(R) = S$ as ψ is onto. Therefore $sSm \subseteq nil_S(M)$ and it concludes that M is weak reduced $S-$module.
(iii) Since the ring S is reduced. So S is reduced S- module (Example 1.3 in [5]). It implies S is weak reduced S-module . By using (i) S is weak reduced R-module.

The above proposition can be extended to weak rigid module in similar way and some examples are given in the following which are direct applications of it.

3.7. Example
M is weak reduced $R-$module if and only if M is weak reduced $\frac{R}{I}-$module where I is any ideal of the ring R. Since $\exists$ a ring homomorphism from R onto $\frac{R}{I}$.

3.8. Example
M is weak reduced R-module iff M is weak reduced $\bar{R}$-module, where $\bar{R}=\frac{R}{ann(M)}$.

Next we start with some important results which will be used later to construct counter examples for the generalizations.

3.9. Lemma
Let ${}_RM$ be a module, then $nil_{M_n(R)}(M_n(M)) = M_n(M)$ for each $n \geq 2$.

Proof. We have to show that each element $A = \begin{pmatrix} m_{11} & m_{12} & \dots & m_{1n} \\ m_{21} & m_{22} & \dots & m_{2n} \\ \vdots & \vdots & \ddots & \vdots \\ m_{n1} & m_{n2} & \dots & m_{nn} \end{pmatrix} \in M_n(M)$ is a nilpotent element over $M_n(R)$.
Case I: If $A = 0$, then there is nothing to prove.
Case II: If $A \neq 0$, then at least one of the entries of A is non zero say m_{ij}, where $1 \leq i, j \leq n$. Let E_{ij} be the matrix in $M_n(R)$ whose $(i,j)^{th}$ entry is 1 and 0's elsewhere. It is clear that, ${E_{ki}}^2 = 0$ for $k \neq i$. Thus ${E_{ki}}^2A = 0$ but $E_{ki}A \neq 0$ as m_{ij} is non zero. It implies A is a nilpotent element of $M_n(M)$.

3.10. Proposition
${}_{M_n(R)}M_n(M)$ is weak reduced module for a module ${}_RM$ and $n \geq 2$.

Proof. Let $A^2L = 0$ for any $A \in M_n(R)$ and $L \in M_n(M)$. Then $ABL \in M_n(M)$ for any $B \in M_n(R)$. This implies $ABL \in nil_{M_n(R)}(M_n(M))$ by using Lemma 3.9. Thus $AM_n(R)L \subseteq nil_{M_n(R)}(M_n(M))$. So ${}_{M_n(R)}M_n(M)$ is weak reduced module.

3.10.1. Corollary ${}_{M_n(R)}M_n(M)$ is weak rigid module for a module ${}_RM$ and $n \geq 2$.

3.11. Example
Let $R = \mathbb{Z}$ and $M = \mathbb{Z}$. From Proposition 3.10 $M_2(\mathbb{Z})$ is weak reduced module over $M_2(\mathbb{Z})$. Now $A^2L = 0$ for $A = E_{12}$ and $L = E_{22}$. But $AL = E_{12} \neq 0$. Thus $M_2(\mathbb{Z})$ is not rigid module over $M_2(\mathbb{Z})$. So it is not reduced module (reduced module $\Rightarrow$ rigid module).

3.11.1. Remark Let $A = E_{12} \in T_2(R)$ and $L = E_{22} \in M_2(M)$. Clearly ${E_{12}}^2E_{22} = 0$. But $E_{12}E_{22} = E_{12}$ is not a nilpotent element in $M_2(M)$ over $T_2(R)$ even when it is a nilpotent element over $M_2(R)$. Since there does not exist any non zero $B \in T_2(R)$ and $k \in \mathbb{N}$ for which $B^kE_{12} = 0$ and $BE_{12} \neq 0$. Therefore $M_n(M)$ is not weak reduced module over $T_n(R)$ as it is not weak rigid. (We can also use Proposition 3.6 that it is not true as there does not exist any onto ring homomorphism from $T_n(R)$ to $M_n(R)$ for any ring R and $n \geq 2$.)

It was given that if M_t is a reduced R_t-module for each $t \in \Gamma$, then Π_tM_t is a reduced Π_tR_t-module (in Example 1.3 of [5]). In the next proposition we will show that it is also true in case of weak reduced module.

3.12. Proposition
If M_i is a weak reduced R_i -module for $i \in I$. Then $M = \Pi_iM_i$ is a weak reduced R=Π_iR_i-module.

Proof. Let us assume that $a^2m = 0$ for $a = (a_i)_{i\in I}$ and $m = (m_i)_{i\in I}$ where $a_i \in R_i$ and $m_i \in M_i$. It implies $a_iR_im_i \subseteq nil_{R_i}(M_i)$ for all $i \in I$. Now to show that $aRm \subseteq nil_R(M)$ where $M = \Pi_{i\in I}M_i$ and $R = \Pi_{i\in I}R_i$. Let $r = (r_i)_{i\in I}$ be any element of $R = \Pi_{i\in I}R_i$. So for $arm = (a_ir_im_i)_{i\in I}$. Here $a_ir_im_i \in nil_{R_i}(M_i)$ for each $i \in I$. By definition $\exists t_i \in R_i$ and $k_i \in \mathbb{N}$ such that ${t_i}^{k_i}a_ir_im_i = 0$ but $t_ia_ir_im_i \neq 0$. Now we consider $t = (t_1, 0, 0, ...) \in R = \Pi_{i\in I}R_i$ and $k = k_1 \in \mathbb{N}$. Then $t^{k_1}arm = 0$ but $tarm = (t_1a_1r_1m_1, 0, 0, ...) \neq 0$. Therefore arm is a nilpotent element of $M = \Pi_{i\in I}M_i$ for any r and so $aRm \subseteq nil_R(M)$.

3.13. Proposition
Let R be an integral domain. Then (1) and (2) are equivalent:
(1) M is weak reduced (respectively weak rigid) R-module.
(2) $Tor(M)$ is weak reduced (respectively weak rigid) R-module.

Proof. (1)⇒(2), Since $Tor(M)$ is a submodule of weak reduced (respectively weak rigid) module M . So clearly $Tor(M)$ is a weak reduced (respectively weak rigid) R-module.
(2)⇒(1), let $a^2m = 0$ for $a \in R$ and $m \in M$. If $a = 0$ then clearly $aRm = 0 \subseteq nil_R(M)$ (respectively $am = 0 \in nil_R(M)$) without using the statement (2). If $a \neq 0$, then $a^2 \neq 0$ as R is domain. Therefore $a^2m = 0$ implies $m \in Tor(M)$. Therefore a^2m implies $aRm \subseteq nil_R(Tor(M))$ (respectively $am \in nil_R(Tor(M))$) by using condition (2). Now using Lemma 2.2, $aRm \subseteq nil_R(M)$ (respectively $am \in nil_R(M)$).

The following example shows that weak rigid modules need not be weak reduced module.

3.13.1. Example In Example 2.20 in [6], it was proved that $M = T/TX$ is a rigid T-module where $T = L < X, Y >=$ the set of all polynomials in X and Y over a field L and both variables are non commuting indeterminates. Hence M is weak rigid T-module by 3.2.(i). Now for $X \in T$ and $\bar{1} = 1 + TX \in M$, $X^2\bar{1} = X^2(1 + TX) = X^2 + TX = TX = \bar{0}$. Let $t = a + bX + cY \in T$ be a non zero element where $a \in L$ and $b, c \in T$. Cleary $t^k = d + eX + fY \in T$ for any $k \in \mathbb{N}$. Therefore $t^kXY\bar{1} = t^kXY(1 + TX) = t^kXY + TX \neq TX = \bar{0}$ because $t^kXY \notin TX$ as X and Y are non commuting indeterminates. So $XY\bar{1} \notin nil_T(M)$ whenever $X^2\bar{1} = \bar{0}$. Hence M is not weak reduced T-module.

4. Conclusion
In this paper the study introduced weak reduced module (3.1) and weak rigid module (3.2), which generalized the concept of reduced module and rigid module respectively. Moreover weak rigid module is itself a generalization of weak reduced module and it is emphasized by the constructed example given in 3.13.1. Finally a number of properties of these generalizations are investigated here.

Acknowledgments. The authors express their thanks to the referee for useful suggestions to improve the presentation.

References
[1] H Kose, B Ungor and S Halicioglu 2012 ,*A generalization of reduced rings*, Hacettepe Journal of Mathematics and Statistics, **41(5)**, 689-696.
[2] L Ouyang 2008, *Extensions of generalized α−rigid rings*, International Electronic Journal of Algebra, **3**, 103-116.
[3] M V Mehrabadi and S Sahebi 2017, *Central α− rigid rings*, Palestine Journal of Mathematics, **6(2)**, 569-572.
[4] J Krempa 2008, *Some examples of reduced rings*, Algebra colloquium,**3**, 289-300.
[5] T K Lee and Y Zhou 1985,*Reduced modules in :Rings, modules,algebras and abelian groups*, Lecture notes in pure Appl. Math, Dekkar, New York,**236**, 147 - 169.
[6] A M Buhphang et al 2013, *On Rigid modules*, East-West journal of mathematics, **15**, 71-85.
[7] M B Rege and A M Buhphang 2002, *Semi-commutative modules and Armendariz modules*, Arab Journal of Mathematical science,**8**, 53-65.
[8] D Ssevviiri and N Groenewald 201), *Generalization of nilpotency of ring elements to module elements*, Communications in Algebra,**42**, 571-777.
[9] M B Rege and A M Buhphang 2008, *On Reduced modules and Rings*, International electronic journal of Algebra,**3**, 58-74.
[10] M Baser and N Agayev 2006, *On reduced and semicommutative modules*, Turkish J. Math.,**30**, 285-291.
[11] P P Nielsen 2006,*Semi-commutativity and the McCoy condition*, Journal of Algebra, **298**, 134–141.
[12] Y Gang 2007, *Semicommutative and Reduced Rings*, Vietnam Journal of Mathematics, **35:3**, 309–315.

[13] A Harmanci and H Kose 2017, *On a Class Of Semicommutative Rings*, New Zealand Journal Of Mathematics, **47**,69-85.
[14] N Agayev and A Harmanci 2007, *On Semicommutative Modules and Rings*, Kyungpoo Math. J.,**47**, 21-30.
[15] A Buhphang and M B Rege 2002, *Semi-commutative modules and Armendariz modules*, Arab Journal of Mathematical Sciences, **8**, pp.53-65.
[16] M Baser and A Harmanci 2007, *Reduced and p.q.-Baer modules*, Taiwanese J.Math., **11(1)**, 267-275.
[17] N J Groenewald and D Ssevviiri 2016, *Classical completely prime submodules*, Hacettepe Journal of Mathematics and Statistics,**45 (3)**,717 – 729.
[18] M Behboodi and H Koohy 200), *Weakly Prime Modules*, Vietnam Journal of Mathematics,*32:2*, 185–195.
[19] M Behboodi, O A S Karamzadeh and H Koohy 2004, *Modules Whose Certain Submodules Are Prime*, Vietnam Journal of Mathematics, **32:3**, 303–317.
[20] N Agayev, A Harmanci and S Halicioglu 2009, *On symmetric modules*, Riv. Mat. Univ. Parma, **(8)2**, 91-99.
[21] E Guner and S Halicioglu 2014, *Generalized Rigid Modules*, Revista Colombiana de Matematicas, *48:1*, 111-123.
[22] N Agayev, S Halicioglu, A Harmanci and B Ungor 2015, *Modules which are Reduced over their Endomorphism Rings*, Thai Journal of Mathematics, **13(1)**, 177–188.
[23] P Kanwar, M Khatkar and RK Sharma 2017, *Idempotents and units of matrix rings over polynomial rings*, International Electronic Journal of Algebra, **22**, 147-169.
[24] D S Dummit and R M Foote 2004, *Abstract algebra*, Wiley Hoboken, **l3**. .
[25] T Y Lam 2013, *A first course in noncommutative rings*, Science and Business Media, **131**.

Improved Reliability of Voice over Internet Protocol(VoIP) using Machine Learning

Vinod Kumar[1], Om Prakash Roy[2]

[1,2]Department of EE, NERIST, Arunachal Pradesh 791109, India

[1]vinodnerist@gmail.com, [2]oproy61@gmail.com

Abstract. Start Voice over Internet Protocol (VoIP) is the communication means inspired with the revolutionizing wireless internet technology to transfer voice signals. However, VoIP over wireless means faces several challenges in terms of data loss resulting in poor voice quality or communication delay. To deal with the practical aspects of voice communication over Internet Protocol(IP) we proposed a highly secure wireless network. The proposed VoIP as a secure wireless network for VoIP combines the advantage of Artificial Bee Colony (ABC) based network optimization based on node property. This is followed by implementation of two classifiers; Support Vector Machine (SVM) that identifies the affected route and Convolutional Neural Network (CNN) that detects the malicious nodes present in the affected route to offer secure data transmission. The quality of voice communication is evaluated in terms of dropped packets, delay and jitter to offer interactive communication service. The simulation analysis over 50 nodes had proved the effectiveness in achieving reliable quality voice calling service with average throughput of 98.65% with comparatively lower jitter, packet loss and latency of 2.325 ms, 1.35% and 1.616 ms, respectively.

Keywords: VoIP, Reliability, Wireless Network, ABC, SVM, CNN

1. Introduction

VoIP can be understood as a voice calling process via data packets. In the recent decade, communication means have been revolutionized from Public Switched Telephone Network (PSTN) to internet based voice calling using VoIP. In VoIP communication the voice conversion data is sent using (Internet Protocol) IP packets over internet [1]. Skype calling has been a commercial application taking advantage of VoIP service has nearly 8 million users [2]. The rising popularity is due to the inherent advantages of VoIP services such as minimal voice calling cost. VoIP has proved to be cost effective means of communication however there exist some reliability issues adjoining this service like quality of voice communication which is highly challenged.

1.1. VoIP components

VoIP essentially consist of three parts, CODEC (COder/DECoder) for coding and decoding voice signal, packetizer and depacketizer that fragments and defragments the encoded signal and playout buffer at the receiver end that smoothen the delay [3][4]. Overview of the VoIP system to offer end-to-end voice communication is shown in Figure 1. Table 1 lists some of the most frequently implemented CODECs for VoIP services comparing their bit rate and frame size [5]. The security aspects of VoIP service majorly rely on the degree of security adjoining the network it has been using as means of transmission. The transmission process first converts audio signals into IP packets and transfers through the network. During transfer instances of data loss and attack has been observed that have negative effect on the Quality-of-Service (QoS) offered by VoIP. Here, QoS is the major concern in reflecting customer satisfaction measured in terms of latency, dropped packets, jitter, echo and throughput.

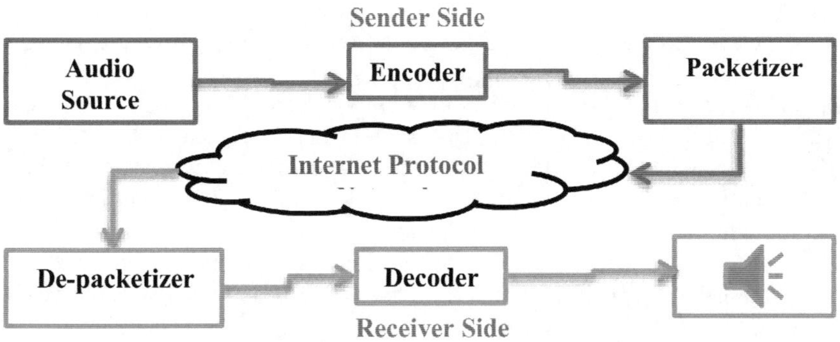

Figure 1. End-to-End VoIP communication process.

It has been observed that rising traffic in a network results in packet drop and even as small as 1% loss of packet could significantly damage VoIP service [6]. In the current study, attempt is made to offer a secure network for VoIP service using ABC based optimization followed by SVM and CNN to decrease the latency and packet drop.

Table 1. Popular Linear Prediction CODECs in VoIP service.

Codec (standard)	Bit Rate (kbps)	Frame Size (ms)	Audio Compression
G.279.1 (ITU-T)	8-32	10	CS-ACELP Algorithm
Speex	8,16,32	30 (NB), 34	CELP algorithm
EVRC (3GPP2)	9.6/4.8/1.2	20	RCELP algorithm
SILK (Skype)	6-40	Variable	LPC method
iLBC	8	30	AAC method

The paper is summarized in five sections while covering the introduction of VoIP service, popular CODECs implemented in the service, components necessary for the deployment of the service. Section 2 is dedicated to discuss the existing approaches that were implemented for the improvement of QoS related adjoining VoIP, section 3 describes the author proposed implementation to improve the reliability issues like QoS whose results and outcomes are discussed in section 4. Lastly, section 5 concludes the paper.

2. Literature Review

This part of paper summarizes the work done by various researchers. In this context researchers had discussed the security challenges faced by VoIP services. They analysed the existing approaches in terms of defence offered for best practice in addition to postulating guidelines for future researchers to secure network [7]. A security enhanced Session Initial Protocol (SIP) system bestowed was developed with protective firewall and proxy server that proved to be effective in defending against SIP flooding attacks [8]. The researchers offered Elliptic Curve Encryption approach to secure VoIP against the man in middle attack. In the process they also performed network modelling to address the DoS attacks. The simulation analysis proved the implemented strategy to successfully defend the network against SIP flooding and Denial of service attack [9]. Ad hoc On-Demand Distance Vector (AODV) routing algorithm is tested in network modelling for QoS parameters like Throughput evaluation, Packet Delivery Ratio (PDR) calculations, and energy usage [10]. The researchers developed VoIP technique to detect DoS attack and to increase security in private communication [11]. The machine learning techniques are applied for the prediction of voice quality in distinct network environment [12]. An intelligent security system has been examined to protect from Distributed Denial-of-Service (DDoS) attack that mainly involves, firstly DDoS attack detection and secondly discriminator to detect individuals with affected areas. This presented approach has been designed over a simulated telephone network. The telephone service providers can change their circuit switches networks to packet switched by using the VoIP security model [13]. The researcher proposed a VoIP by using stream cipher technique and chaotic cryptography to produce secure key. The stream without encryption as well as stream on the basis of cipher and logistic key generator has been examined [14]. Another researcher proposed a secure VOIP model that worked on the basis of hidden Markov model against dynamic voice spammer. The malicious user has been detected based on the voice behaviour and the results have been computed in two scenarios that is one for heavy traffic and other for medium traffic. The TPR (True Positive Rate) for heavy and medium traffic of 95% and 92.5% has been attained [15]. The approach Convolutional Neural Network based Multi-domain Learning Scheme (CNN-MLS), has been used a classifier for arriving at a result based on features outcome from total subnets [16]. The researchers reviewed various research papers in the context of Artificial Bee Colony (ABC) algorithm and found useful to solve the real time engineering problems [17]. A reliable wireless network presented for VoIP based application and examined performance for routing overhead, average delay and average jitter [18].

3. Methodology

This section discusses the proposed solution designed to offer a secure VoIP service. Literature reflects number of contributions made by researchers in this context, however none have been able to offer a desired level of network security required for VoIP service.

3.1. Network Deployment

The very step is to define a simulation area for experimentation and evaluating the performance of VoIP service. Therefore, a simulation area (A) of 1000 m^2 is defined in the proposed work. This is followed by the deployment of the network containing 'N' number of nodes with each node having a coverage limit of 25% of the total simulation area. Among the nodes a source node and a destination node are defined and route is created between the two for data transmission. The performance of deployed network for VoIP service are computed and improved using ABC fitness function for optimizing the network based on route node property. ABC works by evaluating nodes properties based on the packet dropped, latency and inter-node distance. Two machine-learning classifiers namely, SVM and CNN follow this step. In the present work, SVM acts to identify affected route based on the node properties and this information is fed to CNN that further identifies the affected node present in the identified affected route. The performance of the route to offer secure VoIP service is computed in terms of QoS parameters. The flow diagram of proposed design is shown in Figure 2.

3.2. Optimized Node Property using ABC

In response to the unsatisfactory performance of the network routes that were created for deployment of VoIP service few steps are followed to improve the network quality. The very first step is the evaluation of the nodes based on the inherit properties. ABC works by identifying an optimal threshold values based on the fitness function to offer a satisfactory outcome. All the nodes exhibit distinguished qualities or properties, namely, location, energy consumption, latency and packet drop during data transmission that significantly challenges the overall network performance. In the present work, ABC fitness function is used to distinguish nodes into two classes as normal and abnormal nodes based on node property. The steps involved in the process are given in Algorithm 1.

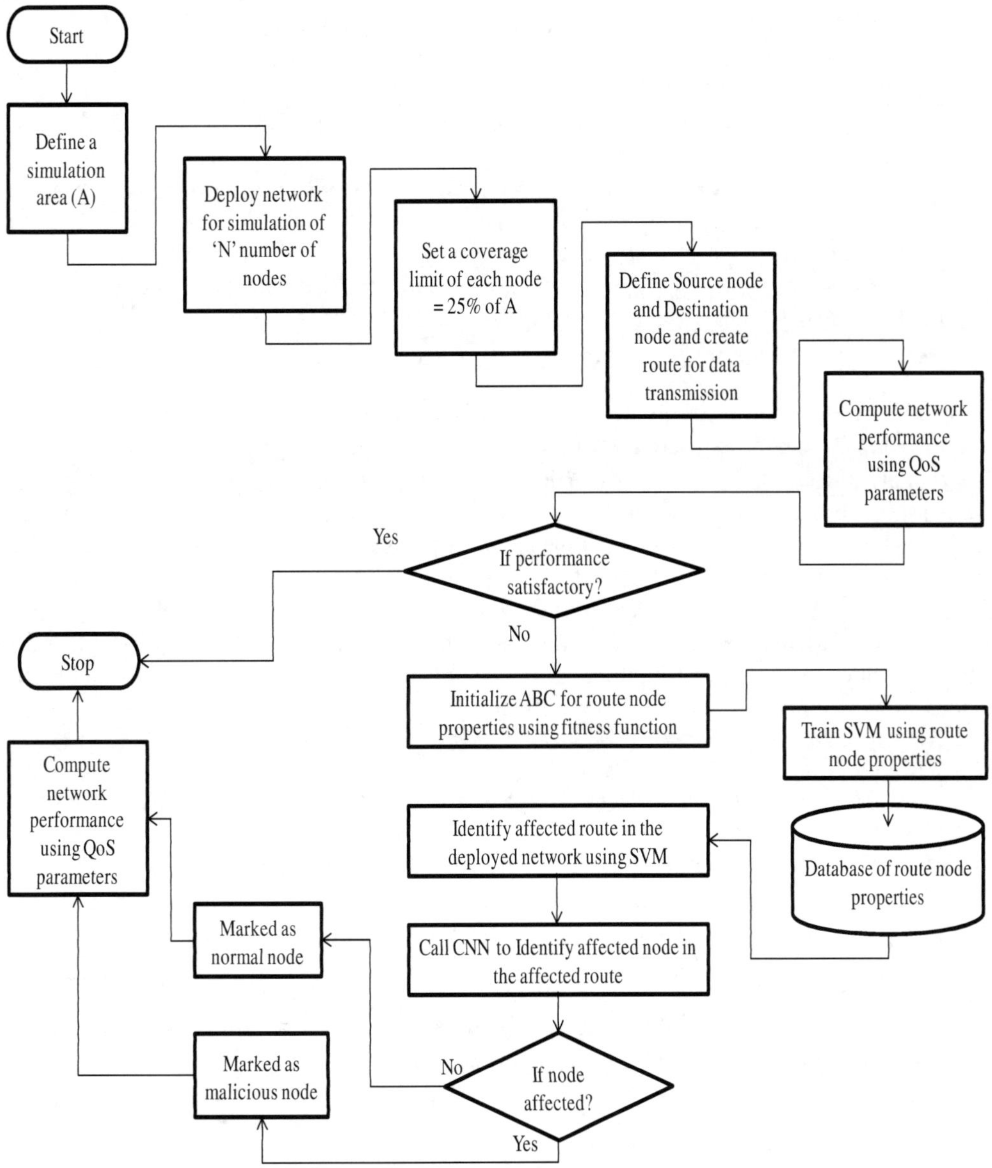

Figure 2. Flowchart of the proposed design.

Algorithm 1: ABC for defining node property
1. Input: $N \rightarrow$ Number of nodes (bee population) deployed in 1000m^2 area
2. Output: $node_a$ and $node_n$ // group of abnormal and normal nodes
3. Initialize parameters Calculate Length of $Node$
4. $N = length(Nodes)$ //represents size of nodes
5. Locate random nodes
6. $Node_{position} = random$ (0,1) //randomly locate next node within network area
7. For_{each} i in N Calculate distance matrix
8. $Node_{dist} = \min(Node_{i+1} - Node_i)$ // finds distance between current and the next node.
9. Calculate fitness score based on all node properties

10. $f_{th} = \sum_{i=1}^{N} Node_{prop}$ // calculate threshold value for nodes based on node property (latency, delay, energy consumption and distance)
11. Calculated the fitness function
12. $if\ Node_{prop} < f_{th}$ // checks for low latency, energy consumption
13. $Assign$ the $node_{current} \rightarrow node_n$
14. $Else,\ assgin\ node_{current} \rightarrow node_a$ // node with high latency, energy consumption and smaller node distance is assigned as affected node
15. Select the one with least fitness value
16. $Best_{value} = min(Node_{prop})$ //select the best one with least $node_{prop}$ value
17. End_{if} End_{for}
18. $Return: node_a\ and\ node_n$ // nodes distinguished as affected and normal nodes

The above algorithm is called to divide the 'N' number of nodes into two groups, namely normal nodes $node_n$ and the abnormal nodes $node_a$ based on the node properties $node_{prop}$ exhibited by them during data transmission. ABC analyze the nodes in the created route in terms of their inter-node distance that is expected to be larger for better network coverage. In addition to this, $node_{prop}$ namely latency, energy consumption and packet drop are also analyzed that are expected to be of smaller magnitude in order to offer better network performance. Based on this information, ABC fitness function is called to divide all the nodes into two groups as $node_n$that contains nodes demonstrating the properties of normal nodes to offer better network data transmission and $node_a$ that contains nodes that exhibited property that challenges the quality of data transmission. This information is further forwarded to computational intelligence techniques.

3.3. Identification of Affected Route using SVM

In the last step, ABC objective function was used to distinguish nodes based on their properties during data transmission from source node to destination node. In the current step, the route is evaluated using SVM based on the information passed by the ABC regarding $node_{prop}$. In the present work, radial basis function (RBF) kennel is used that is considered one of the popular kernelled training algorithms. The steps involved in identifying the route property is given in Algorithm 2.

Algorithm 2: SVM for identifying affected route
1. Input: $T_{data} \rightarrow node_a\ and\ node_n$ // Categories data of optimized nodes based on $node_{prop}$
2. Output: R_a and R_n // affected route and normal route
3. Initialize SVM Set parameters
4. $Kernel_{function} \rightarrow$ RBF polynomial
5. $T_{data} \rightarrow$ For training of the network
6. $RSVMtrain_{st} = SVMTrain\left(T_{data}, Category, Kernel_{function}\right)$ // train SVM
7. $R_{gp_{prop}} = SVM.classify(RSVMtrain_{st}, node_{prop})$ // identifies trusted route based on property of group of nodes involved
8. $if\ R_{gp_{prop}} == Valid$ // if the route node group property gets validated
9. $Assign\ R_{gp_{prop}} \rightarrow R_n$ // normal route
10. $Else,\ R_{gp_{prop}} \rightarrow R_a$ // affected route

The above algorithm uses the categories data T_{data} as training data that was based on the $node_{prop}$. This information along with SVM kernel function is used to train SVM that creates a training database of route node properties $R_{gp_{prop}}$. The trained SVM now proved to successfully predict the affected route R_a among the deployed routes. The information of the affected route is further used in the next step.

3.4. Malicious node prediction using CNN

The current sub-section discusses the implementation of CNN for predicting the affected node within the affected route identified by SVM. CNN architecture is more complicated and considered as a multilayer perceptron. It is one of the classes of deep neural networks that has pooling layer that significantly reduces the dimensionality of the features. It consists of convolutional layers as input layer, pooling layers also known as sub-sampling layers and fully connected layers that are the last few layers in the architecture to share the output. CNN training is considered to be computationally expensive however; overall it speeds up the computation by reducing feature size while reducing the processing time. It is initialized using parameters such as number of epoch, neurons, cross entropy, gradient, and mutations in addition to technique as Scale Conjugate. The steps involved in the prediction of affected node are given in Algorithm 3.

Algorithm 3: CNN to predict affected node

1. Input: $T_{data} \rightarrow R_n\ and\ R_a$ // Categories data of route as training data
2. Output: $N_n\ and\ N_m$ // normal node and malicious node in the affected route
3. Initialize CNN parameters
4. E_{num} // epochs number representing iterations used by CNN
5. N_{num} // number of neurons used
6. $Cross\ Entropy$, Gradient, Mutation, validation
7. Scaled Conjugate// technique used
8. $random$ // data division
9. $RCNNtrain_{st} = CNNTrain\ (T_{data}, Category, N_{num})$ // train CNN
10. $N_{gp_{prop}} = CNN.classify(RCNNtrain_{st}, node_{prop})$ // identifies affected node
11. if $N_{gp_{prop}} == Valid$ // if the route node group property gets validated
12. Assign $N_{gp_{prop}} \rightarrow N_n$ // normal node
13. Else, $N_{gp_{prop}} \rightarrow N_m$ // malicious node

The above algorithm uses the information of affected route R_a passed by SVM classifier. Based on this information, convolutional layer trains the CNN structure to evaluate the nodes present within the affected route. In case the node property get validated, the particular node is marked as normal node N_n and considered to be secure for data transmission otherwise, the node is marked as malicious node N_m.The above discussed steps are designed to achieve a reliable and secure VoIP based communication. The knowledge based selection of ideal and secure nodes no doubt improved the communication system. Hence, using this hybrid approach speeds up the simulation and results in a faster communication system desired for VoIP services.

4. Results and Discussion

The performance of the proposed improved communication process is evaluated in terms of network reliability. VoIP transmission is evaluated under two scenarios, firstly under normal network without attack and secondly VoIP communication under attack. In the following sections QoS is measured in terms of parametric values of Throughput, Packet loss rate, jitter and latency observed for VoIP service under both scenarios.

4.1. Throughput Comparison

Table 2 lists the Throughput observed before and after the attack with variation in the number of nodes from 5 to 50. The values are plotted in Figure 3 for graphical analysis with node count plotted on X-axis against the throughput on Y-axis. It is observed that an average throughput of 96.88% is achieved using ABC with SVM in comparison to 97.7% by ABC with CNN in comparison to 98.65% by the proposed design before attack. The same get lowered to 95.62%, 97.26% and 98.46% due to network attack. The enhanced throughput by the proposed design is due to involvement both SVM and CNN by offering a secure route discovery.

Table 2. Throughput comparison (%).

Node Count	Before Attack			After Attack		
	(ABC+SVM	ABC+CNN	Proposed)	(ABC+SVM	ABC+CNN	Proposed)
5	98.35	98.8	99.5	98.9375	98.579	99.438
10	98.02	98.5	99.3	98.6985	98.2205	99.208
20	97.24	97.9	98.7	97.862	97.5035	98.518
30	96.74	97.6	98.4	97.5035	97.145	98.173
40	95.84	96.9	98.1	96.7865	96.3085	97.828
50	95.14	96.5	97.9	95.95	95.8305	97.598

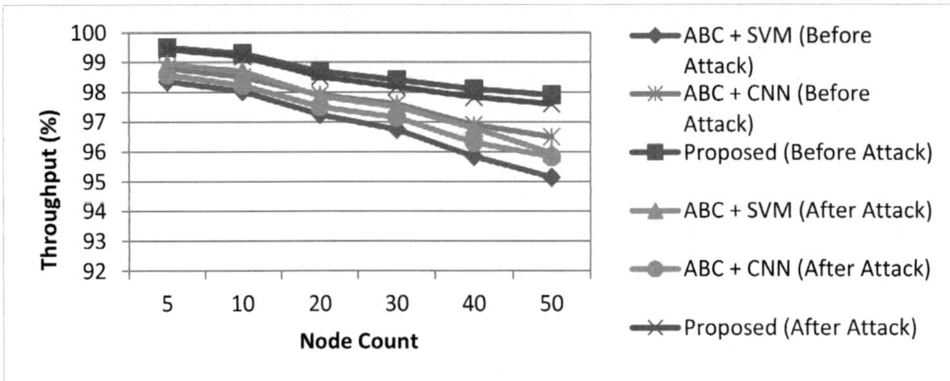

Figure 3. Throughput Comparison (%).

4.2. Packet Loss Comparison

In VoIP packets are broadcasted over the network and any loss in the packets may significantly deteriorate the voice quality. The packet loss observed before and after attack are computed and listed in Table 3. The node counts for the same varies between 5 and 50. The proposed approach has shown relatively lesser packet loss as compared to ABC with SVM.

Table 3. Packet Loss comparison (%).

Node Count	Before Attack			After Attack		
	(ABC+SVM	ABC+CNN	Proposed)	(ABC+SVM	ABC+ CNN	Proposed)
5	0.9	1.2	0.5	1.0625	1.421	0.562
10	1.1	1.5	0.7	1.3015	1.7795	0.792
20	1.8	2.1	1.3	2.138	2.4965	1.482
30	2.1	2.4	1.6	2.4965	2.855	1.827
40	2.7	3.1	1.9	3.2135	3.6915	2.172
50	3.4	3.5	2.1	4.05	4.1695	2.402

Packet loss of the proposed in comparison to ABC with SVM and ABC with CNN is shown in Figure 4. It is observed that proposed approach exhibited lower average packet loss of 2.0% and 2.37% are found using ABC with SVM, 2.3% and 2.7% using ABC with CNN are found before and after attack as compared to the proposed work using ABC with SVM and CNN that exhibited an average packet loss of 1.35% and 1.53% before and after attack, respectively. Overall, lower packet loss has been observed using proposed approach.

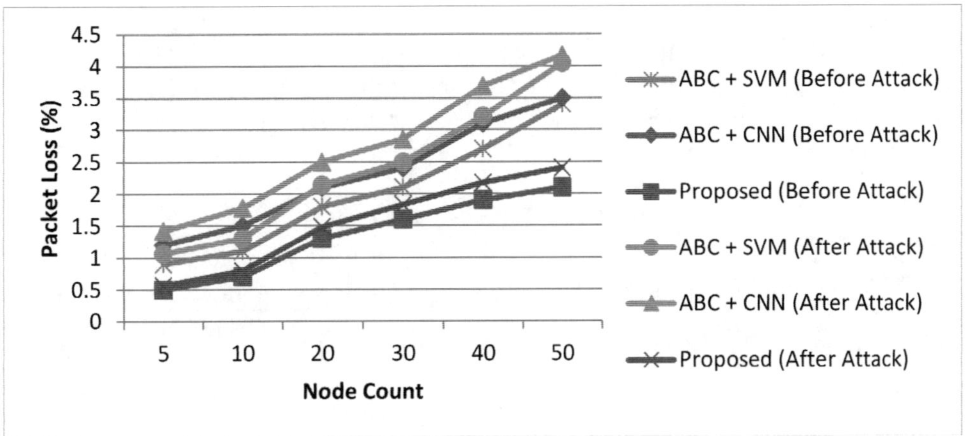

Figure 4. Packet Loss comparison (%).

4.3. Jitter Comparison (ms)

Jitter represents the variations observed in the delay of the packets in milliseconds (ms). The jitter observed before and after attack has been summarized in Table 4 for both ABC with SVM and proposed work. Overall lower jitter has been demonstrated by proposed approach.

Table 4. Jitter comparison (ms).

Node Count	Before Attack			After Attack		
	(ABC+SVM	ABC+CNN)	Proposed)	(ABC+SVM	ABC+ CNN	Proposed)
5	1.4	1.2	0.48	1.653	1.415	0.539
10	1.9	2.1	0.87	2.248	2.486	0.9875
20	2.7	2.6	1.5	3.2	3.081	1.712
30	3.9	3.6	2.6	4.628	4.271	2.977
40	4.8	4.3	3.8	5.699	5.104	4.357
50	5.4	5.1	4.7	6.413	6.056	5.392

Jitter comparison in Figure 5 shows that before attack proposed approach demonstrated a lower average jitter of 2.325 ms as compared to 3.35 ms by ABC with SVM and 3.15 ms by ABC with CNN. The jitter for all the cases increased to 3.973 ms (ABC with SVM), 3.735 ms (ABC with CNN) and 2.66 ms (proposed) as a result of network attack. However, jitter for the proposed work remains lower in comparison to the other combinations.

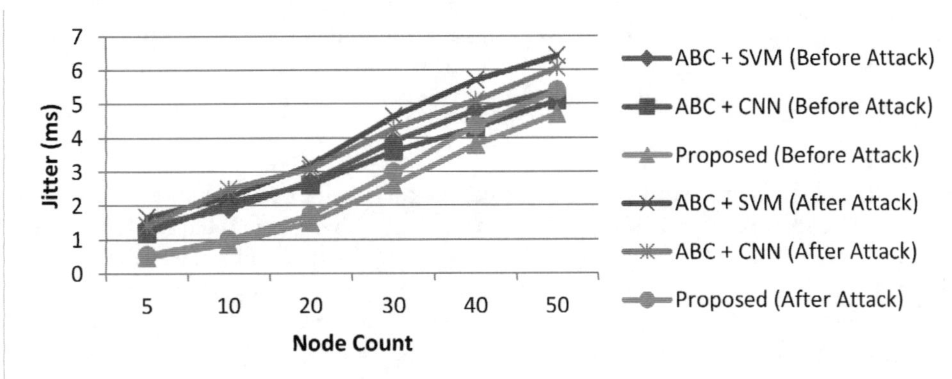

Figure 5. Jitter Comparison (ms).

4.4. Latency Comparison (ms)

Latency is compared to identify the overall delay in the packet delivery over the network. Latency values of the proposed and the ABC with SVM along with ABC with CNN are listed in Table 5 before and following the attack. The tabulated latency values show that overall lesser average delay has been observed using proposed approach.

Table 5. Latency comparison (ms).

Node Count	Before Attack			After Attack		
	ABC+SVM	ABC+CNN	Proposed	ABC+SVM	ABC+ CNN	Proposed
5	1.1	0.8	0.3	1.296	0.939	0.332
10	1.5	1.3	0.9	1.772	1.534	1.022
20	2.6	2.1	1.4	3.081	2.486	1.597
30	3.4	2.6	1.9	4.033	3.081	2.172
40	4.5	3.8	2	5.342	4.509	2.287
50	5.9	4.6	3.2	7.008	5.461	3.667

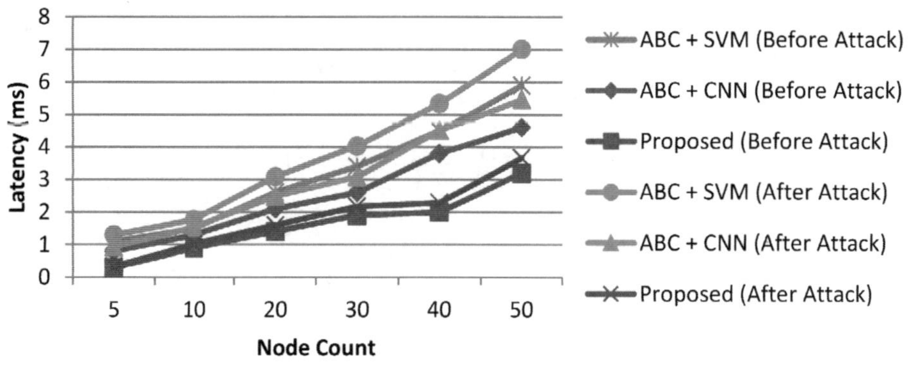

Figure 6. Latency Comparison.

Latency of the three combinations is plotted in Figure 6 for both cases before and after the network attack. An average lower latency of 1.616 ms has been observed using proposed approach in comparison to 3.166 ms and 2.533 ms using ABC with SVM and ABC with CNN, respectively. The latency values get increased to 3.75 ms, 3.001 ms and 1.846 ms following the network attack in case of ABC with SVM, ABC with CNN and the proposed work, respectively. The overall lower latency was demonstrated by the proposed approach. The performance analysis has shown that the proposed work involving three fold strategies, namely optimization of node properties using ABC followed by identification of attacked route and malicious node in the attacked route proved the effectiveness of the design in achieving a reliable and secure VoIP service.

5. Conclusion

A reliable and secure wireless network for VoIP network has been designed in MATLAB simulator. To show the efficiency of the proposed work, the results have been computed in the presence and in the absence of attack. The ABC based optimization is performed to optimize nodes based on node properties. This information was used by SVM classifier to detect affected route followed by CNN to identify the malicious node present in the affected node. The efficiency of the designed approach has been evaluated over 50 nodes in terms of parametric values of throughput, packet loss, latency and jitter to have a reliable and secure VoIP service. The network performance demonstrated the higher average throughput of 98.65% and 98.46% before and after attack with overall lower packet loss, jitter and latency of 1.35%, 2.325 ms, and 1.616 ms before attack, 1.539%, 2.66 ms and 1.846 ms after attack for the proposed design as compared to the existing approach that reflects a better and improved VoIP service over wireless network.

References

[1] Golait D and Hubballi N 2016 Detecting anomalous behavior in VoIP systems: A discrete event system modeling *IEEE Transactions on Information Forensics and Security* **12(3)** 730-745

[2] Kumar D, Ryu Y and Jang H 2008 Quality of service (QoS) of voice over MAC protocol 802.11 using ns-2 *Proceedings of the 1st ACM international workshop on Communicability design and evaluation in cultural and ecological multimedia system* 39-44

[3] Cai L, Xiao Y, Shen X, Cai L and W Mark J 2006 VoIP over WLAN: Voice capacity, admission control, QoS, and MAC *International Journal of communication systems* **19(4)** 491-508

[4] Markopoulou A P, Tobagi F A and Karam M J 2003 Assessing the quality of voice communications over internet backbones *IEEE/ACM transactions on networking* **11(5)** 747-760

[5] Kumar V and Roy O P 2020 Reliability and Security Analysis of VoIP Communication Systems *In Rising Threats in Expert Applications and Solutions* **1187** 687-693

[6] Amin A M 2005 VoIP Performance measurement using QoS parameters *Second International Conference on Innovations in Information Technology* 1-10

[7] Butcher D, Li X and Guo J 2007 Security challenge and defense in VoIPinfrastructures *IEEE Transactions on Systems, Man, and Cybernetics, Part C (Applications and Reviews)* **37(6)** 1152-1162

[8] Deng X 2008 Security of VoIP: Analysis, Testing and Mitigation of SIP-based DDoS attacks on VoIP Networks 31-49

[9] Bommagani G, Gunjan V K, Paidipelli A and Singh P 2013 Security enhancement of voip protocols using ECC *International Journal* **3(12)** 158-164

[10] Kumar V and Roy O P 2020 Ad Hoc Network Based Energy Efficient Reactive Routing Algorithm for Internet of Things and VoIP *SSRN 3516707* **1556-5068** 59-64

[11] Safoine R, Mounir S and Farchi A 2018 Comparative study on DOS attacks Detection Techniques in SIP-based VOIP networks *IEEE 6th International Conference on Multimedia Computing and Systems* 1-5

[12] Hu Z, Yan, H, Yan T, Geng H and Liu G 2020 Evaluating QoE in VoIP networks with QoS mapping and machine learning algorithms *Neurocomputing* **386** 63-83

[13] Semerci M, Cemgil A T and Sankur B 2018 An intelligent cyber security system against DDoS attacks in SIP networks *Computer Networks* **136** 137-154

[14] Ibrahem M K and Kassim H A 2018 VoIP Speech Encryption System Using Stream Cipher with Chaotic Key Generator *Journal of Fundamental and Applied Sciences* **10(6S)** 204-210

[15] Vennila G, Manikandan M S K and Suresh M N 2018 Dynamic voice spammer's detection using Hidden Markov Model for Voice over Internet Protocol network *Computers and Security* **73** 1-16

[16] Huang Y, Li B, Barni M and Huang J 2019 Identification of VoIP Speech With Multiple Domain Deep Features *IEEE Transactions on Information Forensics and Security* **15** 2253-2267

[17] Sharma A, Choudhary S, Pachauri R K, Shrivastava A and Kumar D 2020 A Review on Artificial Bee Colony and it's Engineering Applications *Journal of Critical Reviews* **7(11)** 4097-4107

[18] Kumar V and Roy O P 2020 A Reliable and Secure Wireless Network for VoIP Applications *Electronic System and Intelligent Computing* **686** 1-10

IOP Conf. Series: Materials Science and Engineering 1020 (2021) 012026 doi:10.1088/1757-899X/1020/1/012026

Molecular Identification of Lactic Acid Bacteria as a predominant probiotic microorganism found in indigenous fermented pig fat of Assam, India

Ranjita Yukhaibam[1] and Kimjolly Lhouvum[2]

[1,2] Department of Biotechnology, National Institute of Technology, Arunachal Pradesh, Yupia-791112, India

Email: ranjita.phd@nitap.ac.in / kimjolly@nitap.ac.in

Abstract. The study was focused on identifying probiotic bacteria from traditionally fermented pig fat (Sathu) procured from Karbi Anglong Assam. Two types of bacteria were isolated (KJc8_C8 and KJR2_C9) from the food sample. The isolates were morphologically found to be cocci and rod-shaped. The isolated strains showed resistance to inhibitory compound such as NaCl (1-10%), Bile salt (0.1-1%) that exhibit better growth in acidic condition which were the important characteristics for bacteria to be probiotic. In addition the isolates showed positive results in metabolizing different carbohydrates sources. Molecular identification using 16SrDNA gene sequencing confirmed the presence of probiotic bacteria namely *Bacillus* and *Lactobacillus* species. The experimental results further indicated the absence of any other spoilage bacteria in the food sample which predominated only Lactobacillus and Bacillus species. This strongly supports the food product as a potential probiotic. Further studies of the food product will be a milestone for economic growth of the society beyond its nutritional effects.

1. Introduction

Northeast represents one of the biodiversity hotspots region in India with a huge reserve of traditional fermented foods. Among various fermented food products, the fermented pork fat is one of the ethnic fermented food which is traditionally prepared from animal source. It was consumed by the diverse groups of people and the tribes such as Dimasa, Hmar, Zeme, Naga, Rangmei, Baite, Vaiphei, Hrangkhol, Karbi, Kuki, Khasi, Jayantia and Khelma living in North Cachar district of Assam [1,2]. Since eating of meat is a habitual culture among indigenous people in Northeast. Domestic animals are slaughtered on festivals and special occasion, the meat is cooked and the fat portion are preserved by fermentation for future consumption.

Some of the common traditionally processed fermented meat products which were lesser known and scientifically documented from Assam were Mogonggrain (meat of buffalo /pig, boar/ deer), Honoheingrain(pig/boarmeat), Sathu (pig fat), Gwai-Rou(cow/ buffalo fat and intestine), Saphak(pig meat with fats), Gwag –Ruum (pork fat and meat) Cheu(Semi-boiled pork pieces), Dohklong(boiled pork), Doh Khleik(pig brain), Doh Kpu (Finely minced(pork), smoked dried pork/beef Eg-adin banum(pork pieces mild cooked), Saphak Assam(Pork) [1,2]. These fermented meats are at low cost and use for socio-cultural reasons by different ethnic people of Assam. In making these fermented

IOP Conf. Series: Materials Science and Engineering 1020 (2021) 012026 doi:10.1088/1757-899X/1020/1/012026

food and marketing of products, ethnic women with traditional knowledge play a major role. However preparation methods may vary due to diversities in ethnicity and specific region. The climatic condition, geographical location, food preference and seasonal availability of plants and animals were the important factors responsible in producing varieties of indigenous foods and availability within the North East States [3]. Fermented foods are an essential part in the daily meals for the people of north eastern region. Not only Northeast, traditional fermented products have played an integral part in the diet of developing nation and far East because of health benefits driven by the probiotic organisms in fermented foods. In the world, it has assumed around 50-400g of fermented foods and alcoholic beverages have consumed per day which represented 5-40%of the daily meals [4]. This is because fermented foods harbor major sources of probiotic bacteria that have been associated with human diet from traditional times. By probiotic means live microorganism that confer health benefits to hosts by modulating mucosal and innate immunity when taken in adequate amount thereby enhancing dietary and microbial stability in the intestinal tract [5,6]. Among them Lactic acid bacteria (LAB) including Lactobacillus species have accepted as probiotics and Generally Recognized as Safe (GRAS) organisms above its functional property of fermenting food materials [7,8]. In fact most strains of Bifido bacterium and LAB isolated from fermented food materials increased the survival of compatibility with the human gut micro flora and improved their chances of survival and showed effective in many ways. In general the lactobacillus and Bifido bacterium are predominant bacterial strains used as probiotics microbial flora. Among them Lactic acid bacteria including Lactobacillus Lacto coccus, Enterococcus, Bacillus, Pedi coccus, Saccharomyces were commonly occured microorganisms in raw meat as well as also representative probiotics [9,5,10]. This means of probiotic has used in food and feed industry as starter culture and research has proven their exert beneficial effects to the host that include improving intestinal microbial balance, production of bacteriocins or bactericidal proteins compounds, moreover produces bioactive molecules such as ethanol fatty acids, diacetyl, hydrogen peroxide as lactic acid thus decreasing toxicity level , lowered the luminal pH that has strong antagonistic effects in the inhibition of spoilage and other pathogenic bacteria[10]. Thus occurrence of probiotic bacteria to the fermented meat products can benefit in improving health through balancing activities of intestinal microbial flora in association with LAB. Moreover the connection between probiotic characteristics and the dairy products has scientific evidence. So, probiotic properties from animal sources provided by microorganisms has also becoming a burning research during recent years.

With the present growing concern regarding healthy lifestyles, having of good food becomes the first priority. For this, isolation of probiotic bacteria from potent food sample is the first step for production of probiotic food products. Thus the study aims to identify probiotic microorganisms by 16s rDNA sequencing and analysis of probiotic activities responsible in changing the meat products with the help of advance biotechnological tools. Processing such fermented pig fat can largely contribute to the regional economy as well as can provide good benefits towards the health sector of the society.

2. Materials and Methods

2.1 Collection of sample and isolation of bacteria.

The fermented product of pig fat was collected from local kuki lady Karbi Anglong, Assam. The sample was aseptically transferred to a sterile container and stored at 4°C refrigerator. In 10ml of Mann Rogosa Sharpe (MRS Himedia) broth 0.1 gm of sample was suspended and homogenized well. The broth was incubated anaerobically at 37°C for 24 hrs. at 200 rpm and stored in refrigerator as initial mother stock. From the stock solution, 1ml was taken and inoculated in 9ml of sterilized MRS broth which was defined as 10^{-1} dilution. Appropriate serial dilution up to 10^{-5} were prepared. 100μL each of the broth cultures from different dilution were spread plated on MRS agar plate for 48 hrs. at 37°C. The heterogeneous bacterial colonies were isolated by streaking in different MRS agar plates and incubated 24hrs at 37°C .

2.2 Morphological and biochemical characteristic

From the isolated colonies, colony morphology and Gram nature were studied. The purified isolates were preserved at 4°C for immediate used and in 30% glycerol at -20°C for future studies.

The biochemical tests such as catalase, motility, Methyl Red (MR), Voges Proskanuer(VP) , Indole and Simmons's Citrate Slant test(IMVIC test)and carbohydrates utilization tests were performed.

2.3 Physiochemical characteristics of the isolates

2.3.1 Determination of optimum pH

To determine optimum pH, 100µL of overnight grown cultures were inoculated into each sterilized 10 ml of MRS broth of different pH ranging from 2,3,4,5,6,7, 8 and 9 by adjusting the pH with 1N HCl and 0.5 M NaOH for acid and base. The inoculated broth were incubated anaerobically for 24 hrs. at 37°C. After 24 hrs. incubation, the optical density of the bacterial isolates were recorded using spectrophotometer at 600 nm along with uninoculated controlled broth [11].

2.3.2 Determination of optimum temperature

From overnight grown culture of isolated bacteria, 100µLwas inoculated into different 10ml of sterilized MRS broth and incubated anaerobically at different temperature such as 25° C, 30° C, 35° C and 45° C for 24 hrs. at 200rpm. After incubation, the bacterial growths were recorded by spectrophotometer at 600 nm along with uninoculated controlled broth [12].

2.3.3 NaCl tolerance test

The isolates were investigated for NaCl tolerance by inoculating 100µL of freshly prepared overnight grown cultures into each sterilized MRS broth with different concentration of NaCl (0,2%,4%,6%,8%,10%)and incubated for 24 hrs. at 37°C. After incubated optical density of the bacteria were measured at 600 nm [13].

2.3.4 Bile salt tolerance test

Tolerance to bile salt was performed by measuring the optical density of isolates at 600 nm of sterilized MRS medium containing different concentration (0, 0.1%, 0.3%, 0.6%, 0.8%, and 1%) of bile salts in which 100µL of overnight isolated cultures were inoculated and incubated for 24 hrs. at 37°C [11].

2.4 Molecular Identification

2.4.1 Extraction of genomic DNA form isolates

Bacterial genomic DNA was isolated using HiMedia HiPurA™ Bacterial Genomic DNA purification Kit with spin-column technology. 1.5ml of each bacterium Lactobacillus MRS Broth culture was spun down at 10000rpm for 2min and the supernatant was discarded followed by lysis of cells. The lysis was done according to the protocol based on Gram characteristics. For Gram positive bacteria, 45 mg/ml lysozyme solution was used instead of AL solution (provided with the Kit). RNase A and Proteinase K solution was added every time after cell lysis to avoid RNA ad protein contamination. After addition of Elution Buffer, the spin-columns were incubated for 30 min to obtain greater yields of bacterial genomic DNA, 10 µL of each genomic DNA sample was electrophoresed on 0.6 % Agarose gel and visualized with pro-stained Ethidium bromide under Ultraviolet light at 254nm along with 1kbDNA ladder in Gel Documentation machine. The purified DNA was stored at-20C refrigerator.

2.4.2 PCR Amplification of 16S-rDNA gene

Purified DNA was used as a template to amplify a segment of the 16S-rDNA gene by the PCR techniques using the two universal prokaryotic primer27F (5-AGAGTTTGATCMTGGCTCAG-3) and 1492 R (5-CGGTTACCTTGTTACGACTT-3)100ng of genomic DNA isolated from each bacterial strain was amplified in the PCR reaction. A total volume of 25 ml of each PCR reaction mixture was prepared where 2XPCR Tag mixture (HiMedia) was added and final concentration was made to 1X by half dilution. The program of PCR reaction was set as initial denaturation for 2 minutes at 95°C followed by denaturation step of 35 cycles at 95°C for 30 seconds , annealing step of 30 seconds at 52°C and 2 minutes extension step at 72°C with a final extension step at 72°C for 15 minutes. The amplified PCR products were checked on 1% Agarose gel with Ethidium bromide dye under ultraviolet light in Gel DOC machine and image was captured [7].

Sequencing of 16S rDNA gene of the isolates were carried out to verify the strains within each cluster. The amplified DNA was sequence with primers 785F(5-GGATTAGATACCCTGGTA-3) and 907R(5-CCGTCAATTCMTTTRAGTTT-3) by Aakaar Biotechnologies Private Limited R&D Unit 9/322 Jankipuram Extension, Lucknow, 226031, India. The sequences were aligned by using the sequence alignment and cluster W. Its homology was analyzed through BLAST algorithm to find the most related sequence in the NCBI nucleotide sequence database. After sequencing Phylogenic tree construction was performed with mega X by neighbor-joining analysis

3. Result

3.1 Isolation of Bacteria

Two types of bacteria were isolated from the fermented pork fats named as KJc8_C8 and KJR2_C9.Both the isolates produced round white and small brownish color on MRS agar plate. The isolate KJR2_C9 was found dominated on the MRS agar plate.

3.2 Morphological and Biochemical Characterization

When gram stained and examined under microscope a gram positive cocci, chain and short rod shape were found in the medium (Fig 2 and 4) typically similar to be reference lactobacillus strain MTCC 9748. The isolates were further taken for biochemical characterization, and were found negative for catalase, oxidase, indole MR, VP, Citrate test but positive in case of KJc8_C8 in MR and citrate test. Carbohydrate utilization tests were also performed to investigate the ability to ferment Glucose, sorbitol fructose, maltose, Lactose, Inositol sucrose mannitol, arbitol. The isolate KJR2_C9 was able to ferment the given sugars but in KJc8_C8 except sorbitol all the sugars were fermented (Table 1).

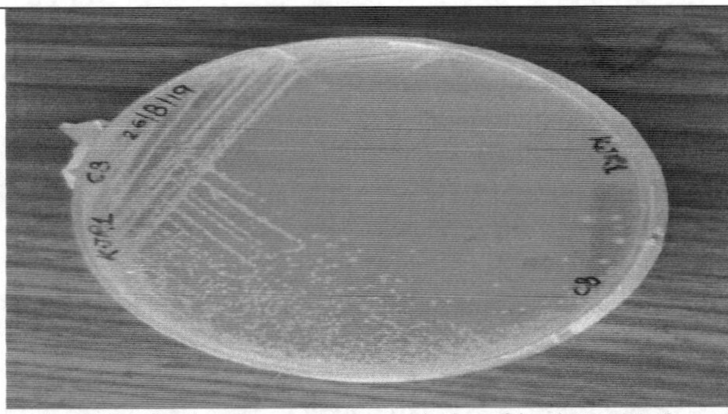

Fig. 1 Typical colony characteristic of isolate KJc8_ C8 on MRS agar

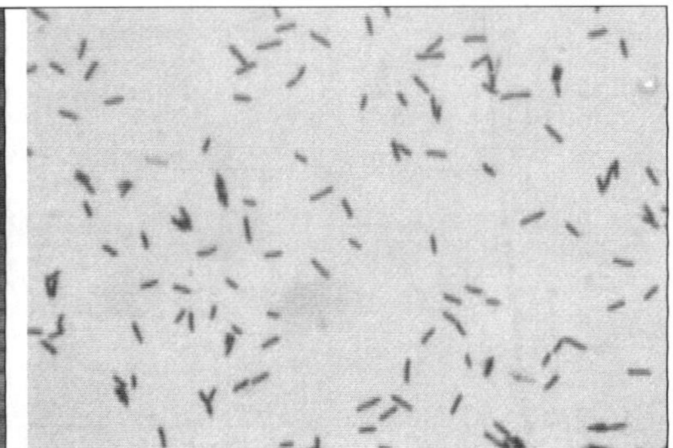

Fig .2 Microscopic view of KJc8_C8 after Gram staining

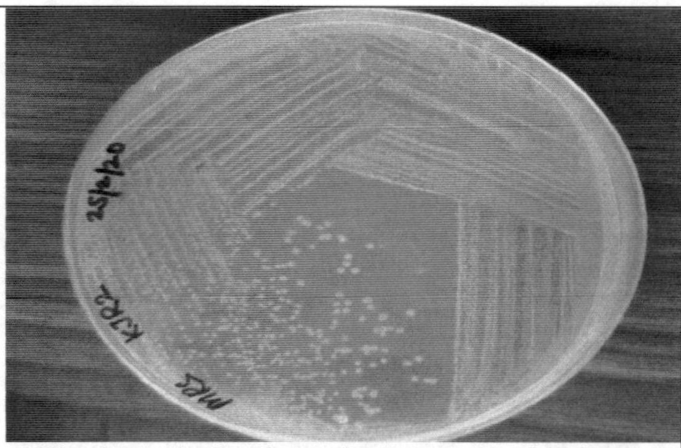

Fig. 3 Typical colony characteristic of isolate KJR2_C9 on MRS agar

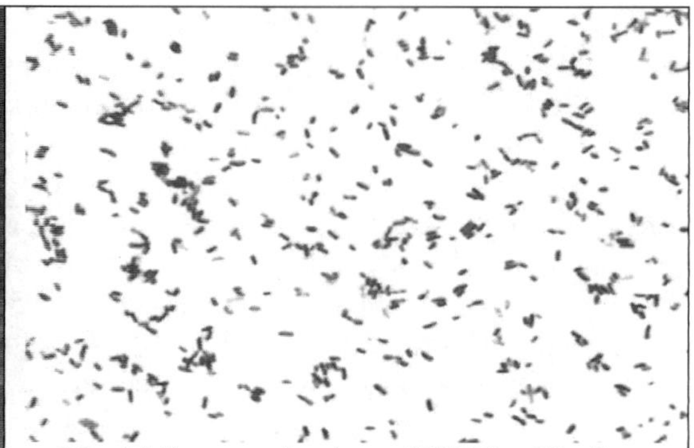

Fig.4 Microscopic view of KJR2_ C9 after Gram staining

Table 1-Morphological and Biochemical characteristic of the isolates

Test parameter	**KJc8_C8**	**KJR2_C9**	**Reference**
Gram stain	+	+	
Colony morphology	Round white color	Small and brownish	Bansal et al 2009
Microscopic view	Cocci chain	Short Single rod	Menconi et.al 2004
Catalase test	+	_	
Methyl Red(MR)Test	+	_	
Indole test	-	-	
Citrate test	+	-	
Voges Proskauer(V) test	-	-	
Glucose fermentation	+/-	+	
Sorbitol Fermentation	-	+	
Fructose Fermentation	+	+	
Maltose Fermentation	+	+	
Lactose Fermentation	+	+	
Inositol Fermentation	+	+	

+: Positive reaction, -: Negative reaction, +/- : weak reaction

3.3 Physio-chemical Characteristics of the isolates

The isolated bacteria were determined for the following physiochemical characteristics.

3.3.1 Determination of optimum pH:

The isolates were able to grow between pH 4 to 8. Maximum growth was found at pH 6 for both the isolates. However growth of KJc8_C8 decreases drastically after pH 6 in MRS broths at 35^{0}C in Fig. 5.The result showed KJR2_ C9 was tolerant to high pH condition. Thus the optimum pH of KJc8_C8 andKJR2_C9 were pH 6 and pH 8 respectively which means both the isolates could tolerate in high pH condition.

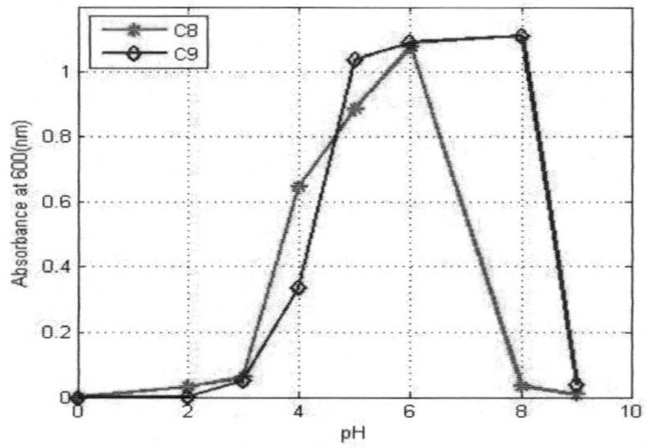

Fig. 5 Effect of pH

3.3.2 Determination of Optimum temperature:

Among the different temperature such as 25^{0}C, 35^{0}C, 40^{0}C and 45^{0}C, the best growth temperature of KJc8_C8 and KJR2_ C9 were found at 35^{0}C. After 35^{0}C growth of both the isolates decreases as in Fig. 6 .

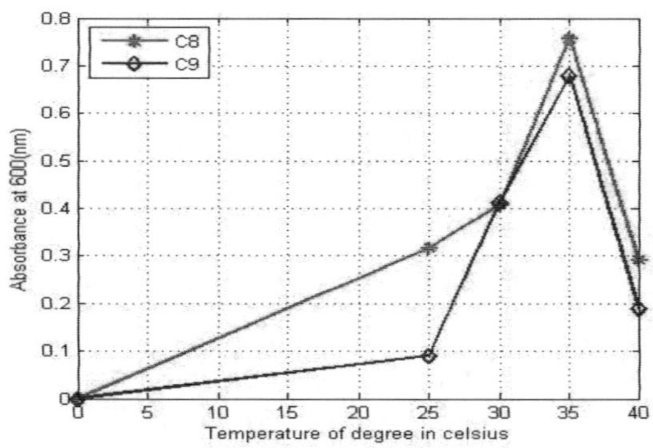

Fig. 6 Effect of temperature

3.3.3 NaCl tolerance test

The isolated species from fermented pig fat grew well in 2% NaCl concentration and started decreases their growth rate with increasing concentration .The result showed KJc8_C8 was more tolerant to NaCl than KJR2_ C9

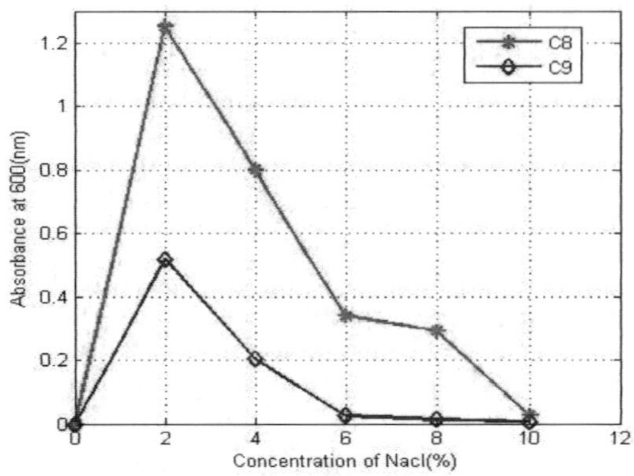

Fig. 7 NaCl tolerance test

3.3.4 Bile salt tolerance result

Both the isolates were able to endure in 0.1 to 1 % bile salt but the growth rate decreases with increase concentration of bile salt. The result showed the isolate KJR2_C9 was more inhibited by bile salt than KJc8_C8 .This indicates that bile salt act as inhibitory effect as illustrate in Fig.8.

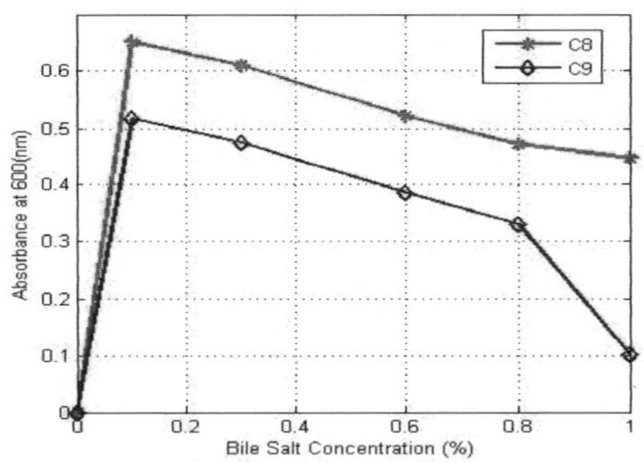

Fig. 8 Bile salt tolerance test

3.4 Molecular identification of the isolates

The amplification products of 16s rDNA sequences were run through 1% agarose gel. A sharp band at around 1.5kb of ladder sequence was observed which the size of the sequence of interest (Fig 9).

After DNA sequencing of the 16s rDNA sequences, the sequences were blasted with known sequences using NCBI nBLAST. The result showed that KJc8_C8 has highest similarity (but not 100% identical) with *Bacillus australimaris* (Table 2) whereas; KJR2_C9 has highest similarity (but not 100% identical) with *Lactobacillus brevis* (Table 2). This concludes that the isolates belong to new strains which are to be named as *Bacillus australimaris* KJc8_C8and *Lactobacillus brevis* KJR2_C9. The new sequences were submitted to NCBI Gene Bank and accession number of MK720731.1 and MK729004.1 were given for *Bacillus australimaris* KJRc8_C8 and *Lactobacillus brevis* KJR2_C9 respectively as shown in table 2.

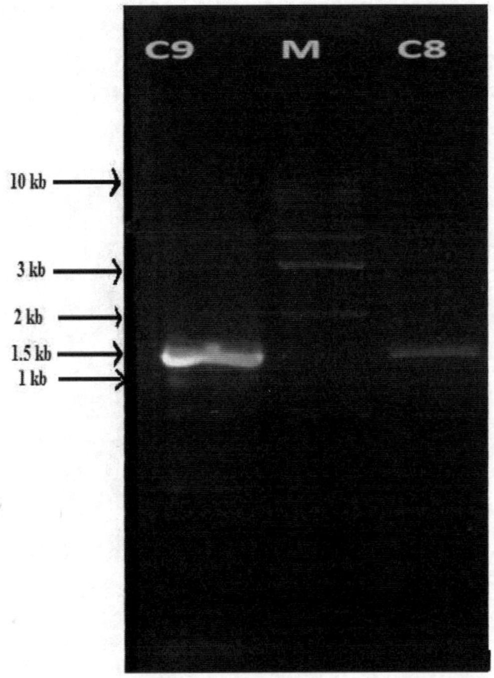

Fig. 9 Agarose gel visualization on UV trans illuminator showing PCR band of C8 and C9. C9 =KJR2_C9, M= DNA ladder ,C8 =KJc8_C8 respectively.

Table 2 Details submission of sequences in NCBI GeneBank of identified bacteria from fermented pig fat based on I6s rDNA sequencing

Name of isolate	Sequence identified	Sequence length	% identity with closest species	Gene Bank Strain accession on no.
KJRc8_C8	*Bacillus australimaris*	979	97%	MK720731.1
KJR2_C9	*Lactobacillus brevis*	1361	99%	MK729004.1

Table 3 Nucleotide BLAST result of KJc8_C8 showing the top sequences with the name of organism

Description	Max score	Total score	Query cover	E value	Ident	Accession
Bacillus australimaris strain MCCC 1A05787	1751	1751	99%	0	99%	NR_148787.1
Bacillus xiamenensis strain MCCC 1A00008	1746	1746	99%	0	99%	NR_148244.1
Bacillus stratosphericus strain 41KF2a	1742	1742	99%	0	99%	NR_042336.1
Bacillus zhangzhouensis strain MCCC 1A08372	1740	1740	99%	0	99%	NR_148786.1
Bacillus aerius strain 24K	1736	1736	98%	0	99%	NR_118439.1
Bacillus safensis strain NBRC 100820	1735	1735	97%	0	99%	NR_113945.1
Bacillus altitudinis strain 41KF2b	1724	1724	97%	0	99%	NR_042337.1
Bacillus atrophaeus strain JCM 9070	1696	1696	99%	0	98%	NR_024689.1
Bacillus subtilis subsp. subtilis strain 168	1668	1668	99%	0	98%	NR_102783.2
Bacillus mojavensis strain NBRC 15718	1657	1657	97%	0	98%	NR_112725.1
Bacillus drentensis strain LMG 21831	1526	1526	98%	0	95%	NR_118438.1

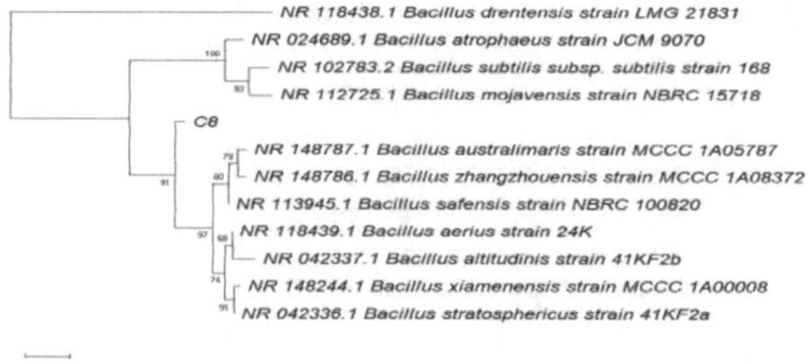

Fig.10 Phylogenetic tree (Neighbor joining method) of KJc8_C8 based on 16s rDNA sequence

Table 4 Nucleotide BLAST results of KJR2_C9 showing the top hits sequences with the name of organism

	Description	Max score	Total score	Query cover	E value	Ident	Accession
1	*Lactobacillus brevis* strain ATCC 14869	2503	2503	100%	0	99%	NR_116238.1
2	*Lactobacillus senmaizukei* strain NBRC 103853	2394	2394	100%	0	98%	NR_114251.1
3	*Lactobacillus yonginensis* strain THK-V8	2390	2390	100%	0	98%	NR_109452.1
4	*Lactobacillus hammesii* strain TMW 1.1236	2375	2375	100%	0	98%	NR_042243.1
5	*Lactobacillus koreensis* strain DCY50	2364	2364	100%	0	98%	NR_116854.1
6	*Lactobacillus cerevisiae* strain TUM BP 140423000-2250	2357	2357	100%	0	98%	NR_158030.1
7	*Lactobacillus zymae* strain R-18615	2329	2329	99%	0	98%	NR_042241.1
8	*Lactobacillus paucivorans* strain TMW 1.1424	2266	2266	99%	0	97%	NR_116943.1
9	*Lactobacillus oryzae* strain SG293	2174	2174	99%	0	96%	NR_114339.1
1	*Lactobacillusparabuchneri* strain LMG 11457	2156	2156	99%	0	95%	NR_114962.1
1	*Pediococcusethanolidurans* strain Z-9	2045	2045	99%	0	94%	NR_043291.2

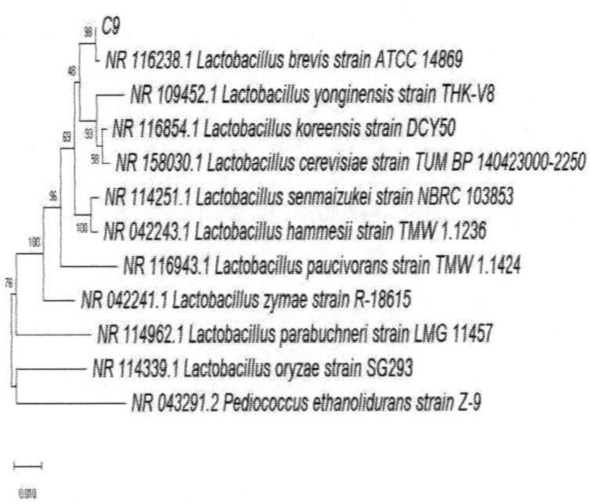

Fig. 11 Phylogenetic tree (Neighbor joining method) of KJR2_C9 based on 16s rDNA sequence

4. Discussion

According to our finding, fermented pig fat contained a rich microbiota mainly composed of two types of bacteria, *lactobacillus Brevis* KJR2_C9 and *Bacillus australimaris* KJc8_C8 without the occurrence of food borne pathogens and spoilage bacteria [1].The identification of KJc8_C8 and KJR2_C9 were carried out through phenotypic, biochemical and genotypic methods to give polyphasic approach. The combination of this method has provided greater accuracy for identification of new bacterial strain. It is also a major criterion during the selection of potential probiotic bacteria. The physiochemical characteristics tests parameters of KJc8_c8 and KJR2_ C9 were listed in Table 1.The result of gram positive reaction with morphological shape of rod and cocci with brownish and whitish colour indicated that the isolates have belonged to Lactobacillus species. The momentous growth of isolates at pH ranging from 6 to 8 at 35°C more confirmed their identification as Lactobacillus along with the results of IMVIC tests and carbohydrates fermentation tests [14]. The tested isolates showed maximum growth at 35°C while KJR2_C9 could grow up to 40°C.The ability to grow at high condition is a good characteristic for lactic acid production which decreases chances of contamination by other microorganism .The ability to grow in acidic and bile salts condition are also an in vitro tests guidelines to evaluate during the selection of potential probiotic bacteria. The strains were also determined for optimum pH. Both isolates exhibited the ability to grow in between pH 4 to 8 but favourable growth was found at pH 6. The results indicated growth rate decreases with increase in pH which has in correlation[15]. However the isolates KJR2_C9 could grow up to pH 8. This indicates Lactobacillus spp. were strain dependant. The isolated strains were able to resist (1-9 %) w/v concentration of NaCl MRS broth. Both isolates maintain good growth at 2% concentration of NaCl and declined sharply with increase of NaCl concentration in the medium. In bile salt tolerance test, the isolates showed good growth upto 0.3% w/v of bile in the MRS broth and continued resistance up to 0.8% concentration of bile salt. Tolerance to lower concentration depicted the isolates has the potential to colonize and balancing the intestinal microbial inside the small intestine [11,16]. Both the strain deceases their growth rate with increased concentration of bile salt and NaCl. In some research articles, tolerant to 0.3% bile concentration was the first property of a probiotic bacterium required for human consumption. Based on the results the isolated strains were able to live through the harsh condition of toxicity [13] therefore can be used as a potential probiotic bacteria. As reported by Lilis Nuraida not only LAB in fermented food some species of bacillus have also potential probiotics [17]. Further, 16s rDNA sequencing revealed both the isolates as a new strains of *lactobacillus Brevis* and *Bacillus australimaris* which were named as *Bacillus australimaris* KJc8_C8and *Lactobacillus brevis* KJR2_C9.

5. Conclusion

Two predominant probiotic bacteria with new strains have found in the ethnic fermented pig fat. The two bacterial strains belong to *Lactobacillus brevis* and *Bacillus australimaris* and are named as *Bacillus australimaris* KJc8_C8 and *Lactobacillus brevis* KJR2_C9. Both the isolates assure the capability and functional properties of probiotic, resistance to harsh conditions such as low pH, high salt and bile concentration. In this regards, fermented pig fat may be a potent sample for the development of probiotic food. Our study has established documented proof of the beneficial effects that may exert health promotion in human by consumption of fermented pig fat. This study confirms the occurrence of potential probiotic bacteria with new strains present in fermented meat pig fat. Further studies will provide a base for the recognition of such products as healthy foods and a supplementary for beneficial bacteria.

Acknowledgement

The authors thank Dr. Shantanu Roy, DST Inspire faculty for profound help and guidance during the course of the works.

References

[1] Chakrabarty J, Sharma GD, Tamang JP 2014 Traditional technology and product characterization of some lesser-known ethnic fermented foods and beverages of North Cachar hills district of Assam *Indian J. Traditional Knowl. 113, 706-715*
[2] Tamang JP et al. 2012 Microorganisms and nutritional value of ethnic fermented foods and alcoholic beverages of North East India. *Indian Journal of Traditional Knowledge, 11,7-25*
[3] Das G et al. 2016 Diversity of traditional and fermented foods of the Seven Sister states of India and their nutritional and nutraceutical potential: a review. *Front Life Sci, 9, 292-312*
[4] Wolkers Roojackers JCM et al. 2016 *Food Fermentation Brewing and Diary*. Elesevier
[5] Bansal S, Singh A, Mangal M, Sharma Sk 2013 Isolation and characterization of lactic acid bacteria from fermented foods. Vegetos, *26, 325-330*
[6] Penner R, Fedorak RN, Madsen KL 2005 Probiotics and nutraceuticals: Non-medical treatments of gastrointestinal diseases. *Currrent Opinion in Pharmacology, 5, 596-603*
[7] Adimpong DB et al. 2012 Genotypic characterization and safety assessment of lactic acid bacteria form indigenous African fermented food products. *BMC Microbiology, 12, 10-12*
[8] Halder D et al. 2017 Indigenous probiotic Lactobacillus isolates presenting antibiotic like activity against human pathogenic bacteria. Biomedicines, *5,1-11*
[9] Amara AA, Shibla 2015 Role of Probiotics in health improvement, infection control and disease treatment and management. *Saudi Pharmaceutical Journal, 23, 107-114*
[10] Nanasombat S et al.2017 Lactic acid bacteria isolated from raw and fermented pork products: Identification and characterization of catalase- producing Pediococuus pentasaceus. *Food Science and Biotechnology, 26,173-179*
[11] Menconi A et al. 2014 Identification and Characterization of Lactic Acid Bacteria in a Commercial Probiotic Culture. *Bioscience of Microbiota Food and Health, 33, 25-30*
[12] Pyar H & Peh KK 2014 Characterization and identification of Lactobacillus Acidophilus using biolog rapid identification system. *International Journal of Pharmacy and Pharmaceutical Sciences, 6, 189-193*
[13] Ahmed MM 2012 Screening of Lactobacillus spp. From Buffalo Yoghurt for Probiotic and Antibacterial Activity. *Journal of Bacteriology and Parasitolog, 3, 156*
[14] Islam KN et al. 2016 Characterization and Confirmation of Lactobacillus spp. from Selective Regional Yoghurts for Probiotic and Interference with Pathogenic Bacterial Growth. *Asian Journal of Biological Sciences, 9, 1-9*
[15] Chowdhury et al. 2012 Screening Lactobacillus spp. From Buffalo Yoghurt for Probiotic and Antibacterial Activity. *Journal of Bacteriology and Parasitology,* 3, 8
[16] Nawaz ASN, Jagadeesh KS, Krishnaraj PU 2017 Isolation and Screening of Lactic Acid Bacteria for Acidic pH and Bile Tolerance. *International Journal of Current Microbiology and Applied Sciences, 6, 3975-3980*
[17] Nuraida L 2015 A review: Health promoting lactic acid bacteria in traditional Indonesian fermented foods. *Food Science and Human Wellness, 4, 47-55*

Word Level Language Identification on Code-Mixed English-Bodo Text

Nayan Jyoti Kalita, Ankita Goyal Agarwala, Jayprakash Das
Department of Computer Science and Engineering,
The Assam Royal Global University
E-mail: nayan.jk.123@gmail.com, ankita89.goyal@gmail.com, dasjay084@gmail.com

Abstract. Since social media has become an active part of one's life, people express their views freely in mixed informal languages on such platforms. So, in a multi-lingual country like India, it becomes really difficult for conventional language detectors to identify such languages. This paper mainly aims to detect the language at word level where the code mixed text can be in English-Bodo-Assamese. The data for the same is collected from some related Facebook pages and various classification algorithms are used to predict and compare the accuracy with which the detection is done.

Keywords: Code Mixing, Classification Algorithm, English, Assamese, Bodo

1. Introduction

With the increasing use of social media like Facebook, WhatsApp, etc., people often tend to express their ideas, feelings, statements, etc. in mixed languages like English-Hindi or Hindi-English-any Regional language together. Identifying the languages of the texts extracted from such platforms becomes very difficult. Social media facilities the sharing of ideas, thoughts and information and the building of the virtual networks and communities.However, it has adversely affected the language identification task as social media users writes texts in short or with spelling mistakes or with different phonetic typing, acronyms, or mixed with different languages, etc. Also, instead of using only Unicode, social media users makes use of phonetic typing, Roman script or transliteration for writing texts and have developed the habit of mixing different languages more frequently to express heir views and thoughts.

The main purpose of Language Identification is to examine the extracted text of each document to determine the primary language and any secondary languages present in that document.

2. Motivation

The text or data collected from any social media platform is highly unstructured with a lot of non-standard abbreviations. Moreover, these data can also include texts written in some regional languages and the traditional language detectors cannot determine the language behind these texts because of their multilinguality. And Assam being rich for its diverse set of languages, the problem of language identification becomes all the more severe. So, the main objective of this piece of work is to identify the language of the code mixed text which will be written in English-Bodo-Assamese language.

3. Organisation of the Paper

The paper is organized as follows: Section 4 depicts a summary of the various work done on code mixing using social media text. Section 5 then describes the corpus that has been built from posts collected from Facebook groups and pages and which is code mixed among English, Bodo and Hindi language. Section 6 describes the different classification algorithms used along with the methods used for word level language detection, based on character n-grams, dictionaries. The results and discussion are reported in Section 7, while Section 8 concludes this piece of work and proposes some future line of work.

4. Background Study

A lot of work has been done so far on the language identification of text collected from social media platforms out of which some of them are summarized as follows:

In [1], Ted Dunning in 1994 had built a language identification model based on Markov model and it got 92% accuracy using 20 bytes size of test text data and 50 thousand training data set.

Arjen Poutsama in [2] introduced a language identification method which was based on Monte Carlo sampling in 2002. The proposed technique performed somewhat less than best- performing classification techniques, but it worked much faster.

Heba Elfardy et al. [3] in 2013 presented a model which could perform sentence level identification between modern standard Arabic and Egyptian Dialectal Arabic. They used token level to derive sentence level features. Their system achieved an accuracy of 85.5% on online - community dataset.

William B. Cavnar et al. in [4] proposed an N-Gram- Based Text Categorisation system. This system was based on calculating and comparing profiles of N-gram frequencies. They achieved as high as an 80% correct classification rate and concluded that N-gram frequency profiles provides a simple and reliable way to categorise documents in a wide range of classification tasks.

In [5] , Utsab Barman et al. proposed a model at word- level classification in 2014 using a simple dictionary-based method, linear kernel support vector machines (SVMs) and a k-nearest neighbour approach. Based on these experiments, finally they selected SVM- based system and presented results for the Nepali-English and Spanish-English datasets.

In 2014, Amitava Das et al.in [6] proposed a model using n-gram with weights and support vector machine(SVM) along with additional features such as dictionary based for the English Bengali Hindi code mixing corpus and they achieved 94% accuracy on their system.

Nayan Jyoti Kalita et al.in [7] in 2018 proposed a model on social media code mixed corpus on English Assamese language and they found 89% accuracy using Support Vector Machine with additional features such as n-gram with weights, dictionary based and word position checking.

With this detailed study, it was noted that apart from the classification algorithms, the accuracy of identification also greatly depends on the language itself, the source from where the data is taken and the size of the corpus. Therefore, same classifier may have different accuracies with respect to these factors. In this work, the identification will be done using three different classifiers and the results will then be compared with each other.

5. Corpus Design

A lot of work has been done so far under language identification of texts collected from Social Media. Most of these have central languages like English,Hindi, some Arabic languages, etc. However, not much has been done with regional languages. Here, the code mixing using English-Bodo-Assamese language. The Bodo language is the official language of the Bodoland Autonomous region and co-official language of the state of Assam. It is also one of the 22 official language of India.

The English Bodo code mixing language was collected from two Bodo pages of Facebook which contains 15,022 tokens and 6,602 unique words. Since Hindi is the official language of

Tag	Description	Example	Total	%
EN	English	The Like	4612	30.30
BD	Bodo	Mwjang, Ma	7,055	46.96
HI	Hindi	Guru, Bhawan	160	1.06
ACRO	Acronym	OMG, RIP	214	1.42
UNIV	Universal	!,&,()	1495	9.95
NE	Named Entity	Assam,Delhi	840	5.59
UN	Undefined	Ned,Cograts	416	2.76
WM	Word Mix	Publickho	230	1.53

Table 1. Code Mix Tag Set

India so it acts as a bridge for communication between different tribes. In the corpus collected, a small percentage of Hindi of about 1% was also found.

5.1. Annotation
The collected corpus was annotated at word level with 8 different tags. The annotation is shown below in the table 1.

5.2. Level of Code Mixing in the Corpus
The code mixing occurs at different level. The distribution of different code switching present in the corpus is as follows: 31.59% from intra sentential, 65.15% from inter senenial and 3.36% from intra word. The examples of sentences present in the corpus in the stated three forms are as follows:

- Inter-sentential: Mwjang/BD jadwang/BD Sir/EN.
- Intra-sentential: Banlgadeshi/NE Sir/EN Jwb/BD nama/BD.
- Intra-word: Publickho/WM aakhai/BD phwnaghwnai/BD nonga/DB.

In the intra-word case, *publickho* is the mixing of two words from two different languages, public(English) and kho(Bodo) .

6. Classification Algorithms Used
The language identification here is done using three classification algorithms namely the Decision Tree, Naive Bayes and the Multilayer Perceptron. These three classifiers are described as follows:

6.1. Decision Tree Classifier
Decision trees [8] tries to evaluate the instances of data by creating a tree in a greedy manner and traverses as deep as possible until a prediction can be made. Once the tree is made, it is pruned in order to improve the model's ability to generalize to new data.

6.2. Naive Bayes Classifier
Naive Bayes [9] is a simple classifier that used the Bayes Theorem as its base. For the implementation, this method calculates the the posterior probability for each of the seven mentioned classes and makes a prediction for the class with the highest probability.

Classifiers	Precision	Recall	F-Score	RMS
Decision Tree	0.708	0.544	0.471	0.269
Naive Bayes	0.762	0.693	0.659	0.299
Multilayer Perceptron	0.335	0.425	0.372	0.288

Table 2. Detailed accuracy with different features

6.3. Multi Layer Perceptron Classifier
A multilayer perceptron (MLP) [10] is a classifier that generally works with three layers namely: the input layer, hidden layer and output layer. A technique called the backpropagation is used for training the network and some activation functons are used for the hidden layers.

For the said study, English Bodo code mixing language were selected through various post and comment from some Facebook pages. The collected corpus was annotated at the word level with different tags. Weka V3.9[8] tool was used to implement three different classifiers with the default parameters and it was trained with the features like:- n-gram with weight, dictionary based, Acronym check, etc. The reason behind choosing three different classifiers was mainly to test their accuracy. Also the chosen classifiers show better results in comparison to others across multiple domains including language identification. Following are some of the features used to train the model:

- **N-gram with weights:** This feature is carried out using the bag-of-words principle. If there are n unique n-grams for a language pair, then consider them as n unique features. For example, that in is the ith bigram in the list. In a given word w (e.g., training), a particular n-gram occurs k times (twice for in in raining). If the pre-calculated weight of in is t_w^i the feature vector is 1,2,...,(t_w^i * k),...,(n-2),(n-1),n
- **Dictionary Based:** There are two English dictionaries, so there are two binary features. 1 is used to mark the presence of a word in a specific dictionary and 0 is used to mark its its absence. Lexical Normalisation Dictionary and ABC (Australian Broadcasting Commission 2006)3 are used.
- **Acronym Check:** For an acronym and named entity, we use there Boolean features. In general, named entity begins with a capital letter. Again, acronyms are written in all capital letters. Therefore, we have checked whether all the letters are capitalized, only the first letter is capitalized or any letter of the word is capitalized.

7. Observations and Results
The model was trained by three different classifiers with n-gram and dictionary-based features. In word level evaluation, different accuracies were obtained from the code-mixed corpus. Table 2 gives a detailed accuracy of the model with respect to the differnt classifiers and also demonstrates their respective precision, recall, F-score and Root Mean Square (RMS) error values.

The confusion matrix of the experimental setup for all the tags is reported in the Table 3. Form the experimental set-up, it was observed that decision tree classifier correctly classified 60% of the original instances, Naive Bayes Classifier identified 78% of the instances correctly and MLP classified 65% instances correctly. The overall accuracy score was found a bit low in all the three classifiers as the size of the corpus was comparatively small because of the limited availability of structured English-Bodo code mixed data. Moreover, some of the words of the Bodo language happens to be same in either English, Assamese or Hindi language. This property again contributes to the wrong classification of the instances thereby reducing the overall accuracy.

	ACRO	UNIV	NE	EN	HI	UN	BD	WM
ACRO	0	0	0	12	0	0	0	0
UNIV	0	20	0	87	0	0	0	0
NE	0	0	18	12	0	0	7	0
EN	0	0	0	344	0	0	8	0
HI	0	0	0	6	0	0	1	0
UN	0	0	0	1	0	0	0	0
BD	0	0	3	113	0	0	188	0
WM	0	0	1	2	0	0	0	0

Table 3. Confusion Matrix for 8 class in Naive Bayes Classifier

8. Conclusion and Future Scope

The scope and dimension of language processing has progressed a lot with the more prevalent use of social media where culture and language are getting mixed more often. This paper presented a study on the detection of code-mixing in the context of social media texts. Since social media texts and contain words in different languages, so identification of language becomes a challenging task. So this identification needs to be done at word level.

In this work, the corpus has been built with only the Facebook posts written in mixed English and Bodo language together. In the future, this model can be applied and experiment with other languages and other types of social media text, such as tweets. And there may be other features which can increase the accuracy. Facebook posts are generally restricted to 140 characters and tweets are even shorter. It would be interesting to investigate how this total character restriction of Facebook affects the level of code-mixing in tweets. It would be challenging to perform language identification with respect to tweets owing to its small size.

Also, this work has emphasized on code-mixing in romanized Indian social media texts, but there are other possible code-mixing cases such as Unicode and romanized Indian language text plus English.

References

[1] Dunning T 1994 *Statistical identification of language* (Computing Research Laboratory, New Mexico State University Las Cruces, NM, USA)
[2] Poutsma A 2002 *Computational Linguistics in the Netherlands 2001* (Brill Rodopi) pp 179–189
[3] Elfardy H and Diab M 2013 *Proceedings of the 51st Annual Meeting of the Association for Computational Linguistics (Volume 2: Short Papers)* pp 456–461
[4] Cavnar W B, Trenkle J M *et al.* 1994 *Proceedings of SDAIR-94, 3rd annual symposium on document analysis and information retrieval* vol 161175 (Citeseer)
[5] Barman U, Wagner J, Chrupała G and Foster J 2014 *Proceedings of The First Workshop on Computational Approaches to Code Switching* pp 127–132
[6] Das A and Gambäck B 2014 *Proceedings of The 11th International Conference of Natural Language Processing*
[7] Kalita N J and Saharia N 2018 *Proceedings of the International Conference on Computing and Communication Systems* (Springer) pp 433–440
[8] Quinlan J R 1986 *Machine learning* **1** 81–106
[9] Rish I *et al.* 2001 *IJCAI 2001 workshop on empirical methods in artificial intelligence* vol 3 pp 41–46
[10] Rumelhart D E, Hinton G E and Williams R J 1985 Learning internal representations by error propagation Tech. rep. California Univ San Diego La Jolla Inst for Cognitive Science

IOP Conf. Series: Materials Science and Engineering 1020 (2021) 012028 doi:10.1088/1757-899X/1020/1/012028

Object Detection in Live Streaming Video Using Deep Learning Approach

Jeberson Retna Raj, Senduru Srinivasulu

Department of Information Technology, Sathyabama Institute of Science and Technology, Chennai, India

Abstract. Object detection in a complex scene is of challenge as the image contains numerous features. Object detection is important in surveillance, healthcare, recognition, manufacturing, military and agriculture etc. The traditional approaches are limited in accuracy and mainly focussing on image datasets. With the application of deep learning algorithm, the accuracy can be increased. In this paper, the Convolutional Neural Network (CNN) is introduced to detect the objects in the image frames. The input of video is decomposed into image frame and the particular image frame is considered for processing. The image is preprocessed and the noises are removed for further processing. In the next step, feature extraction process contains edge detection and extracting the meaningful shape and texture in the image. The shape which contains the higher level feature is then feed into convolutional layer. The convolutional layer consist of four parts such as number of filter, filter size, pooling layer and max pooling layer. With the application of activation function Relu, the objects can be identified. Furthermore, the Gradient descent algorithm is applied for optimization to improve the accuracy. The system has been tested with ImageNet and the results are very promising.

1. Introduction

Data Science is a broad area which employs Artificial Intelligence, Machine Learning and Deep learning. Artificial intelligence means that to make intelligent externally for any kind of the system. Furthermore, it taking decision based on insight of the available data. Machine learning allows machine to crunch data and construct knowledge and intelligence. It combines applied statistics and optimization theory. Deep learning is the sub topic of machine learning which focuses on building, training and deploying large and deep neural networks. The neural network works based on the data in parallel. The journey of the problem can be defined as data science is a path, Machine Learning is a tool and AI is the Goal. The term analytics can be defined as the combination of discern and analysis made from a given data source. There are three types of analytics defined in literature. First descriptive analytics focus on what happened to the system or change of state after applying the algorithm. The fundamental is database and what we seen as a dashboard. The second is diagnosis analysis deals about root cause analysis of the problem. Data mining is the fundamental source to discover the facts in the data. The data mining techniques aims to discovering facts from wealth of information. Diagnosis analysis is a level 2 which comprising of data mining and ETL. The third is predictive analytics deals with historical things. The prediction is based on the historical data. The mathematical model comprising AI and ML allows for solving the problem. The final one is prescriptive analytics deals with decision making process. It focuses on the term 'what I do next' phrase. It is a antology or knowledge representation techniques. For performing this task, a toolkit is required with computational form. Data science employs with the techniques data handling and harnessing of data with the system. It focuses on creation, manipulation, transformation and diffusion of row archived data into information turned knowledge with support of decision.

The need for deep learning is to address the N-gram feature representation of data. The existing techniques are limited in accuracy and performance. In essence, the deep learning focuses on reusing the existing trained models. The convolutional Neural Network is one such algorithm to solve the n-gram feature vector of the image or data. The word convolution is derived from the image processing applications. The canny filter is used to identifying the edges and corners from n layers and detects the edges and boundaries of the given image. The first layer is feed with the raw data and all intermediate each subsequent layer are used to extract the data. The n-th layer is for displaying the final output. The classification is divided into binary classification and multiclassification. The binary classification techniques are used to classify the output based on one neuron at the output layer. Whereas, the multiclass classification used to classify the data with multiple class labels. In essence, the neural network passes the input value along with the weight value. The activation function used to process the data with the weight values. Many activation functions are available in literature which includes sigmoid,Tannet and Relu functions for processing hidden layer neurons. The softmax is the activation function mainly used for multiclass classification problems.

For multilayer perceptron, the following key functions are required. First, we need to define the cost function, back propagation, learning rate alpha, optimizer and regularizer parameters. The optimizer may be gradient descent or any other similar algorithm can be used. The epoch used to perform one forward and backward pass of all training examples. In essence, there are two problems can occur in these kind of processing. The over-fitting and under-fitting problems are exist in classification. The gap between the training accuracy and validation accuracy is defined as over-fitting. The under-fitting is categorized as the training itself the model not provide accuracy. To solve this over-fitting problem the drop out mechanism has been used. It randomly removes connection in hidden layer. The second important solution to the over-fitting problem is based on data augmentation technique. It is based on increase the amount of data based on available data. The parameter epoch is used to fine tune the model for early stopping of iteration.

1.1. Literature Survey

Lex et al introduced a dilated kernel based CNN algorithm is used to classify the earth surfaces of satellite images. The method replaces the traditional kernel with the dilated kernel to increase the classification accuracy. The hybrid dilated CNN model is used to address the information loss problem [1]. Lin Yuan et al proposed a CNN model for extracting the image feature of MRI brain images to classify tumor cells. The authors use multiple convolutional layers to extract the gradient information in images and combine the spatial features [2]. In [3], an ensemble based CNN method is applied with the medical images for classification. The method is applied with various dataset like Googlenet and Alexnet for measuring the performance of the method. Hou et al proposed an object detection method in panchromatic images using spatial template matching method. The approach making use of single shot multibox detector (SSD) method to extract the multi scale features used to classify the image and identify location [4]. Tang et al proposed an object detection mechanism to linking the objects in the same frame. Furthermore, cuboid network proposal method used to extract the spatio temporal features in the image [5]. Fang et al proposed a faster R-CNN algorithm for object detection. The R-CNN experiencing low object experiences in bounding boxes, to increase this Faster R-CNN algorithm is introduced [6]. Haijun et al compared the performance of different state-of the art algorithms for object detection in UAV satellite images. The algorithms successfully detect buildings, trucks, car, building collapse etc. The algorithm detects the bigger objects and unable to detect smaller objects [7]. Vipul et al presented a detailed survey on object detection using deep learning algorithms. They are focusing on image dataset and not applicable for video data [8]. Yong Tian et al introduced a metal object detection system using SVM algorithm for wireless charging technology for electric vehicles [9]. In [10], object detection in UAV images is presented. Several deep learning algorithms

IOP Conf. Series: Materials Science and Engineering 1020 (2021) 012028 doi:10.1088/1757-899X/1020/1/012028

includes Faster RCNN, cascade RCNN, R-FCNN, YOLO and its variants are discussed. These algorithms are applied for the static images and not applied for video data.

2. Proposed Methodology

The proposed method using video as the source of input and the object of interest is detected as output. The video is decomposed into image frames and the selected frame is then taken for detecting objects. In the first step, the features of the images are extracted in the image. Not all the pixel is feed into the neural network, the shape and structure is enough to identify the meaningful features in a given image. Using feature extraction process, the edges of the image can be detected. With the help of the edges, the shapes can be determined. Then the shape used to detect the higher level features such as vehicle, people, sign board, etc. The first layer is convolution layer which consist of four parts such as number of filter, filter size, pooling layer and max pooling layer. The sequence is important in CNN method. First starts with convolutional layer by perform the feature extraction process. The low pass filter is used to smooth the edges and high pass filter used to identify the edges in the image. Basically, the filter used to train the image. In our approach, sobel operator is used to train the image. For corner pixel feature, we use the 0 padding concept.

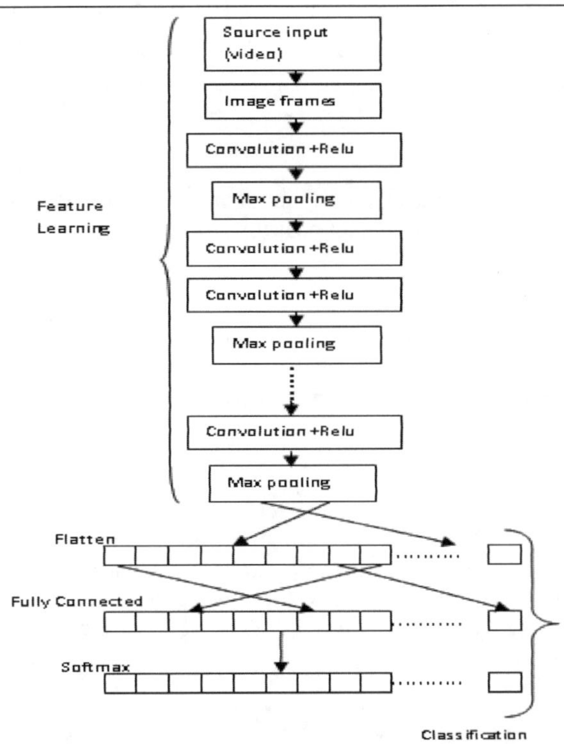

Figure 1. Flow of object detection using CNN Algorithm

The kernel value is defined using the formula. If H*W and F*F are defined as 5*5 and 3 * 3, then K can be computed as

$$K = \frac{H - F + 2P}{S} + 1$$

Where s is a stride value. For example H=5, F=3 and P=0, the filter size is odd. This means that the filter size is odd which used to fitting center pixel. Padding can be defined using the formula

IOP Conf. Series: Materials Science and Engineering 1020 (2021) 012028 doi:10.1088/1757-899X/1020/1/012028

$$p = \frac{F-1}{2}$$

where p=1

Next we have to define the max pooling layer. The maxpooling layer is used to reduce the spatial pattern. There are two pooling is available, one is max pooling and another is average pooling. Max poling layer is used to reduce the pixel window size. The maximum pixel value is considered for reducing the window size. For example 4*4 window size is reduce into 2*2 window. In our system, the max pool is reduced into 2*2 and the stride value which makes the filter is go with a non-overlapping fashion. Furthermore, maxpooling provides better results.

The convolutional neural network is an important algorithm in deep learning algorithms. The applications of CNN which includes object identification, face recognition, classification, etc. in essence, the computer process the image as array of pixels of height, width and dimension. For example 6*6*3 as RGB color image and 6*6*1 as grayscale image. Generally, CNN algorithm works with train and test data. The process involves of convolutional filter, that is kernel, pooling, fully connected layer and finally apply softmax function to classify the objects with the probabilistic values between 0 and 1.

2.1. *Convolutional layer*

Convolution is the first layer used to train the image by extracting the image features from an input image. It requires two input, one is sample window h*w*d and kernel input fh*fw*d and the output is convolutional featured 2D image.

2.2. *Stride*

Stride used to shift the number of pixels over the input image. stride equals 1 means one pixel shift and stride value 2 means shifting with two pixel positions in the image. in our work, the stride value is set as 2.

2.3. *Padding*

Padding is used to fit the kernel in the image window. There are two paddings used to fit the kernel in the image region. To cover the boundary region of the image zero padding mechanism used. otherwise, the part of the image can be apply with the kernel filter which does't cover corner pixels.

2.4. *Relu*

The softmax function Relu stands for Rectified Linear Unit.

$$f(x) = \max(0, x)$$

The Relu function used to apply the input data for getting non linearity in the convolutional network. This provides the CNN with non-negative linear values. The other functions Tanh and Sigmoid can be used instead of Relu. However, the Relu provides higher performance comparing with others.

2.5. *Pooling layer*

Pooling layer used to reduce the dimension in the feature map. There are three pooling are used for dimensionality reduction of the image feature. They are max pooling, average pooling and sum pooling. The max pooling takes the maximum pixel value in the image grid of (2*2) with the stride value of 2 and the average pooling is to sum of all pixel in the image grid and take average. The sum pooling is to sum of all elements in the selected image grid.

2.6. *Fully connected layer*

The fully connected layer used to flatten the matrix into vector data. The input vector data is feed into neural network. We create a model with the help of these fully connected layers to combine the features. With the help of softmax function, we can classify the objects. Finally, the Gradient descent algorithm is used for optimization to increase the accuracy.

3. Experimental Result

The proposed system is implemented using the CNN algorithm. The video input is processed and decomposed into image frame and the image is used for further processing. The image is preprocessed and the noise is removed. The sobel operator is used to detect the edges and corner pixels. The shape is identified and based on the shape the higher level features are identified. These higher level features are feed into CNN layer. The CNN network consist of sequence of steps containing convolutional filter with the activation function, maxpooling, Figure 1 shows the input image of road view is considered and the CNN algorithm is applied. The image consists of many objects like car, person, bus, auto etc. The system is trained such a way that it can able to detect the objects separately. The image is divided into meaningful segments. The region of proposal is extracted from the image.

Figure 2. Sample image for detecting the objects

Figure 3. Output image with objects detected using the CNN algorithm

Figure 2 shows that the objects which are detected in a video frame using CNN algorithm. The objects car, man tree are detected based on the training of data. The output clearly shows that the algorithm successfully detecting the objects.

Table 1 performance comparison

Algorithm	Accuracy (%)
Random forest	85
CNN with Gradient descent	92

Table 1 shows that the accuracy of the proposed method of object detection with gradient descent algorithm. With this algorithm, the accuracy is increased from 85% to 92%.

4. Conclusion

Detecting objects in a streaming video is complex due to the depth of information available in each frame. The video is sliced into image frames and the image of interest will be identified for processing. The images are divided into training and testing set for experiment. The system successfully processed with the images. The CNN algorithm with the Gradient descent optimization algorithm is used and the results are compared. The results are very promising when the gradient method is integrated with the CNN. The system can be extended with Fast RCN method for similar data sets.

References

[1] Lei, X., Pan, H., & Huang, X. (2019). A dilated CNN model for image classification. IEEE

Access, PP:124087-124094 1–1. doi:10.1109/access. 2019.2927169
[2] Yuan, L., Wei, X., Shen, H., Zeng, L.-L., & Hu, D. (2018). Multi-center Brain Imaging Classification Using A Novel 3D CNN Approach. IEEE Access, PP: 124082-124095, 1–1. doi:10.1109/access.2018.2868813
[3] Kumar, A., Kim, J., Lyndon, D., Fulham, M., & Feng, D. (2017). An Ensemble of Fine-Tuned Convolutional Neural Networks for Medical Image Classification. IEEE Journal of Biomedical and Health Informatics, 21(1), 31–40. doi:10.1109/jbhi.2016.2635663
[4] B. Hou, Z. Ren, W. Zhao, Q. Wu and L. Jiao, "Object Detection in High-Resolution Panchromatic Images Using Deep Models and Spatial Template Matching," in IEEE Transactions on Geoscience and Remote Sensing, vol. 58, no. 2, pp. 956-970, Feb. 2020, doi: 10.1109/TGRS.2019.2942103.
[5] P. Tang, C. Wang, X. Wang, W. Liu, W. Zeng and J. Wang, "Object Detection in Videos by High Quality Object Linking," in IEEE Transactions on Pattern Analysis and Machine Intelligence, vol. 42, no. 5, pp. 1272-1278, 1 May 2020, doi: 10.1109/TPAMI.2019.2910529.
[6] F. Fang, L. Li, H. Zhu and J. Lim, "Combining Faster R-CNN and Model-Driven Clustering for Elongated Object Detection," in IEEE Transactions on Image Processing, vol. 29, pp. 2052-2065, 2020, doi: 10.1109/TIP.2019.2947792.
[7] Haijun Zhang, Mingshan Sun, Qun Li, Linlin Liu, Ming Liu, Yuzhu Ji, “An empirical study of multi-scale object detection in high resolutionUAV images”, Neuro Computing 421(20201), PP:173-182
[8] Vipul Sharma Roohie Naaz Mir, A comprehensive and systematic look up into deep learning based object detection techniques: A review, Computer Science Review 38 (2020), 100301
[9] YongTian, ZhengLi, Yawen Lin, LijuanXiang, Xiaoyu Li, Yonghong Shao, JindongTian, Metal object detection for electric vehicle inductive power transfer systems based on hyperspectral imaging, Journal of Measurement 168 (2021) 108493
[10] Payal Mittal, Raman Singh, Akashdeep Sharma, Deep learning-based object detection in low-altitude UAV datasets:A survey, Image and Vision Computing, 104 (2020) 104046

IOP Conf. Series: Materials Science and Engineering 1020 (2021) 012029 doi:10.1088/1757-899X/1020/1/012029

Dielectric Material Selection for High Capacitance Ratio and Low Loss in MEMS Capacitive Switch using Ashby's Methodology

Kurmendra[1], Rajesh Kumar[2]
[1]Department of Electronics and Communication Engineering, Rajiv Gandhi University (A Central University), Doimukh, Arunachal Pradesh.
[2]Department of Electronics and Communication Engineering, North Eastern Regional Institute of Science and Technology, Nirjuli, Arunachal Pradesh.

Abstract. The performance of the RF MEMS capacitive switch is majorly decided on capacitance ratio and loss in the dielectric layer as they decide the isolation (in dBs) in off condition. The capacitances in a MEMS capacitive switch in pull-up and pull-down states are used for deciding the capacitance ratio. In this article, we have presented a method to find out the best dielectric material using Ashby's methodology for high capacitance ratio and low loss thus high isolation in switch off condition. Firstly, a database of 10 mostly employed dielectric materials have been created, and then, material indices are derived using material properties such as dielectric constant, resistivity, conductivity, and loss tangent which have been used in the selection of dielectric material. Through the study it is observed that to have a high capacitance ratio and low loss in dielectrics, the down capacitance should always be much greater than pull-up capacitance and tangent loss must be kept at possible minimum. This method will be highly useful in the selection of the best dielectric material by switch designers.

1. Introduction

Integrated circuit (IC) industries have seen revolutionary developments in the RF components in the last few decades due to growth in MEMS [1]. The switches are one of these components which have witnessed lots of new technological development in recent years. The MEMS technology has tremendously acquired an important position in development of these switches [2]. In general, if a switch has been fabricated using micromachining process it possess various advantages such as very small size, low voltage requirement, low weight and most importantly a greater RF characteristics even at higher frequencies [3-5]. Taye et al. (2013) [6] presented a capacitive switch using MEMS where silicon nitride was used as dielectric material to have better isolation and high capacitance ratio. Use of silicon nitride as dielectric material had resulted in better down state capacitance and thus provided an improved capacitance ratio in the range of 83.20-177 for air gap variation of 2-3 μm. Reddy et al.[7] have proposed a novel capacitive RF MEMS switch by using a beam of butterfly shape in which aluminium nitride was used as dielectric material. The results were obtained for capacitance ratio and it is found that the capacitance ratio is 44-125 for air gap variation of 1-3 μm. Fouladi et al. [8] have presented a RF MEMS switch structure in which a standard 0.35 μm CMOS technology was used to create a wrapped metal- dielectric membrane. In their result it was found that the capacitance ratio was enhanced to 91 using standard CMOS process. Yu Liu et al. [9] introduced a new MEMS switch based on (BaSrTiO) thin film technology, through their results it is found that the capacitance ratio was improved to 175. A.B. Yu et al. [10] demonstrated a membrane polarisation method for enhancing isolation of MEMS switch. The capacitance ratio was calculated as 38 for a range of air gap of 2.5-3 μm. Jae Y. Park et al. [11] experimentally tested the Strantium titanate oxide (STO) material for fabricating a MEMS switch. The thin films of STOs were fabricated and tested for dielectric constant at different temperatures and it was found that STO can work even at higher temperature and provide a good dielectric constant for having a capacitance ratio of 600 or more. V. Thakur et al. [12] designed a three metal switch structure using MEMS for energy harvesting where the capacitance ratio was observed as 72 for an aluminium oxide thin membrane of dielectric material and X Rottenberg et al. [13] had used three structures for their study as standard, top metal and boosted. All three structure uses tantalum pent-oxide (Ta_2O_5) as dielectric material where it was found that the boosted structure provides the capacitance ratio value as 459 which was way better that the top metal structure. This increase in the capacitance ratio was achieved because the pull up capacitance was lowered to approximately 5 times compare to top metal.

Therefore from above introduction, it can be said that the selection of a dielectric material should be done carefully to obtain good performance in RF MEMS switches. The Dielectric material used and capacitance ratio achieved by using them is listed in Table. 1.

Table.1 Dielectric materials used in MEMS Switch structure for enhancing capacitance ratio

Sr. no.	Dielectric material used in the switch structure	Capacitance ratio obtained	References
1	Silicon nitride (Si_3N_4)	83.20-177 for varying air gap of 2-3 μm	Taye et al. [6]
2	Aluminium nitride (AlN) (er-9.5)	44-125 for varying air gap of 1-3 μm	Reddy et al. [7]
3	Dielectric developed under standard 0.35 μm CMOS process	91 for a wrapped metal dielectric membrane	Fouladi et al. [8]
4	BST-MEMS Switch	175	Yu Liu et al. [9]
5	Silicon nitride (Si_3N_4)	38	A.B. Yu et al. [10]
6	Strantium titanate oxide (STO) with high dielectric constant 110-120 at 300 degree celcius	600	Jae Y. Park et al. [11]
7	Al_2O_3	72.7	V. Thakur et al. [12]
8	Ta_2O_5	459	X Rottenberg et al. [13]

2. Methodology for Selection of Materials

This section presents methodology used for selection of dielectric material for an RF MEMS capacitive switch. To select a material for dielectric layers from available materials, Ashby's Methodology [14] is used which is considered as one of most employed methods for selection of material for microsystem and MEMS devices previously. As per Ashby's Methodology, any systems performance can be characterised by three requirements namely functional requirements F, geometric requirements G and materials requirements M.

$$P = f(F,G,M)$$

The three parameters F, G and M are independent. If functional and geometric requirements are met, the material requirement can be obtained easily. The material selection process requires material indices to select best available material from all materials. The material indices are found using material properties of a material. Then, these material indices are plotted to select best available material from all materials. This can be understood from flowchart provided below.

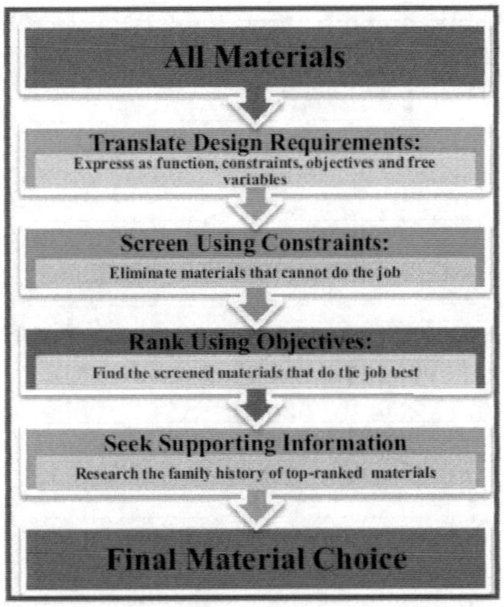

Figure.1 Ashby's material selection methodology [14]

To select a dielectric material from all available dielectric materials, the material indices are deduced from available information in the literatures. This study aims to suggest a material to improve capacitance ratio of a capacitive MEMS switches. Thus, equations for capacitance will be utilized in up and down capacitance. Later, equations for ratio of down-state capacitance to upstate capacitance will be used to see which material will provide good capacitance ratio.

The up-state capacitance in a MEMS capacitive switch is given by [15],

$$C_u = \frac{\varepsilon_0(A_o - A_h)}{g_0 + \frac{t_d}{\varepsilon_r}} + C_f \quad (1)$$

Where, C_u is upstate capacitance, $\boldsymbol{\varepsilon}_0$ is permittivity of free space, A_o is overlapping contact area, A_h is area of one perforation made on beam bridge, t_d is thickness of dielectric materials and C_f is fringe capacitance.

The down state capacitance in an RF MEMS switch is expressed as [15]

$$C_d = \frac{\varepsilon_0 \varepsilon_r (A_o - A_h)}{t_d} \quad (2)$$

where, C_d is downstate capacitance.

Dielectric constant

Here, t_d is the only one parameter which affects material properties and will be considered as first material index.

$$C_u, C_d \ \alpha \ \frac{1}{t_d} \quad (3)$$

The t_d is related to dielectric constant 'k'' as [16]

$$k' = \frac{C.t_d}{\varepsilon_0 A} \quad (4)$$

Where, k is the dielectric constant of dielectric material. It can be seen that k is proportional to dielectric thickness $'t_d'$. So first material index is

$$MI_1 = k' \quad (5)$$

Resistivity

The resistivity of a dielectric material should be high to offer good isolation property. The capacitance in down state should be high to offer high capacitance ratio. Thus, capacitance and resistivity are proportional to each other. Thus, next material index would be

$$MI_2 = \rho \quad (6)$$

Loss tangent

The losses in the MEMS capacitive switch are not only dependent of metal electrodes but it is also dependent on dielectric loss tangent (tan δ). The loss tangent (tan δ) is a quantitative parameter which is used to see the losses in dielectrics due to their material properties such as electrical conductivity and polarisation of dielectrics ($\boldsymbol{\varepsilon}_r$). The dietetic loss can be expressed as [22]

$$Dielectric\ loss = \frac{\sigma}{2}\sqrt{\frac{\mu_0}{\varepsilon_r}} \quad (7)$$

Where, 'σ' is electrical conductivity of the dielectric material, and ε_r is polarisation of the dielectric material. From equation (8), it can be observed that the loss in dielectric layer of the switch is related to two material properties electrical resistivity (σ) and is polarisation of the dielectric material (ε_r). So, next two material indices are

$$MI_3 = \sigma \text{ and } MI_4 = \sqrt{\frac{1}{\varepsilon_r}} \quad (8)$$

The above four material indices will be used in the selection of dielectric material for a MEMS capacitive switch using Ashby's methodology.

3. **Result and Discussions**

The material properties as per material indices requirement have been listed in Table-2. As per methodology discussed in previous section, the plots have been presented below.

Table-2 Dielectric material and their properties used [17, 18, 19, 20].

Dielectric Material	Dielectric constant k'	Resistivity(Ω- m)	Electrical Conductivity [S/m]
Silicon nitride (Si_3N_4)	9.5	10^{12}	10^{-12}
Aluminium nitride (AlN)	9.3	10^{11}	10^{-11}
SiO_2	3.9	10^{15}	10^{-15}
HFO_2	23.4	10^{7}	10^{-7}
Strantium titanate oxide (STO)	120	10^{8}	10^{-8}
Al_2O_3	9.8	10^{12}	10^{-12}

Ta_2O_5	25	10^{11}	10-^{11}
TiO_2	86	10^{10}	10-^{10}
ZnO	8.5	10^{13}	10-^{13}
ZrO_2	23	10^{10}	10-^{10}

Ashby's methodology of material selection uses material indices to find out best available material. In MEMS capacitive switches, the down capacitance is required to be as high as possible. Thus to obtain a high downstate capacitance in the switch, the dielectric material needs to have large dielectric constant value. The dielectric in MEMS switches must exhibit good resistance to electrical signals in off state and thus, to maintain good isolation in the switch. The isolation property in the switch is related to dialectic material resistivity (Ω- m), high resistivity results in good isolation in the switch and vice-versa is true. It is desired that a good dielectric material must have good resistivity in off state and high dielectric constant.

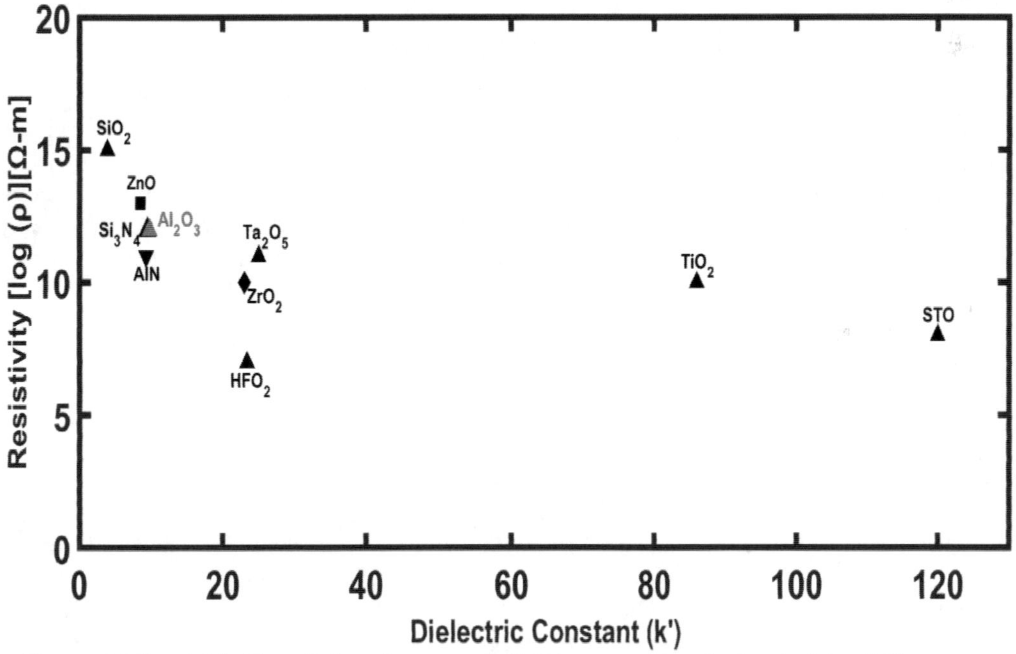

Figure.2 Material indices dielectric constant (MI_1) versus electrical resistivity (MI_2)

The plot for dielectric constant and resistivity of dielectric material has been presented in Figure.2 where the resistivity is taken in logarithmic due to simplifying a larger range of resistivity of dielectric materials. Thus, from Figure.2 it is found that the dielectric material TiO_2 and STO have high dielectric constant and also moderate resistivity compared to other available materials. So, these two materials can be a good choice for obtaining good capacitance in downstate conditions. However, a material with a very high dielectric constant may cause stiction issues in the switch and reduces the reliability of the switch [21]. HFO_2 is also used in the MEMS capacitive switches but its resistivity is comparatively low from other dielectric material which is not desired for good isolation in the switch. The dielectrics such as Si_3N_4, Al_2O_3, ZnO, and AlN would be preferred choice for MEMS switches since they have good resistivity and moderate dielectric constant which will also improve the reliability of the switch. SiO_2 shows excellent resistivity among all available materials in this study but it is not preferable since it has the lowest dielectric constant.

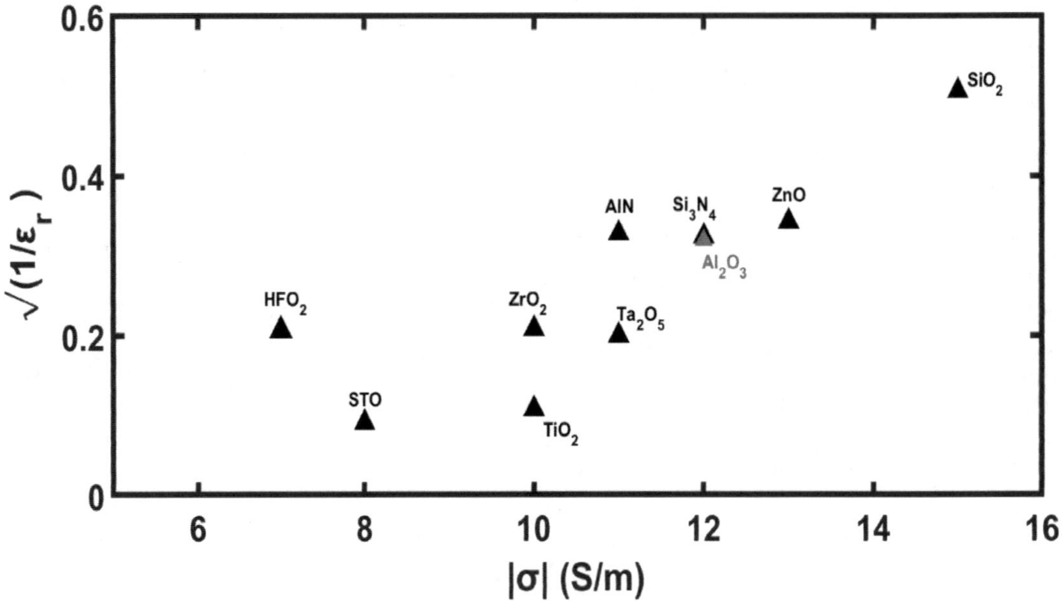

Figure.3 Material indices MI_3 versus electrical conductivity (MI_4).

The loss of signals in dielectric also reduces switch performance and therefore, while selecting a dielectric material a MEMS switch requires prominent attention. The electrical signal conductivity and reciprocal of dielectric polarization are considered to be two very important material indices for the selection of a dielectric material which results in the low loss while switching operation. The plot for logarithmic of electrical conductivity and reciprocal of dielectric polarization is depicted in Figure. 3. The modulus and logarithmic functions in electrical conductivity are used for positive value and to cover a larger range respectively. As discussed above, the material which exhibits good resistance to electrical signals that is bad electrical signal conduction and extremely low value of $\sqrt{\frac{1}{\varepsilon_r}}$ will be best choice of material for low loss in the dielectrics. The dielectric materials HFO_2 and STO are the best choices of materials for the low loss but the reliability issues due to these two make them less utilized in MEMS switches. SiO_2 is not recommended at all since it will allow more loss in the dielectric layer. The TiO_2 is the most suitable candidate for low loss and however, the stiction and dielectric charging is more compared to other dielectrics such as AlN, Si_3N_4, Al_2O_3, ZnO, and Ta_2O_5 will result in a moderate loss but the advantage is that the switch will be free from stiction and improved reliability is achieved.

4. **Conclusion**

The selection of material in any device demands utmost care since it improves the performance of that device. The method for selection of dielectric material has been proposed on the basis of Ashby's methodology. The dielectric constant, electrical resistivity, electrical conductivity, and tangent loss of the dielectric material are used as material indices to get a material that improves performance in the switch. The capacitance ratio and dielectric loss are considered a key parameter for the selection of dielectric material in this study since good capacitance ratio and minimum loss results in high isolation. The good choices of materials for the dielectric layer are Si_3N_4, Al_2O_3, ZnO, and AlN since they offer good isolation in the switch. These materials are preferred since they improve reliability in the switch, unlike TiO_2 and STO. Further, Ashby's methodology of material selection proved to be most accurate among all available material selection methods. This study also concludes that a trade-off is needed between performance and reliability in MEMS switches while selecting a material for these switches.

References

[1]. Nagod, S., & Halse, S. V. (2017). Evolution of MEMS Technology. *Evolution*, *4*(12).

[2]. Cao, T., Hu, T., & Zhao, Y. (2020). Research Status and Development Trend of MEMS Switches: A Review. *Micromachines*, *11*(7), 694.

[3]. Iannacci, J. (2018). RF-MEMS technology as an enabler of 5G: Low-loss ohmic switch tested up to 110 GHz. *Sensors and Actuators A: Physical*, *279*, 624-629.

[4]. Shekhar, S., Vinoy, K. J., & Ananthasuresh, G. K. (2017). Surface-micromachined capacitive RF switches with low actuation voltage and steady contact. *Journal of Microelectromechanical Systems*, *26*(3), 643-652.

[5]. Kageyama, T., Shinozaki, K., Zhang, L., Lu, J., Takaki, H., & Lee, S. S. (2018). Fabrication of an Au–Au/carbon nanotube-composite contacts RF-MEMS switch. *Micro and Nano Systems Letters*, *6*(1), 1-8.

[6]. Taye, J., Guha, K., & Baishya, S. (2014). Design and analysis of RF MEMS shunt capacitive switch for low actuation voltage and high capacitance ratio. In *Physics of Semiconductor Devices* (pp. 445-448). Springer, Cham.

[7]. Reddy, B. L., & Shanmuganantham, T. (2014). Design of novel capacitive RF MEMS shunt switch with aluminum nitride (AlN) dielectric. *Procedia materials science*, *6*, 692-700.

[8]. Fouladi, S., & Mansour, R. R. (2010). Capacitive RF MEMS Switches Fabricated in Standard 0.35-μm CMOS Technology. *IEEE transactions on microwave theory and techniques*, *58*(2), 478-486.

[9]. Liu, Y., Taylor, T. R., Speck, J. S., & York, R. A. (2002, June). High-isolation BST-MEMS switches. In *2002 IEEE MTT-S International Microwave Symposium Digest (Cat. No. 02CH37278)* (Vol. 1, pp. 227-230). IEEE.

[10]. Yu, A. B., Liu, A. Q., Zhang, Q. X., Alphones, A., Zhu, L., & Shacklock, A. P. (2005). Improvement of isolation for MEMS capacitive switch via membrane planarization. *Sensors and actuators A: Physical*, *119*(1), 206-213.

[11]. Park, J. Y., Kim, G. H., Chung, K. W., & Bu, J. U. (2000, January). Electroplated RF MEMS capacitive switches. In *Proceedings IEEE Thirteenth Annual International Conference on Micro Electro Mechanical Systems (Cat. No. 00CH36308)* (pp. 639-644). IEEE.

[12]. Thakar, V., Wu, Z., & Rais-Zadeh, M. (2012). A high ON/OFF ratio MEMS capacitive switch with applications in solar energy harvesting. In *Solid-State Sensors, Actuators and Microsystems Workshop* (pp. 385-88).

[13]. Rottenberg, X., Jansen, H., Nauwelaers, B., Fiorini, P., De Raedt, W., & Tilmans, H. A. C. (2002, November). Boosted RF-MEMS capacitive shunt switches. In *Proceedings of Workshop on Semiconductor Sensor and Actuator (SeSens, Veldhoven, The Netherlands, 2003)* (pp. 667-671).

[14]. Ashby, M. F., & Cebon, D. (1993). Materials selection in mechanical design. *Le Journal de Physique IV*, *3*(C7), C7-1.

[15]. Mehmood, Z., Haneef, I., & Udrea, F. (2018). Material selection for Micro-Electro-Mechanical-Systems (MEMS) using Ashby's approach. *Materials & Design*, *157*, 412-430.[15]

[16]. Yakut, S., Ulutas, K., & Deger, D. (2019). Effect of thickness on the dielectric properties and glass transition of plasma poly (ethylene oxide) thin films. *Materials Science and Engineering: C*, *104*, 109962.

[17]. Muhamad, N. F., Osman, R. A. M., Idris, M. S., & Yasin, M. N. M. (2017). Physical and electrical properties of SrTiO3 and SrZrO3. In *EPJ Web of Conferences* (Vol. 162, p. 01052). EDP Sciences.

[18]. Patra, P., & Angira, M. (2019). Investigation on Dielectric Material Selection for RF-MEMS Shunt Capacitive Switches Using Ashby, TOPSIS and VIKOR. *Transactions on Electrical and Electronic Materials*, 1-8.

[19]. Todorova, Z., Donkov, N., Ristić, Z., Bundaleski, N., Petrović, S., & Petkov, M. (2006). Electrical and Optical Characteristics of Ta2O5 Thin Films Deposited by Electron-Beam Vapor Deposition. *Plasma processes and polymers*, *3*(2), 174-178.

[20]. Raimondi, D. L., & Kay, E. (1970). High resistivity transparent ZnO thin films. *Journal of Vacuum Science and Technology*, *7*(1), 96-99.

[21]. Bansal, D., Kumar, A., Sharma, A., & Rangra, K. J. (2015). Design of compact and wide bandwidth SPDT with anti-stiction torsional RF MEMS series capacitive switch. *Microsystem Technologies*, *21*(5), 1047-1052.

[22]. https://rb.gy/ec78op

Novel design and performance analysis of WSN node using NRF and ATmega328

Ashish Gupta[1], Rajesh Kumar[2]

1 Asst. Prof, ABES Engineering College, Ghaziabad (UP) & Ph.D. Scholar Department of Electronics & Communication Engineering, North Eastern Regional Institute of Science and Technology, Nirjuli, Itanagar, Arunachal Pradesh, India:er.ashish.miet@gmail.com

2 Associate Professor, Department of ECE, North Eastern Regional Institute of Science and Technology, Nirjuli, Itanagar, Arunachal Pradesh, India: rk@nerist.ac.in, itsrk2006@gmail.com

***Abstract*—Designing an efficient wireless sensing network (WSN) node over a remote location is a challenging task in the heterogeneous environment. The two implementation challenges are – first, large distance with comparatively signal coverage and second, power consumption for devices. It is observed that the use of energy harnessing techniques by the research communityhas proven successfully redemption in power consumption for circuits involving antenna, microcontroller and sensor. In present work, a novel schematic design of a low power wireless sensing node has been discussed which is used for connecting different sensors such as moisture and temperature sensor. The design considered the wireless sensor node with a power module, a NRF, an Atmega-328 microcontroller and a GSM module to upload data over the cloud. NRF facilitates long-range communication up to a range of 1000m by using a maximum current consumption of 15mA. The wireless sensor node are key elements in implementing Internet of Things.**

The performance in terms of power consumption is analyzed for long range communication. Experiment has been designed to transmit with an interval of 15 sec and interval of 60 sec. At the receiver end, the central node GSM sim800A module is used to upload data over the cloud platform Thingspeak. In comparative study with different microcontroller development boards & wireless methods it has been found that the designed WSN node is effective in saving power consumption. The node can act as master or slave and can communicate with several other nodes also.

*Keywords:*Wireless sensor network(WSN) node, NRF, Design, microcontroller, GSM, Internet if things (IoT).

I. INTRODUCTION

The wireless communication has helped the world to transmit data from remote locations. After development of amplitude modulation, frequency modulation and digital modulation techniques, it is possible to achieve a greater bandwidth at a faster data transfer rate. Advancement in communication technologies such as 4G and 5G has made easy the implementation of the Internet of things (IoT) and machine to machine interface (M2M) [1]. These technologies use frequency bands which need to get licenses, are expensive to use and devices are costly to transmit data. It has drawn the interest of researchers to develop wireless sensor nodes and networks [2, 3] which may not require licensing or purchasing the transmission on these bands. In WSN major functional block is sensor node which senses a physical property, collect data, and then transmit the information wirelessly [4, 5]. The major problem with the WSN systems is power. It is powered by a battery which needs to be charged or replaced from time to time. WSN nodes are placed in a harsh environment where it is not possible to change batteries [6-8]. Energy can be harvested using suitable techniques, from the surrounding environment [9]. Since the energy harvested is quite small, the power consumption of the WSN must also be maintained at a lower value. This highlights a need to come up with a newer method of designing a WSN node having low power consumption [10]. The factor of power consumption is important for a WSN node designer.

Sensing, processing, and transmission are three major activities of WSN node [11, 12]. A sensor node for monitoring air quality has been designed using the Arduino Yun development board in which the power consumption at the node was also observed [13]. The node relied on power harvested by a solar panel which was stored in batteries. The set-up was stipulated to run for days but due to adequate power consumption, a need to recharge again and again was highlighted. Rohit et al [14] used Raspberry Pi for designing an IoT based home automation WSN model. A critical limitation in their model was the need for a regular power supply thereby proving its inefficacy in running on battery sources. In similar lines, authors have also developed a sensor node for deployment in an infant incubator, however, since it was not wireless it consumed a significant amount of power [15]. Al-Haija et al. developed a WSN based system using arduino and gas sensors to monitor the air pollution. This system provide real time data of the concentration of various gas contaminants exist in air [17]. They have used wifi technology to transmit the observed data by the sensors over a long distance. The usage of Wi-Fi has been increased but due to power consumption during wifi transmission, the node life time has been affected [18]. Verma et al. has made use of Raspberry pi and wifi technology to develop home automation systems using a home network through which the home appliances can be accessed from any place the world. A script has been written in php for web portal in which the the user controls the appliances by giving some input. Python language has been used to configure the Raspberry pi [19]. Literature survey reveals that the most of the WSN node are designed on development boards with libraries available for programming. The power consumption performance parameter is ignored in most of the designed nodes. A comparative analysis of different systems has been done with the proposed system as represented in Table 1.

Table1: comparison with existing system

Sl. no.	System existing/proposed	Features	Remarks
1	Existing system[19]	Iot for home automation, raspberry pi	Cot of hardware is high, wifi power consumption not studied, range is very low in few meters.
2	Existing system[18]	System for ambient monitoring, a low power node has been designed. Based on Psoc.	The usage of Wi-Fi has been increased. The node life time has been affected with increase in power consumption. Low maintenance with self calibration.
3	Existing system[16]	Vehicle pollution monitoring system developed using ESP 8266 with wifi.	Power consumption for sensor node is not studied, consumes a lot of power due to wifi and no use of sleep mode.
4	Existing system[14]	System designed for home automation using raspberry pi connected to wifi.	No power consumption studied. Cost of one node is high
5	Existing system[15]	Infant incubator system developed using arduino. Sensor node for temperature and pulse detection.	a sensor node for deployment in an infant incubator, however, since it was not wireless it consumed a significant amount of power
6	Proposed system	WSN node, can be used for IoT, novel design, Range upto 1000m and remote data access.	Data sent over server and action taken based on analysis, can be used with many types of sensors. Node life has been increased, NRF communication & GSM in sleep mode help to reduce power.

II. PROPOSED SYSTEM

We have designed a new WSN node with low power consumption and small size. The designed node can act as transmitter or receiver. The circuit can easily be connected to sensor giving output on single pin in analog or digital form. Fig. 1 shows power consumption takes place during transmission. The consumption, however, varies directly with the amount and the distance up to which data must be transferred. Out of all units, transmission unit require maximum power. We observe the power consumption in two cases. First the sensing data is sampled after every minute and then data is sampled after 15 seconds, the consumption changes quite drastically. Power consumption is also affected by the communication protocol used to transmit data.

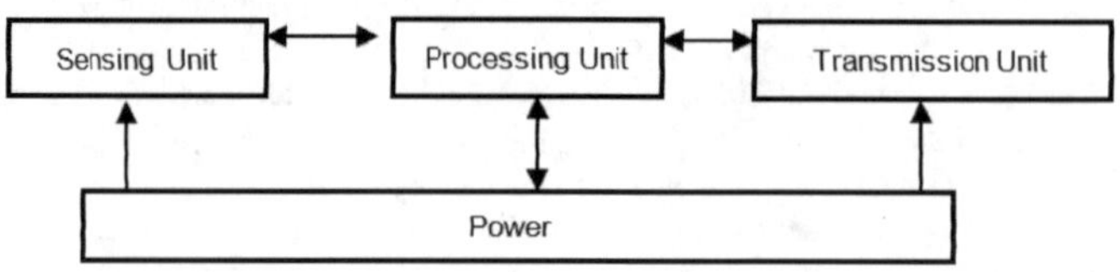

IOP Conf. Series: Materials Science and Engineering 1020 (2021) 012030 doi:10.1088/1757-899X/1020/1/012030

Fig.1: Power usage in a Wireless sensor network node

HARDWARE & SOFTWARE DESCRIPTION

Atmega-328 is an AVR family microcontroller with 28 pin DIP. It has 6 analogue inputs and 13 digital input/output pins. A high-performance Microchip power in pico-watts 8-bit AVR RISC-based microcontroller combines 32KB ISP flash memory with read-while-write capabilities. It has a 1024B EEPROM, 2KB SRAM, 23 general-purpose I/O lines, 32 general purpose working registers, three flexible timer/counters with compare modes, internal and external interrupts. It also has a serial programmable USART, a byte-oriented 2-wire serial interface, SPI serial port, a 6-channel 10-bit A/D converter (8-channels in TQFP and QFN/MLF packages), programmable watchdog timer with internal oscillator, and five software selectable power saving modes. The device operates between 1.8-5.5 volts.

NRF24l01 is a single chip trans-receiver working in frequency of 2.4-2.4835 GHz ISM band. It is embedded with a baseband protocol engine known as Enhanced ShockBurst. It is generally used for low power wireless applications. Its maximum current consumption is 15mA. It can be operated using SPI (serial peripheral interface) using Master Out and Slave In(MOSI) and Master In and Slave Out(MISO) pins as shown in figure-2. The frequency channel, output power and the data rate can be configured depending upon the need. It supports a data rate of 250kbps, 1Mbps and 2Mbps. Overlapping channel spacing is 1 MHz and 2 MHz. A single NRF module can connect with six (6) others modules as the transceiver. We can create a network of more than 3000 nodes using tree topology of 5 levels deep as shown in figure-8. Each node is assigned with address in the octal format defined by 15-bit address.

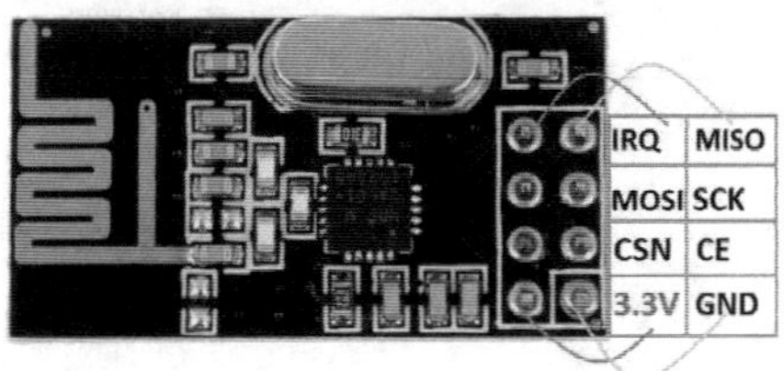

Fig. 2: NRF24l01 module pin diagram.[30]

The SIM800A is a Quad-Band GSM/GPRS Module working on RS232 interface. It is a GSM/GPRS solution in an LGA(Land grid array) type which can be included in applications. SIM800A support Quad-band 850/900/1800/1900 MHz, and it can transmit voice, SMS and data information with low power consumption. It is very small in size of about 100 x 53 x 15 mm and it can fit into slim and compact pcb designs for customized products. It features AT commands and also have embedded module to process the same. It allows total cost savings and fast time-to-market for customer applications. RS232 interface enables easy connection with the computer or laptop by using a USB to the Serial connector, or to the microcontroller using the RS232 to TTL converter. When the SIM800A modem is connected using the USB to RS232 connector, it is assigned a COM port number from the Device Manager of the USB to Serial Adapter.

ThingSpeak is an IoT analytics platform service that allows users to aggregate, visualize, and analyze live data streams in the cloud. Additionally, users can send data to ThingSpeak from their devices, create instant visualization of live data, and send alerts. It provides RESTful and MQTT APIs

CIRCUIT DESCRIPTION

The system is composed of a transmitter and a receiver section. At the transmitter, the sensed value is converted into digital form using microcontroller which is further transmitted by using the NRF module. The system is powered by 3.6volt 2600mAh Li-ion battery cell which is rechargeable. Single transmission of data uses 30mA current and 3.3V to 5 volt supply. Maximum power dissipation by the circuit is 45mW+75mW. The current work focusses on the design of the sensor node. At the terminal, any type of sensor which can give output using only a single pin can be connected. Therefore, temperature and moisture sensors are connected in this terminal.

Block diagram of the system is shown in figure-3. It consists of two section transmitter and receiver. Transmitter node of the system is shown in figure 3a and the receiver node is shown in figure 3b.

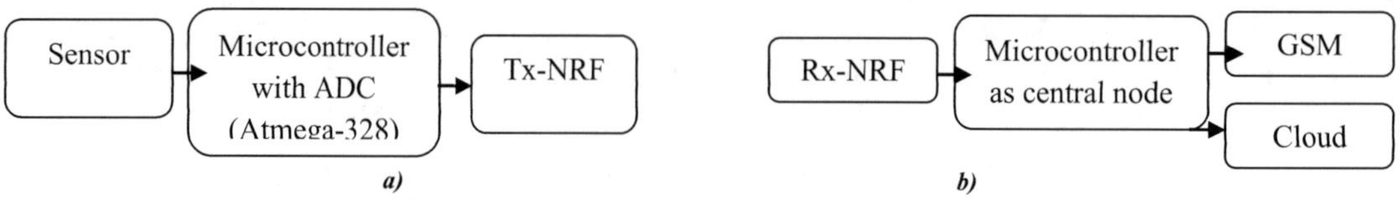

Figure 3: Block diagram of the system a) Transmitter node b) receiver node

Circuit design and PCB schematics are obtained as shown in Fig.4 - 5. Eagle software has been used for PCB schematic and CNC routing & drilling machine has been used for PCB printing. Eagle for board design helped to stack components in a compact space as per preferable dimension. It traces the connections though copper and idealized the design onto the PCB. Fig. 6 shows the PCB after having the components soldered. The size of the PCB is approximately 5cm in length and 4 cm in breadth. Height is around 3cm.

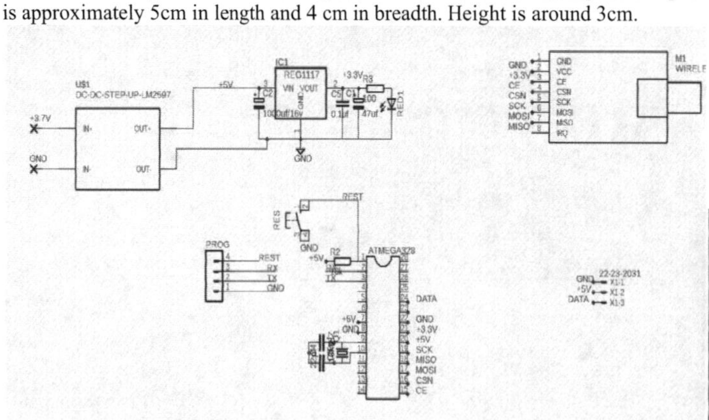

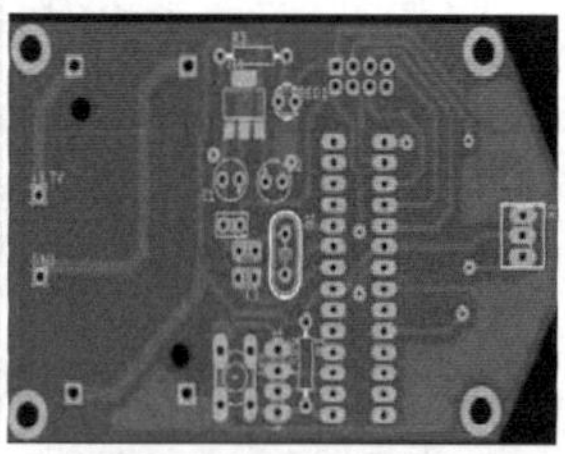

Figure-4: Schematic of circuit designed in Eagle software. Figure-5: PCB schematic before printing.

IOP Conf. Series: Materials Science and Engineering 1020 (2021) 012030 doi:10.1088/1757-899X/1020/1/012030

Figure-6: Actual image of the sensing node after mounting components, attaching power module and NRF antenna.

III. WORKING OF PROPOSED SYSTEM

We have designed two systems, firt the sensing node with transmitter NRF and second the receiving central node which collects the data from all sensing nodes and can upload the data on the cloud. The central node will also process the parameter. Whenever the microcontroller is powered up the program which is loaded in the memory is fetched by the processing unit and executed sequentially according to the algorithm displayed in Fig. 7. After data gets captured using the sensors, the value converted by microcontroller ADC(analogue to digital converter) is pushed onto the radio network using NRF24L01 module. NRF uses the SPI communication protocol to get and send the data from the microcontroller. The network for 2 or more sensing nodes may be arranged using the tree topology. Each NRF node can communicate with maximum 5 nodes in the hierarchy. The address of nodes is defined in the octal form with base address 00 given to central receiver and 01-05 for the first layer nodes as shown in Fig.8. The addressing is continued for the successive nodes.

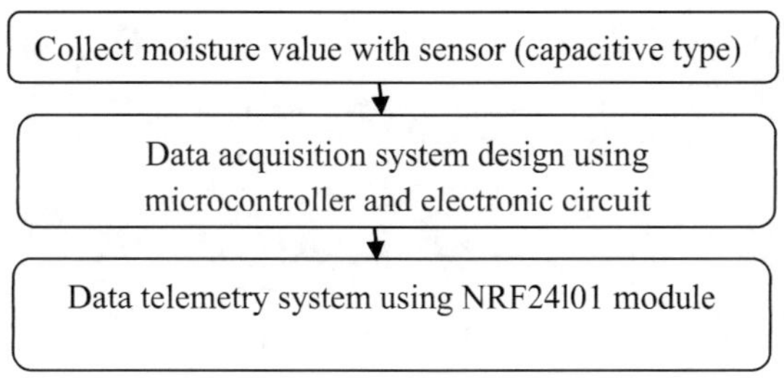

Figure 7: Algorithm for the sensing node

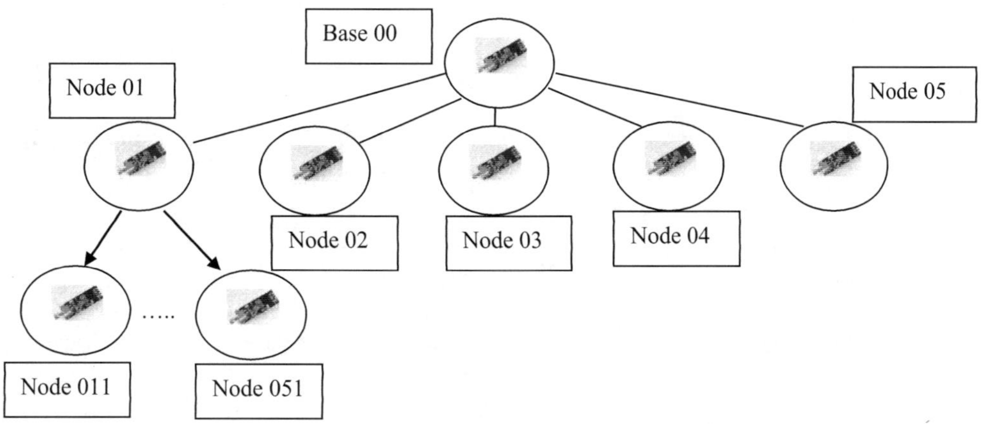

Figure-8: tree topology and the octal address given to nodes

Results:

The experiment for transmission of one sample of data is conducted using the designed WSN node and the observations are recorded in the table2 below.

Table2: power consumption details of the components

Power components in terms of current & voltage									
Mode of operation	Atmega-328p			NRF24l01			GSM-SIM800A		
	current	voltage	power	current	voltage	power	current	voltage	power
Sleep	23μA	5V	115μW	26 μA/ 900nA	3.6V	93.6μW	2mA	3.3V	6.6mW
Active	45mA	5V	225mW	13.5mA	3.6V	48.6mW	350mA	3.3V	1155mW

The sensing node is powered by using rechargeable Li-ion battery of 2600mAH. Ampere hour is the amount of charge stored in the battery. Since a battery changes voltage during the discharge, it isn't a perfect measure of how much energy is stored and we would need watt hour. Multiplying the average battery voltage times the battery capacity in amp-hoursgives you an estimate of how many watt-hours the battery contains.

$E=CxV_{avg}$ E is for energy in watt- hours, C is for charge in amp-hours, and V_{avg} is the average of voltage.

If the current drawn is x amps, the time is T hours then the capacity C in amp-hours is

$$C = xT$$

Current required by sensing nodeduring one transmission = 30mA
Current required by the central node with GSM= 330mA.
Total number of hours battery cell can power sensing node= 2600mAH/30mA= 86Hrs
Total number of hours battery cellcan power central node with GSM= 2600mAH/330mA= 7.8Hrs
We will be using 2 battery cell for central node , one to power the NRF and microcontroller and other to power the GSM.

Comparison of sensing node power consumption using Atmega328P on customized PCB, arduino and NODEMCU.

Parameter	Proposed design (with NRF)	Arduino(with NRF)	Node-MCU(with wifi)
Operating voltage	3.3 to 5 V	3.3 V to 5 V	3.3 to 5V
Current consumption	30mA	50 mA	80mA
Range	1000m	1000m	100m

Another comparison of power consumption by PSoC-BLE during active & sleep mode is also analyzed and shown in Fig. 9 with the help of CRO. It is found that in the active state the current consumption peak is 100mA, whereas, in the deep-sleep state the current consumption is in few μA. In comparison with the NRF module, the range of BLE is in few meters which restrict the usage of BLE over a long-range.

Fig. 10 show the noise with the signal. The markers are added in CRO and the minimum value of the signal is 3.0436 μA and the maximum value of current in the signal is 6.552 μA. The noise signal range

IOP Conf. Series: Materials Science and Engineering 1020 (2021) 012030 doi:10.1088/1757-899X/1020/1/012030

span is about 3 μA. the figure also shows the average, RMS, peak to peak and other parameters in the measurement panel in the lower part of the screen.

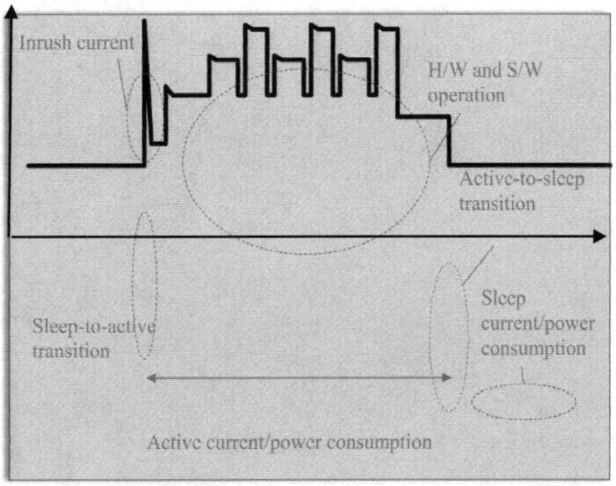

Figure 9.Both sleep current & active current measurement are required for power consumption reduction.

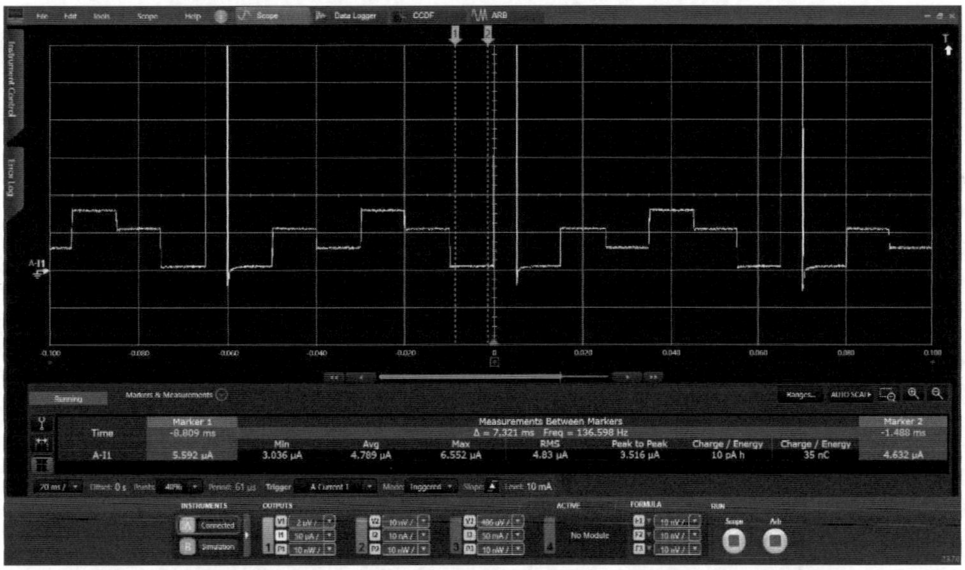

Figure 10.Example of signal hidden with the noise with peak to peak value of 3.516 μA

The analysis is continued further for observing the fast Fourier transform of the noise signal, the result of which is shown in figure. 11. The hardware setup for PSoC-BLE is also shown for the reference in figure 12.

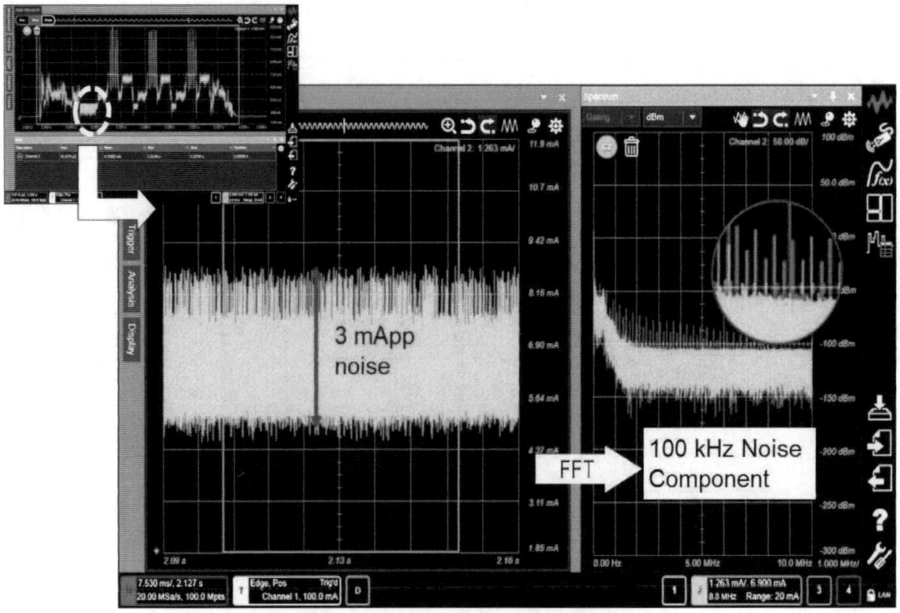

Figure 11.Noise analysis using FFT

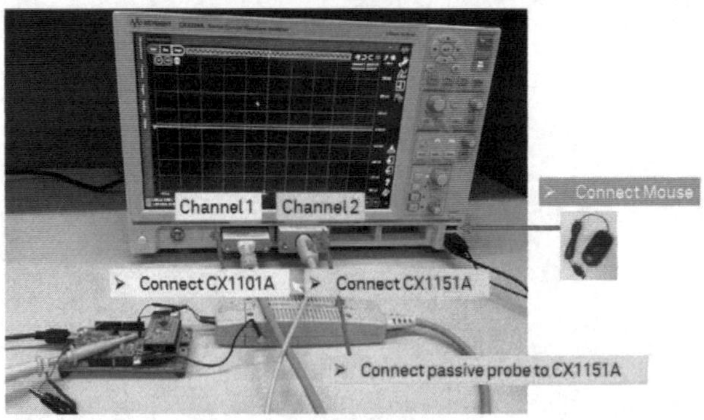

Figure 12: hardware setup of PSoC-BLE with DSO.

Conclusion:

A low power WSN node has been designed using Atmega328P and NRF technology to develop a network of wireless nodes for monitoring data. The nodes can be deployed for a long duration of time and with solar can sustain for days and weeks. The experimental result shows that power consumption is low. The paper is restricted to the design of WSN node and avoids the discussion on the communication protocol. The network is expedient, uncomplicated, and consumed less time to set up. The experiment justifies that power consumption of the whole network has been reduced significantly, and additionally verifies the effectiveness and energy-saving capability of the design. Comparing different module for IoT wireless sensor node for long-range communication, the developed WSN node is most suitable in power consumption and can go in sleep mode as per the design. The size of the PCB developed is also quite compact & of low cost in comparison to development boards. The designed node discussed in the paper can be installed at a remote location with connectivity in a range of kilometres. The range can be extended by using the tree topology. The nodes can be deployed for a long duration of time while maintaining power consumption to a low value.

The power consumption is tested and verified experimentally. Experiments show that the designed wireless sensor node has the following characteristics.

1. Low power consumption, good stability, battery power can work for 86 hours without charging at transmitter side and 7.8 hours on single cell at receiver side with GSM module.
2. Transmission need not to be directional and distance can be increased (typical single node range is 1000m), any placement detection location, to achieve any time and space scale detection.
3. Can achieve multi-hop routing.
4. Scalable according to the specific need of sensor mounted.

REFERENCES

[1] Lai CH, Wang JH, Yen TA, inventors; Brocere Electronics Co Ltd, assignee. Ultra low power sub-wireless sensor network (sub-wsn) for Internet of Things (IoT) system. United States patent application US 15/701,460 (Patent). 2019.
[2] Kocakulak M, Butun I. An overview of Wireless Sensor Networks towards internet of things.2017 IEEE 7th Annual Computing and Communication Workshop and Conference (CCWC). 2017: 1-6.
[3] Christin D, Reinhardt A, Mogre PS, Steinmetz R. Wireless sensor networks and the internet of things: selected challenges. Proceedings of the 8th GI/ITG KuVSFachgesprächDrahtlosesensornetze. 2009: 31-34.
[4] Mahgoub I, Ilyas M. Sensor network protocols. CRC press. 2018.
[5] Khan I, Belqasmi F, Glitho R, Crespi N, Morrow M, Polakos P. Wireless sensor network virtualization: A survey. IEEE Communications Surveys & Tutorials. 2015; 18(1): 553-576.
[6] Mahmood MA, Seah WK, Welch I. Reliability in wireless sensor networks: A survey and challenges ahead. Computer Networks.2015; 79: 166-187.
[7] Jabeen F, Nawaz S. In-network wireless sensor network query processors: State of the art, challenges and future directions. Information Fusion.2015; 25: 1-5.
[8] Gungor VC, Lu B, Hancke GP. Opportunities and challenges of wireless sensor networks in smart grid.IEEE transactions on industrial electronics. 2010; 57(10): 3557-3564.
[9] Meddour F, Dibi Z. An efficient small size electromagnetic energy harvesting sensor for low-DC-power applications.IET Microwaves, Antennas & Propagation. 2016; 11(4): 483-489.
[10] Zaman N, Tang Jung L, Yasin MM. Enhancing energy efficiency of wireless sensor network through the design of energy efficient routing protocol. Journal of Sensors.2016; 2016.
[11] Suryadevara NK, Mukhopadhyay SC, Kelly SD, Gill SP. WSN-based smart sensors and actuator for power management in intelligent buildings. IEEE/ASME transactions on mechatronics. 2014; 20(2): 564-571.
[12] Bouabdallah F, Bouabdallah N, Boutaba R. On balancing energy consumption in wireless sensor networks.IEEE Transactions on Vehicular Technology. 2008; 58(6): 2909-2924
[13] Ashish Gupta, Rajesh kumar. An IOT Enabled Air Quality Measurement, Indian Journal of Science and Technology, *Vol 11(46), DOI: 10.17485/ijst/2018/v11i46/139720,* December 2018
[14] Rohit kumar, Pradeep Jain, Ashishgupta. Internet of Things Enable Home Appliances Control and Monitoring System. Smart Home Control and Monitoring System based on Internet of Things with AI, ICAITCT, 2016.
[15] Design and Development of an Infant Incubator for Controlling Multiple Parameters" Hitu Mittal, Dr. (Mrs.) Lini Mathew, Ashish Gupta. International Journal of Emerging Trends in Electrical and Electronics (IJETEE-ISSN: 2320 - 9569) Vol.11, Issue.5,Sep-2015

[16] Vehicle Pollution Monitoring, Control and challan System using MQ2 sensor based on Internet of Things, Aayushi Gautam · Gaurav Verma · Shamimul Qamar · Sushant Shekhar. Wireless Personal Communications, https://doi.org/10.1007/s11277-019-06936-4, 18nov 2019.
[17] Al-Haija, Q. A., Al-Qadeeb, H., & Al-Lwaimi, A. (2013). Case study: Monitoring of air quality in King Faisal University using a microcontroller and WSN. *Procedia Computer Science, 21,* 517–521.
[18] Folea, S. C., & Mois, G. (2015). A low-power wireless sensor for online ambient monitoring. *IEEE Sensors Journal, 15*(2), 742–749.
[19] Verma, H., Jain, M., Goel, K., Vikram, A., & Verma, G. (2016). Smart home system based on Internet of Things. In *International conference on computing for sustainable global development, New Delhi* (pp. 2073–2075).
[20] https://electronicshobbyists.com/nrf24l01-interfacing-with-arduino-wireless-communication

AUTHOR INDEX